NO TRUTH EXCEPT IN THE DETAILS

BOSTON STUDIES IN THE PHILOSOPHY OF SCIENCE

VOLUME 167

MARTIN KLEIN

NO TRUTH EXCEPT IN THE DETAILS

Essays in Honor of Martin J. Klein

Edited By

A. J. KOX

University of Amsterdam, The Netherlands

and

Einstein Papers Project, Boston University

and

DANIEL M. SIEGEL

University of Wisconsin, Madison

Springer-Science+Business Media, B.V.

Library of Congress Cataloging in Publication Data

No truth except in the details : essays in honor of Martin J. Klein /
  edited by A.J. Kox and Daniel M. Siegel.
      p.   cm. -- (Boston studies in in the Philosophy of science :
  v. 167)
    Includes index.
    ISBN 978-94-010-4097-6 (alk. paper)
    1. Physics. 2. Physics--History. 3. Klein, Martin J.   I. Klein,
  Martin J. II. Kox, Anne J. III. Siegel, Daniel M. IV. Series.
  QC21.2.N64 1995
  530--dc20
                                                              94-38026

ISBN 978-94-010-4097-6   ISBN 978-94-011-0217-9 (eBook)
DOI 10.1007/978-94-011-0217-9

Printed on acid-free paper

TABLE OF CONTENTS

IV. EINSTEIN

V. FURTHER PERSPECTIVES

INTRODUCTION

Plus de détails, plus de détails, disait-il à son fils, il
n'y a d'originalité et de vérité que dans les détails . . .
Stendhal, Lucien Leuwen

I

This book of essays in honor of Martin Klein represents an effort on
the part of his students and colleagues to pay tribute to a lifetime of
dedication to physics and the history of physics. The papers contained
in this volume mirror Martin Klein's contributions to the history of
physics, reflecting his influence as scholar, teacher, and colleague. The
title of this book stems from a quotation used by Martin as an epigraph
in one of his papers. It characterizes Martin's work and the aspirations
of all of us in its call for meticulous attention to historical detail.

Of the themes of this book, more later; it behooves us, at the outset, to
try to give some representation of the career to which this tribute refers.
This effort labors under a major adversity: A significant portion of the
impact of Martin Klein's work comes from the sheer pleasure of reading
it, so that any representation of it by another hand necessarily falls short.
The pleasure in the reading flows, in part, from the skill with which the
words and equations are deployed, so as to convey the message both
clearly and vividly. Standing behind this skill, however, controlling and
commanding it, is something more basic and more important: an attitude
of respect toward both the subject and the reader.

The subject, in Martin Klein's historical writing, is multiple: it is
the science; it is the scientist; it is the history. The science is respected
through the meticulous care with which it is treated; it is treated not
as some epiphenomenon of social history or mere impediment to the
smooth flow of historical narrative, but rather as a thing of beauty and
significance, important in its own right. Similarly, the scientist is not
treated as mere feed for the historian's mill, but rather as a person,
with whom the historian interacts at a distance, but who must still

ix

A.J. Kox and D.M. Siegel (eds.), No Truth Except in the Details, ix–xxi.
© 1995 *Kluwer Academic Publishers.*

be respected as a person. Respect for persons is integral, embracing the direct experience of present individuals as well as the vicarious experience of historical figures. The third subject, the history, is treated with respect as well, through scrupulous honesty, careful construction of narrative, and avoidance of ideological slant.

Beyond this, the reader's situation is kept in mind, to is needs are perceived and attended to. Definitions of words, concepts, symbols are always given. Issues that will be puzzling to the reader, questions that will arise, are anticipated and dealt with; what this requires on the part of the writer is sensitivity and insight into the situation of the reader, as well as a commitment to minister to his needs. In a word, it is respect for the reader, as a person, as a fellow intellect, as a fellow student of the history of science, that drives this process. Beyond all of this, however – to return to the theme with which we began – it is the supreme skill and artistry with which Martin Klein's work is presented that provides the vehicle through which his positive sentiments toward his subjects and his readers are realized in historical narratives that give the highest kind of pleasure and insight.

Let us begin at the beginning, with Martin Klein's work in physics. This work was in the area of statistical mechanics, and in this field he made important theoretical contributions pertaining to the magnetization of thin films. Taking an interest also in questions of the foundations of statistical mechanics and the work of Paul Ehrenfest on this subject, Klein's concern then began to turn toward the history of Ehrenfest's contributions and their broader context in both the history of statistical mechanics and Ehrenfest's own life history. This phase of Klein's career resulted in two important books: the definitive edition of Ehrenfest's collected scientific papers; and a biographically framed study of Ehrenfest and the history of statistical mechanics – including quantum statistical mechanics – that constituted a milestone contribution both to the broader history and to our understanding of Ehrenfest himself.

Martin Klein's work in the history of statistical mechanics then branched out in two directions. First, in the direction of classical thermodynamics and statistical mechanics, as exemplified in the work especially of Rudolf Clausius, James Clerk Maxwell, and Josiah Willard Gibbs. Threading through Klein's work in this area and unifying it is the question of the nature and foundation of the second law of thermodynamics; this work promises to stand as definitive on the history of the second law in the second half of the nineteenth century. To single out just one aspect, Klein's account of the role of Maxwell's demon in his reasoning on this subject has shed clear light on one of the most perplexed issues in the history and foundations of thermodynamics and statistical mechanics.

The other branch in Klein's work led on to the history of quantum statistical mechanics and quantum theory in general. His studies of the origin of the quantum hypothesis in the work of Max Planck were pioneering, showing that analysis in depth can shed light on even the most singular and puzzling events in the history of science; strong continuity with Klein's other work was maintained, as Planck's thinking on the foundations of the second law played a crucial role in his framing of the quantum hypothesis. Most important for Klein's continuing work on the history of the quantum theory, however, was his growing interest in Albert Einstein's contributions in that area. It was Einstein, as Klein has shown, who clarified the nature of Planck's quantum hypothesis, applied it to light itself, and first saw through to the depths of the changes of our view of physical reality that would be required to come to grips with the conundrums generated by quantum theory, especially the wave-particle duality. Martin Klein's intense involvement with the legacy of Albert Einstein has continued in his Senior Editorship of *The Collected Papers of Albert Einstein*, where his direction of that major enterprise is ensuring a product worthy of its subject.

Martin Klein's contributions to the history of physics and the history of science are attributable no less to his personal and institutional activities than to his writing. As Professor of Physics at Case Institute of Technology to 1967, and as Professor of the History of Physics at Yale University thereafter (becoming Eugene Higgins Professor in 1974 and Bass Professor in 1991), Klein has trained a generation of students to go forth and practice the discipline with the clarity, integrity, and skill that characterize his own work. Beyond this, owing to the respect in which he is held in both the physics and the history of science communities, Martin Klein has been an effective facilitator of positive interaction between those communities in the history of physics enterprise. Perhaps the greatest problem of the modern university – and the greatest problem of the intellectual enterprise that is housed in the university – is the fractionalization of learning, the breaking down into departmental and disciplinary feudal domains. Martin Klein is one of those few broad and gifted individuals who is able to bridge the disciplinary gap, producing scholarship that will travel well across disciplinary boundary lines and convince those on each side that the other side has something of value to offer. Especially in bringing history of physics to the physics community, Klein has been an indefatigable ambassador. In numerous lectures and articles directed especially toward physicists, Klein has presented the history of physics as a discipline for which the physicist can have respect. As Chair of the Division of the History of Physics of the American Physical Society, in two separate tours of duty, Klein has done much to further the relationship between the history of physics discipline and the physics community. In a long series of distinguished lectureships, including the George Sarton Memorial Lecture to the American Association for the Advancement of Science, the Morris Loeb Lectures at Harvard, and others, Klein has brought his message to a broad audience of scientists, science teachers, and a variety other constituencies. Not least in his work with the Einstein Papers, Klein has made, and is making, a major contribution to the cooperation of these communities. Serving him in this endeavor have been the respect in which his scholarly contributions are held, as well as the respect in which he is held as a person.

The honors that have come Martin's way reflect the catholicity of his intellectual approach and activity: He is a Fellow of the American Physical Society as well as a full member of the Académie Internationale d'Histoire des Sciences and of the American Academy of Arts and Sciences; his election to the National Academy of Science in 1977 was noteworthy as designating one of a very small number of memberships allocated to individuals who are not primarily scientists. Martin's activity in bringing communities together is manifested also in international ties. There has been a strong connection with the Netherlands, first in his studies of Ehrenfest, which involved his working in Leiden (where Ehrenfest had succeeded H. A. Lorentz), and then through his appointments at the University of Amsterdam, first as Van der Waals Visiting Professor and later as the first Pieter Zeeman Visiting Professor. Martin Klein thus has been, throughout, a unifier of knowledge and of the disciplinary and national communities that generate knowledge. Honors, acknowledgment, and respect for his work have in turn come from many quarters and in profusion.

II

Martin Klein's work in the history of physics has been primarily focused on the late nineteenth and early twentieth centuries, but his influence, through students and colleagues, has touched on the history of physics and related disciplines since Newton – that is, from the seventeenth through the twentieth centuries. The present volume, in like manner, has its primary focus in the late nineteenth and early twentieth centuries, with extensions both earlier and later.

Part I presents a pair of studies dealing respectively with the mathematical and experimental foundations of the physics discipline, as they developed in the late eighteenth and early nineteenth centuries. Elizabeth Garber begins with the French mathematical tradition, embodied in the rational mechanics of the eighteenth century and its generalization in the early nineteenth century, as exemplified especially in work of Joseph Fourier; she then goes on to discuss the transformation of this tradition, especially by the British followers of Fourier and his Continental colleagues, into what may be properly designated mathematical physics. Garber's paper not only advances our understanding of the emergence of a flourishing physics discipline in the course of the nineteenth century, her essay also furnishes background for an understanding of continuing issues, through the nineteenth and twentieth centuries, of the demarcation of physics from mathematics and of the interaction of the two at their borderline. The themes broached in Garber's paper continue into Part II of this volume, in which the work of William Thomson and James Clerk Maxwell on the mathematization of electromagnetic theory are discussed.

The companion paper to Garber's in Part I is by Russell McCormmach, on Henry Cavendish's weighing of the earth. The tradition of precision experiment is the other leg on which the developing nineteenth-century physics discipline stood, and no better example of the roots of this than Cavendish's experiment may be found. In recent decades, among historians of science, experiment has been given short shrift: either experiment has been regarded as unproblematic, and hence unworthy of probing historical analysis; or experiment has been deemed to be so theory-laden, so derivative from theory, as to have no independent significance worthy of study. Most recently, however, there has been a dawning realization among historians that the malleability of experiment – the tendency of its results to be influenced by theoretical expectations –

is accompanied also by a certain recalcitrance of experiment, a tendency to resist manipulation and confound theoretical expectation in certain instances – as working scientists have known all along. Questions of the independence of experiment and the interaction of theory and experiment are important in several of the subsequent articles, especially those by Hiebert, Buchwald, Holmes, Kox, and Brush. McCormmach's article also provides acute insight into aspects of Cavendish's personal life and how it interacted with his science – a theme that continues most conspicuously in Part V of this volume, dealing with Albert Einstein.

Following the discussion, in Part I, of the eighteenth- and early nineteenth-century foundations of the physics discipline, the following two sections of the volume then go on to consider the blossoming of the physics discipline in the nineteenth century, on the basis of those mathematical and experimental foundations. Parts II and III consider respectively the two broad areas of investigation that constituted the central foci of the development of the physics discipline in the nineteenth century: electricity and magnetism; and thermodynamics and matter theory. To separate these two areas is in fact a bit artificial, as there were very important interactions between them, both conceptually and with respect to individual scientists who participated in both. In fact, most of the central figures to be treated in Parts II and III, including Hermann von Helmholtz, William Thomson, James Clerk Maxwell, Heinrich Hertz, and J. Willard Gibbs, made central contributions in both areas.

The first article in Part II, by Ole Knudsen, shows the depth of the connections between energy physics and electromagnetic theory from the outset, that is, from the years around 1850, when the energy law was being formulated. William Thomson started out in electromagnetic theory as described by Elizabeth Garber in her article, namely, as a follower of Fourier's mathematics and methodology. However, as has been shown in recent work, especially by Knudsen himself and by Norton Wise, energy concerns very soon began to play a central role in Thomson's thinking about the electromagnetic field. Knudsen here argues, in particular, that the concept of potential energy in electromagnetic theory was won with much more difficulty than has been realized hitherto, as shown by the fact that both Helmholtz and Maxwell made fundamental errors in their treatments of the relationship between electromagnetic induction and the potential energies of magnetically interacting circuits. It was only Thomson who was able to get this straight, and that, Knudsen argues, was owing to his familiarity with the concept of internal energy from his work with that concept in connection with the energy law and thermodynamics.

Following on directly from Thomson, both in the history of electromagnetic theory and in the current volume, is the work of Maxwell. It was Maxwell who established, in enduring form, the relationship between electromagnetism and optics, in what emerged as the electromagnetic theory of light. Peter Harman, in his article, presents a synthetic overview of the development of Maxwell's thinking on this subject, drawing upon his own previous work in this area as well as that of his colleagues in the Maxwell industry; added to this in Harman's article is the particular insight he brings on the basis of his expert knowledge of archival Maxwell materials, as brought to light in the course of his continuing work on a three-volume edition of Maxwell's scientific letters and papers.

Last in Part II, and bringing the story up toward the end of the nineteenth century, is Erwin Hiebert's article on the history of electrical discharge in rarefied gases, from Faraday onward, to the period just before J. J. Thomson's work in that area, leading to the discovery of the electron. Taken in conjunction with the previous two articles, Hiebert's article completes the treatment, in the present volume, of the Faraday–Thomson–Maxwell triumvirate in the history of field theory. The article itself, however, is rather more concerned with the experimental aspect of the work of Faraday and his successors in this field of research, and in this sense follows on most directly from McCormmach's article in Part I. Hiebert argues for the primacy of experiment in directing research with electrical discharge tubes in the course of the nineteenth century. Those carrying out the research admitted that they knew little about the physical processes taking place inside the tube, and they did not undertake experiments in order to confirm or refute fundamental theories of the phenomenon. Instead, they relied on largely unarticulated feelings, hunches, and low-level generalizations concerning the observed phenomena to direct their attempts to manipulate situations and generate new, and perhaps more revealing phenomena. Technical advances in dealing with the tubes – as in sealing electrodes, pumping to high vacuum, introducing various gases, shaping the tubes, and introducing various objects into the tubes – had more to do with the development of the research than any theory. Part II thus ends on a note of considerable current historiographical interest.

Part III, dealing with thermodynamics and matter theory, parallels Part II chronologically in dealing with nineteenth-century develop-ments, and, as we have seen, there was in fact much interaction between thermodynamics and electricity and magnetism during that period. It is in Part III that we begin to move directly into the part of the history of physics that Martin Klein has made his own. In his article on "Gibbs and the Energeticists," Robert Deltete employs a historical gambit that is familiar to readers of Klein's work: using what one historical fig-ure or set of figures had to say about another figure or set of figures, in order to illuminate the world views and presuppositions of both. In illuminating the energeticist movement and its views concerning the history, nature, and significance of thermodynamics, Deltete's work is important for our understanding both of the beginnings of the energy law and thermodynamics – especially in the work of Robert Mayer, seen by the energeticists as the founder of their movement – and of the role of thermodynamics in the decline of the mechanical worldview toward the end of the nineteenth century. The philosophical orientation implied in the energeticist movement, especially in the form developed and trans-mitted by Ernst Mach, was to be of great importance for Einstein, whose work is discussed in Part IV below.

Jed Buchwald's paper on Heinrich Hertz deals with a scientist per-haps best known for his experimental work in electricity and magnetism, namely, in the generation and detection of electromagnetic waves. It is Hertz's experimental work relevant to issues in thermodynamics and kinetic theory, however, that Buchwald explores here. In thematic con-tent, Buchwald's article continues the emphasis on the integrity of the experimental tradition that was broached in McCormmach's article on Cavendish and further developed in Hiebert's article on gas discharge studies. Buchwald sees Hertz's experiments on electromagnetic waves and his experiments on evaporation as both stemming from the same scientific impulse, the same methodology: the attempt to produce new phenomena in the laboratory, using hints and suggestions stemming in part directly from experience in the laboratory and partly from theoreti-cal ideas. Even when there was theoretical input, however, the primary and overriding object was not to verify or falsify the theory, but rather to use the theory for what it was worth in trying to produce new phe-nomena. Hertz's work on evaporation was hardly as successful as the work on electromagnetic waves, but that circumstance is perhaps all for the best in helping to illuminate Hertz's experimental methodology.

Last in Part III is an article by Frederic L. Holmes, which begins with a problem posed by Antoine Lavoisier in the later eighteenth century, relating to the site of oxidation reactions in the body and the attendant production of heat; the story ends with the answer furnished by the work of Felix Seyler-Hoppe in the 1860s and 1870s, in terms of the transport of oxygen through the bloodstream by hemoglobin, to support oxidation in the individual cells of the body tissues. From Lavoisier's work onward, and especially as manifested in his collaboration with Pierre Simon de Laplace on the subject, the question of animal heat was closely bound up with the physical sciences, and this matter was central for both Mayer and Helmholtz in their pioneering work on the conservation of energy. More directly relevant thematically for the present volume, however, is Holmes's treatment of the interaction between theory and experiment. On the one hand, Holmes shows how the posing of the question of oxidation in the body influenced both the program of experimentation that was undertaken and the conceptualization of the results that were obtained; this tends to diminish the independence of experiment. On the other hand, Holmes's account of the influence of developments in experimental technique on the experimental outcomes and their interpretation, and his detailing of how these in turn brought about changes in the phrasing of the questions that were asked of experiment, tends to enhance the element of independence in the role of experiment. It is, above all, a balanced account of the interaction of theory and experiment that Holmes presents.

Moving on to the early twentieth century is Part IV, on Einstein, a topic central to Klein's oeuvre, which has become even more central since he assumed the Senior Editorship of the Einstein Edition. It is thus not surprising that two of the contributors in this section are involved in the editing of Einstein's papers. Robert Schulmann's paper draws heavily on new archival material, to a large extent unearthed by himself, that sheds new light on Einstein's Swiss years. Schulmann argues convincingly that the old myth of Einstein as an outsider, who suddenly, through his revolutionary work on relativity, statistical physics, and quantum theory, was called back into academia from his exile in the Bern patent office, is indeed a myth. Throughout the patent office years Einstein kept in touch with some of his academic colleagues, and his return to academic life was carefully prepared and executed. The article by Kox shows, through the example of an episode in Einstein's life, the importance to historians of having an edition such as the Einstein edition available.

The article concerns the writing, submission, and eventual retraction, in 1911–1912, of a paper by Einstein on the theory of residual rays. By putting together the various pieces of evidence that can be culled from Einstein's correspondence of that period, now conveniently collected in Volume 5 of the Einstein Papers, the background to the paper, most of its contents, and the reason for its retraction can be reconstructed. As in earlier articles in this volume, in Kox's article, too, experiment and its interaction with theory take a central place, as it was Einstein's disbelief in certain experimental results that prompted his paper, and it was his eventual acceptance of those same results that made him retract it. In the meantime he had argued for a different interpretation of the data, an interpretation based on theoretical considerations. In the article by the renowned Einstein scholar Gerald Holton, finally, we encounter another aspect of Einstein's personality: his love of books. Holton traces Einstein's intellectual development through the books he read and outlines an important future research project, namely, to study the influence of various books on Einstein's thinking through an analysis of the detailed inventory of Einstein's library that is currently being prepared.

In a variety of ways, Einstein's work is a culminating point in the history of science; so also in the composite narrative that emerges from Martin Klein's contributions to the history of physics, and so also in the present volume, where the threads collected in their bearing on Einstein diverge in various directions in Part V, Further Perspectives. Stephen Brush's paper on prediction and theory evaluation continues the theme of the interaction of theory and experiment. It has been widely believed that predictive power furnishes an especially acute test of theory, so that a theory which makes successful predictions – especially if the phenomena predicted run counter to existing belief – will command assent. What Brush demonstrates, in brief, is that this is not true historically. Other issues, such as whether the predicted experimental result is seen as of core relevance to the theory, and whether the theory itself has the kind of internal coherence and structure that will make it generally acceptable, are more important, and will overweigh a given incident of successful prediction in molding the opinion of the scientific community. Experimental tests of the general theory of relativity furnish one set of Brush's examples, and there is, it might be suggested, another Einstein connection as well, having to do with his delineation of two aspects of theory evaluation, respectively the "internal" and the "external." The internal criterion has to do with the internal coherence, harmony, etc. of the theory; the external criterion refers to the fit with experiment. As Brush's investigation shows, especially in the case of Hannes Alfvén's astrophysical and cosmological theories, if the internal criterion is not satisfied, no amount of successful prediction will be convincing.

Two of the articles in Part V, by Daniel Siegel and Abraham Pais respectively, address historiographical issues bearing on the location of the history of physics enterprise at the borderline between physics and history of science, where technical material characteristic of the physics discipline is treated, but with the methods and perspectives of the historian. Martin Klein's work has been at this borderline, furnishing paradigmatic examples of what can be accomplished in this area that will command the respect of both historians and physicists. The history of science discipline was institutionalized in the universities in the period after World War II, partly in response to the feeling that, in the aftermath of the development and use of nuclear weapons in that war, it was necessary for society and, in particular, the intellectual community, to come to grips with the interactions between science, technology, and society, in part through studying the history, philosophy, sociology, and politics of science. The history of science enterprise was envisioned, in this context, as a highly interdisciplinary kind of undertaking, having relationships with science, history, philosophy, sociology, political science, etc. In recent decades, however, history of science has become more of a subdiscipline of history than a truly interdisciplinary undertaking, and relationships with the sciences have been de-emphasized as a result. Pais and Siegel, in their respective articles, voice some discontents with this state of affairs.

Pais's concern in this is primarily with the writing of history, while Siegel is concerned with the reading of historical documents. Pais discusses the primacy of narrative in historical writing, from the journalistic to the scholarly: it is the telling of the tale that matters, that bears the magic, as Pais illustrates with a Chassidic legend. And when the tale is at the borderline between physics and history of science, the teller should be one who understands the physics – otherwise the magic disappears. Siegel, in a parallel argument, suggests that the reader of a historical text that has technical content must be one who is prepared to read that technical material as it was meant to be read, that is, with pen or pencil in hand, poised to fill in the missing steps in the argument, as the reader of mathematical material is almost always called upon to do. Siegel as well makes use of a literary example: a poem by Maxwell that must be read as it was meant to be – out loud – in order to get its message. The concern with reading, writing, and narrative brings us full circle: back to the perspective of the historian, no matter how technical the historical materials may be.

Finally, in the paper by Roger Stuewer on the seventh Solvay Conference, held in Brussels in 1933, we come to the chronological end of the volume. Martin Klein has made good use of conferences, including the Solvay Conferences, as microcosms for the study of historical situations in science; Stuewer here applies the approach to a later time and a different topic, namely, nuclear physics. The seventh Solvay Conference represented both a coming to an end and a new beginning. With the intellectual migration of that period (illustrated by Stuewer as relating to the participants in the conference), the old order characteristic of the period covered by this volume, in which the history of physics was a story of European, especially British, French, and German science, was breaking down, thus beginning the transition to a Post-World-War II situation in which science and technology in the Americas and in Asia play central roles. Also, in physics itself, there was what appeared, briefly, to be an ending, which then very quickly gave way to a new beginning: The quest for the ultimate building-blocks of the universe had been taken down to the molecular level in nineteenth-century kinetic theory, then down to the atomic and subatomic levels in the decades after 1890, and finally to the nuclear level in the second and third decades of the twentieth century. For a moment in the 1920s the quest appeared to have ended with the identification of the proton and the electron as the elementary positive and negative charges and the ultimate elementary particles. However, with James Chadwick and the discovery of the neutron; P. A. M. Dirac, Carl Anderson, and the positron; and Hideki Yukawa and the meson (as variously discussed in the articles of both Stuewer and Brush), this paradise turned out to be, if not exactly a fool's paradise, then perhaps an Eden lost.

With Roger Stuewer's article, then, we write finis to the period in the history of physics characterized by Cavendish, Fourier, Faraday, William Thomson, Maxwell, Gibbs, Hertz, Einstein, and Rutherford, and anticipate the nuclear zoo and other novelties of the end of the twentieth century. Our survey of the period from Cavendish to Rutherford, especially by way of Maxwell, Gibbs, and Einstein, constitutes our tribute to Martin Klein on his seventieth birthday. May he flourish!

A. J. Kox, Daniel Siegel
June 25, 1994

MARTIN KLEIN AT YALE

I first met Martin Klein to talk to in 1964 and in curious circumstances: a few miles above the North Atlantic aboard a KLM airplane bound for Amsterdam. We were, I seem to recall, only three passengers: the third was Derek Price. We talked of many things, and the flight seemed short, at least to me. It may have seemed short to Martin, too, and certainly was when compared to another flight of his, also connected with Amsterdam, where he had stopped to buy cigars on his way home from Denmark. His plane out was hijacked and flown via Beirut to Cairo where it was blown up while he was running away from it through the desert in someone else's shoes. Since then his travel luck has been better, though not of the best.

Martin stayed in Amsterdam, while Derek and I flew on to Hamburg and began plotting ways and means of inducing him to join us at Yale – it took three years, but we succeeded.

In the interval he came to New Haven to give several memorable colloquia which caused the formation of the nucleus of a growing band of faithful followers who do not willingly miss a lecture of his. We were not surprised that he became a popular teacher among Yale's undergraduates and that he achieves his popularity sacrificing neither his subject's dignity nor his own.

His excellence as a lecturer and expositor rests, of course, first on his intimate and detailed knowledge of the warp and weft in the fabric of physics of the late nineteenth and early twentieth century: he can in an instant trace the threads that connect any two related events. His style at the lectern is as far removed from the histrionic as one can imagine, though not without art. He plays his cards quietly and cunningly from a carefully arranged deck and addresses his audience directly, without the intervention of a manuscript, always keenly aware of his listeners' limitations and needs.

I saw a striking example of this last in a lecture to the Connecticut Academy of Arts and Sciences whose chief claim to glory is the publication in 1875–1878 of Gibbs's "On the Equilibrium of Heterogeneous Substances" in its *Transactions*. A celebration of this event's centenary seemed in order, and Martin was the obvious choice for celebrant. He

A.J. Kox and D.M. Siegel (eds.), No Truth Except in the Details, xxiii–xxv.
© 1995 *Kluwer Academic Publishers.*

had prepared himself with the promised presence of most of Yale's physicists in mind. However, at the very last moment we learned that a departmental emergency prevented them from attending, so he faced an audience void of expertise in physics (though not of intelligence). Without so much as a blink – at least as far as this observer could see – he proceeded to give one of the best lectures I have heard, elegant, with few technicalities, yet rich in substance, obviously not the one he had planned, but perfectly attuned to the unforeseen situation.

As said, Martin came to Yale in 1967. Because of an administrative tangle, he was given a presidential appointment, and it became my pleasant task to announce his arrival to Yale's Board of Permanent Officers. To prepare myself I looked carefully at his curriculum vitae and was startled to learn that he had been a child prodigy, for he exhibited none of the conventional tell-tale signs. Indeed, he had graduated from Columbia University in his native New York City with a bachelor's degree in physics at age eighteen, and obtained his master's degree there two years later. He was all of 24 years old when he became a doctor at MIT, but he had by then spent two years in war research for the US Navy – it had to do with the propagation and detection of sound under water (he is an excellent swimmer).

Martin began his tenure at Yale on leave so he could finish, and see through the press, his book on Paul Ehrenfest. I remember well, and with pleasure, an expedition toward New York City in search of an early copy of the *New Yorker* magazine that had in it Jeremy Bernstein's review of it – an informed and enthusiastic long essay. We all basked in his glory. It is, of course, a remarkable book, beautifully written as is all his published work, with great clarity and insight, and ample, but not too many technical details. It can be, and has been read with profit and pleasure by professional and lay alike. I have talked with several scientists who had known Ehrenfest, and all thought it astonishing that Martin, who had not, was still able to capture his essence to perfection.

Martin's principal field of work and mine are separated in time by some two millennia, yet we joined forces in a seminar on Newton's *Principia* and slogged our way through it proposition by proposition. It was heavy going – Newton did not feel kindly toward his reader – but it was great fun and I, for one, learned much.

We held the seminar at the end of the sixties, the golden decade of academic life in America. Never before or since have academic endeavours been so eagerly and generously encouraged and supported, and never before or since and, for that matter, nowhere else, has academic life been so exciting and flourishing. They were heady days, and we were fortunate to have excellent graduate students and visitors, also from abroad, many attracted to Yale by Martin's presence. They all left with his imprint on their standards and style of work.

It behooves old men to count their blessings, but all too many must, alas, find it all too simple. I am not among them for I have been very fortunate and my blessings are many and great. Not least among them is having had a profession and been in circumstances that brought me Martin Klein as a colleague and friend.

Asger Aaboe
Emeritus Professor of Mathematics, History of Science, and
Near Eastern Languages and Literatures
Yale University
U.S.A.

RUSSELL MCCORMMACH

THE LAST EXPERIMENT OF HENRY CAVENDISH*

Martin Klein, who has a long-standing interest in the life of the physicist Josiah Willard Gibbs, gave an address to the History of Science Society in 1982 in which he contrasted Gibbs with another physicist, Paul Ehrenfest. No two personalities in the same field could be more unlike than the austere and reticent Gibbs and the irrepressible Ehrenfest. Klein found Ehrenfest the more accessible of the two. In Klein's reading of Ehrenfest's writings in physics, Ehrenfest's personality comes across on every page. Ehrenfest's forte was criticism, which makes for vivid biography, as Klein's biography of Ehrenfest certainly is. By contrast, the characteristics of Gibbs's most important work are generality and logical simplicity, which are decidedly not characteristics of the biographer's subject, that most complex of things, the unique human personality.

In his search for Gibbs in Gibbs's work, Klein turned to one of Gibbs's early self-proclaimed apostles, Pierre Duhem. Duhem referred to Gibbs's "retiring disposition."[1] Gibbs had no need or desire to try out his scientific ideas on colleagues; he revealed his work to the world only after it was fully worked out. A man of regular, almost monkish habits, Gibbs rarely ventured outside the society of his university, Yale. His powers of scientific concentration were extraordinary. Gibbs's retiring disposition can be seen in his scientific work, Duhem believed, and so does Klein. "Retiring" does not conjure up dramatic scenes, but they are not what Gibbs's life is about. To Gibbs's biographer, a single expression, "retiring disposition," can serve as the starting point for understanding the person.

The subject of this paper is Henry Cavendish, who poses the same problem to his biographers: to bring substance to the shadow, to recreate a thinking, feeling person, in large part from scientific writings. Charles Blagden, the colleague who knew Cavendish best, described Cavendish's lifelong habits as "retired." It is Duhem's and Klein's word for Gibbs. Blagden's choice of the word for Cavendish would seem equally well chosen. Like Gibbs, Cavendish had intense powers of scientific concentration, and he was reluctant to take part in the ordinary affairs of society. Born to the high aristocracy, extremely rich, Cavendish was not drawn to the places of public drama open to him, such as politics

1

A.J. Kox and D.M. Siegel (eds.), No Truth Except in the Details, 1–30.
© 1995 *Kluwer Academic Publishers.*

and war. Taciturn when not mute, guarded and shy in the extreme in the presence of strangers, Cavendish lived all of his adult life in and around London in solid houses with servants to protect his privacy. These houses he turned into places of science. That was where the drama of his life was staged, unseen, internal, and profound.

When Cavendish died in 1810, an official, anonymous biographical notice appeared.[2] The authorship is established by a fragment of the notice in Blagden's handwriting; the fragment breaks off abruptly with the word "secluded." The circumstances of Blagden's notice are explained in two letters to Blagden from Lord George Cavendish, who along with his sons inherited the bulk of Henry Cavendish's estate. The first letter informed Blagden that the Duke of Devonshire, head of the Cavendish clan, had approved Blagden's sketch of Cavendish's "character" for the "Publick Papers." The second letter, written the next day, informed Blagden that the corrections Blagden meanwhile had sent had arrived too late: anxious that nothing about Cavendish appear in public before Blagden's notice, Lord George had already sent it to press. At the bottom of Lord George's letter to him, Blagden wrote out again the three corrections he had requested. They are brief and two of them of no consequence here. The third correction indicated that Blagden wanted Cavendish's habits to be called not "retired" but "secluded."[3] Perhaps the substituted word sounded better to Blagden's ear, but as a writer he was more exacting than elegant, and I think that it was a nuance of meaning he wanted. Weighing the alternative characterizations of Cavendish, "retired" versus "secluded," each conveying much the same impression, Blagden preferred the one that better matched the impression Cavendish had made on his contemporaries. "Retired" suggests withdrawn or inactive, "secluded" shut up.[4] The second word, Blagden thought, is definitely the better word for Cavendish.

The best word for describing the condition of Cavendish biographers is *bewilderment*. The wealth of scientific manuscripts Cavendish left behind confronts them with studies on every topic in the physical sciences, carried out independently of one another, without rhyme or reason other than with the implicit goal of totality. That is a first impression. If the biographers persist, they see that the studies fall into groups, connected by large goals, which belong to the science of Cavendish's time. One extended group of papers has to do with his researches on the earth, including its gaseous envelope and its location and orientation in the solar system. Researches on the earth that were most significant in the eighteenth century tended to involve numbers of investigators working together, in contrast to those on general laws of nature, which tended to be done by individuals on their own, at least in the first instance. Thus, in the several organized researches on the earth that Cavendish took part in, he worked with others, while preserving his measure of essential privacy. However in his last published experiment, – the determination of the mean density of the earth – Cavendish worked in seclusion in the ordinary sense of the word: He brought the earth into his place of seclusion, his home, where he experimented on it virtually alone. Then, because it was science he was doing, he submitted his results to the Royal Society for publication. This experiment came to be known to scientists as *the Cavendish experiment*. It was well named.

I. THE ROUTE TO CAVENDISH'S LAST EXPERIMENT

In 1760, at age twenty-nine, Cavendish was elected Fellow of the Royal Society. The society was, just then, preoccupied with one of the great eighteenth-century scientific projects. In conjunction with societies and academies in other countries, it was planning expeditions to observe the transit of Venus across the sun in 1761. These transits are periodic events, occurring in pairs eight years apart and then not again for 113 years. Their great interest in the eighteenth century was in providing an opportunity to determine the mean distance of the earth from the sun. The newcomer Cavendish was not formally brought into the preparations for the transit of 1761, as he would be into those for the paired transit of Venus eight years later, in 1769. The first evidence of Cavendish's involvement in the transit of 1769 is a letter by him to the president of the society, in 1766; later, the next year, he was appointed to a committee to consider the proper places, methods, and persons for observing the transit.[5] Here is the earliest known participation by Cavendish in a measurement pertaining to the earth, and it is the beginning of his service as a committeeman of the Royal Society, possibly its most called upon and certainly its most versatile.

Newton had concluded, and Huygens had too, that owing to the attraction of the earth and to the centrifugal force of its rotation, its shape ought to be an ellipsoid of revolution, a spheroid flattened at the poles.[6] This theoretical conclusion was disputed by others. On the grounds of previous French measurements, the Cartesian astronomer

Jacques Cassini held the opposite opinion, that the earth is a prolate spheroid, elongated at the poles, like an egg. The implication was clear: if Newton was right, the length of a degree of latitude should increase as one moves from the equator toward a pole, but if the other opinion was right, the length of a degree should decrease. To settle this dispute, two expeditions were sent out, one under P. L. M. de Maupertuis to Lapland, in the direction of the north pole, and the other under Pierre Bouguer and Charles Marie de la Condamine to a place in Peru (now in Equador) near the equator. The question was answered in favor of Newton and his supporters, the "earth flatteners."

Peru is a land of high mountains. If gravitation is a universal law, as Newton reasoned it is, then a plumb bob in the vicinity of a mountain should be affected. Newton calculated the attraction: a hemispherical mountain of earth matter with a radius of three miles would deflect a plumb-line by a minute or two of arc. He thought that the effect was too small to measure, which judgment was received by his eighteenth-century followers as a challenge. Since astronomical instruments depended on a plumb-line to establish the vertical, observations taken with them could be sensibly distorted, and Bouguer and La Condamine took precautions in their determination of the length of a degree of latitude. But since the attraction of mountains had not actually been observed, they did an experiment to see if it really did exist. With a quadrant oriented by a plumb-line, they measured stars directly overhead in two places, one beside the 20,000 foot extinct volcano Chimborazo, the other on a plateau far removed from the mountain. They did see a deflection of the plumb-line in the expected direction, but quantitative measurement was too difficult with the instrument at hand. Returning from the expedition in 1744, Bouguer said that he would like to see the experiment on the attraction of mountains repeated under proper conditions in Europe. His *La figure de la terre, determinée par les observations de Messieurs De la Condamine et Bouguer* ... , published in 1749, would be Cavendish's starting point in his work on the problem.[7]

The figure, density, and internal structure of the earth are connected properties, which in turn are connected to a seemingly remote phenomenon, the precession of the equinoxes. This precession is the slow motion of the earth's axis of rotation relative to the stars caused by the attraction of the sun and the moon on the earth's equatorial bulge. In an unpublished study of the precession of the equinoxes, Cavendish tried to reconcile Bouguer's result for the figure of the earth (which was Newtonian, in general), as obtained by mensuration, with the figure that agreed with the variation of gravity with latitude, as determined by theory and tested by pendulums. He could not,

without assuming some very improbable hypothesis of the density of the earth (if there is any hypothesis which will answer that purpose) or else without denying the theory[,] which seems too well founded to be shaken by these observations[;] & as the irregularity of the surface of the earth particularly in the high mountains of Peru where one of these observ. were made may cause an alteration in the direction of gravity & thereby disturb the accuracy of the experiment . . . we may fairly reject this experiment[al] mensuration & assume that diff. of axes which agrees with the differences of gravity . . . [8]

Cavendish was inclined to favor theory over measurement in this case: the gravitational theory was solid, and the French observations were subject to question, especially in view of the attraction of mountains. He proceeded to calculate the part of the precession caused by the sun using A. C. Clairaut's more probable hypothesis of the earth's interior. (In 1743 Clairaut had published a celebrated theorem relating gravity to latitude, assuming that the interior of the earth consists of concentric strata of uniform density.) Cavendish's result was much larger than what Newton had given, "yet it may perhaps agree full as well with exp. as his[,] as neither the force of the moon to move the equinoxes nor the form and internal struc. of the earth is known." Cavendish did not, as we will see, think that Clairaut's assumption was correct either. The figure and interior structure of the earth remained unknown, and the latter might be unknowable.

Throughout his research on the earth, Cavendish kept in close touch with the astronomer Nevil Maskelyne. With the approach of the first transit of Venus and on the recommendation of James Bradley, the Astronomer Royal, Maskelyne was sent by the Royal Society to St. Helena to make observations. The passage of Venus across the sun was clouded over, so the main point of the expedition was lost. Maskelyne, however, proposed to do another experiment, while on St. Helena, to measure the parallax of the brightest and supposedly closest star, Sirius, using the earth's orbit as base line. That measurement would give the distance of the earth not from the sun but from a fixed star. The parallax of stars is implied by the moving earth of the Copernican system, and astronomers had looked hard for it. Maskelyne had to make reliable observations if his experiment was to stand a chance. Heeding the warning to astronomers contained in Newton's calculation, Maskelyne took into account the possible influence of the attraction of the mountainous island on the plumb-line of his zenith sector. He planned to make corrective observations from the north and south sides of the island, but his instrument proved defective and nothing came of this attempt either.[9]

Yet all was not lost. It had long been known that a pendulum beating seconds is shorter near the equator than at higher latitudes. Newton and Huygens and those who came after them recognized that comparative measurements of the lengths of a seconds pendulum at different latitudes could serve as an experimental means of determining the shape of the earth. Experiments with pendulums had been made at various places around the world, and Maskelyne made another at St. Helena. Using a pendulum clock, he compared the (lessened) gravity on St. Helena with that at Greenwich. He did not, however, draw conclusions about the law of the variation of gravity with latitude or about the figure of the earth. He explained why in his paper reporting on the pendulum observations:

If the body of the Earth was homogeneous throughout, not only the figure of the Earth, but also the law of the variations of gravity in different latitudes would be given, and would be the same as Sir Isaac Newton has described them. But if the Earth be not homogeneous, and there seems great reason, from late experiments, to doubt if it be so, we can form no certain conclusions concerning the figure of the Earth, from knowing the force of gravity in different latitudes; as this force must depend not only on the external figure, but also in the internal constitution and density of the Earth . . . [10]

It is an "intricate" subject, Maskelyne concluded.

The goals of observations of the transits of Venus and of experiments on the density of the earth were similar in that they were both about the earth in relation to the solar system; also, measurements of both remained uncertain in the eighteenth century. The distances of the planets were expressed in terms of the distance of the earth from the sun; likewise, the densities of the sun and some planets were known only relatively, so that the density of the earth had first to be determined to know the density of the other bodies.[11] As one would expect, the same persons worked on the transits of Venus and the density of the earth, among them Cavendish, Maskelyne, and another of Bradley's assistants, the English astronomer Charles Mason.

Upon returning from the Cape of Good Hope where they had gone to observe the transit of Venus for the Royal Society, Charles Mason and his associate Jeremiah Dixon were hired in 1763 to settle the old boundary dispute between the colonies of Pennsylvania and Maryland. This painstaking job took them nearly five years. While they were at it and with the consent of the Royal Society, they measured the length of a degree of latitude. The question was then raised whether Mason and Dixon's measurement could be flawed by the attraction of any mountain and Maskelyne thought not.[12] Reviewing the measurement, however, Cavendish disagreed with Maskelyne. Taking into consideration the attraction of the Allegheny Mountains to the northwest and the deficiency of mountains in the Atlantic Ocean to the southeast, Cavendish calculated that Mason and Dixon's degree could fall short by sixty to one hundred Paris toises. One toise equaling about two metres, this was a considerable error. Cavendish made similar criticisms of the measurements of the length of a degree by R. J. Boscovich between Rome and Rimini and by N. L. de Lacaille at the Cape of Good Hope. Cavendish's study of the length of a degree as measured by Mason and Dixon and the others concluded as follows: "No regular figure can be assigned to the earth which will agree with all these observations so that either the figure of the earth is irregular or the observations have been influenced by the attraction of mountains or some of the observations were not sufficiently accurate."[13]

The problem of the length of a degree was a tangle of several problems; in particular, the form of the earth and the attraction of the earth, taking into account its mountains and its subsurface irregularities, were closely connected, and Cavendish titled his comprehensive discussion of them "Paper Given to Maskelyne Relating to Attraction & Form of Earth." This paper also drew on Cavendish's study of the precession of the equinoxes, and he told Maskelyne that the best way to determine the form of the earth was by gravity, not by mensuration, giving as a reason the better fit with the precession of the equinoxes.[14]

Maskelyne became Astronomer Royal in 1765 and was now in a position to initiate projects of his own. The next in line was the attraction of mountains. The issue was broached in 1771, in a letter from Maskelyne to Cavendish containing two theorems for calculating the attraction of a hyperbolic wedge and an elliptic cuneus; on the back of the letter Cavendish rewrote Maskelyne's two formulas. The analysis of the attracting mountain was underway.[15]

"Paper Given to Maskelyne" reads like a continuation of the letter Cavendish had received from Maskelyne; it gives Cavendish's rules for finding the attraction of a particle at the foot of and at a distance from geometrical solids generated by lines and planes and obeying the law of universal gravitation. After these mathematical preliminaries on the attraction of slabs, wedges, and cones, Cavendish turned to the subject of scientific interest, the real world of attracting bodies. These included the great irregular masses that the earth actually throws up, which distort astronomical observations but which also provide a means for measuring the density of the earth.

Cavendish told Maskelyne, "I know but 2 practicable ways of finding the density of the earth," by the seconds pendulum and by the plumb-line. He began with the first way, using observations from "Bouguer figure de la terre." To judge the effect on the length of a pendulum of the great masses of the Cordillera, Cavendish constructed an approximation to the mountain Pinchincha by joining his Platonic mountains, two half cones, and placing them on an infinite slab, by which he represented, as Bouguer had, the rest of the Cordillera. Assuming that the mean density of the Cordillera is the same as that of the earth, Cavendish calculated the increase in the length of a seconds pendulum placed at the top of the mountain and compared his value with the French observations. There was a difference, and from this difference Cavendish inferred that the mean density of the earth is 2.72 times the density of the surface layer, whatever that should turn out to be. He next did the same calculation as Bouguer had done for an observational site lower down the mountain, at Quito. This time the mean density of the earth came out to be 4.27 times the surface density. He did a further calculation with a different representation of the mountain, this time as a segment of a sphere, and he arrived at the value of 4.44, which was close to Bouguer's 4.7.[16] These were Cavendish's first estimates of the mean density of the earth.

In practice, pendulum lengths depend not only on latitude and on surface masses like mountains but also on the internal structure of the earth. Even if the simplest assumption is made about the earth's interior, it can be shown that the mean density of the earth is much greater than its surface density. To give Maskelyne an idea of what they might expect, Cavendish drew on an entirely different kind of evidence, John Canton's experiment on the compressibility of water. Supposing, Cavendish said, that even if the surface and the interior of the earth are of the same substance, the internal parts will be compressed and therefore be denser, the more so the closer the parts are to the center. Beginning with Canton's demonstration that the density of water is increased 44/1,000,000 by the pressure of one atmosphere, and making a quantitative assumption about the compressibility of earth relative to that of water, Cavendish constructed a table for the densities of the earth at different distances from its center. From these assumptions, he deduced that the mean density of the earth should be more than eleven times the surface density.[17] This value was much higher than the French.

Cavendish did not comment on it, since the interior of the earth was an unknown quantity. Only this far could Cavendish go with theoretical reasoning and observations made by others with the seconds pendulum. What were needed were new observations from a new experiment.

The second practicable way, Cavendish said, was "by finding the deviation of the plumb line at the bottom of a mountain by taking the meridian altitudes of stars." By comparing the acceleration of a pendulum at the top of a conical hill with the deviation of a plumb-line at its foot, he concluded that although the pendulum method is easiest, the plumb-line one is the "more exact." His main point was that a plumb-line seemed "much less affected by any irregularity in the density of the internal parts of the earth." From experiments on gravity, there was good reason to believe that such irregularity exists, and Cavendish made a drawing for Maskelyne of a possible interior of the earth.[18] The method of plumb-lines was the one that Cavendish and Maskelyne would pursue in the Royal Society's experiment on the attraction of mountains.[19]

In 1772 Maskelyne proposed an experiment on the attraction of mountains, which would make the "universal gravitation of matter palpable."[20] In July of that year the council of the Royal Society appointed a committee to consider the experiment and to draw on the society's treasurer as needed.[21] In a paper written for his fellow committee member Benjamin Franklin, Cavendish described what kinds of mountains were best, the main consideration being that the mountain be big and that the observing stations to the north and south be close together. He told Franklin how to estimate the sum of the deviations of the plumb-line on the two sides of the mountain by sectioning the mountain, fitting geometrical solids to it, and then consulting Cavendish's enclosed table of deviations. The want of attraction of a valley, he told Franklin, was as good as the attraction of a mountain and perhaps better; so his "correspondent" should watch out for valleys too.[22]

Maskelyne wrote to Cavendish in January 1773, returning his "Rules for Computing the Attractions of Hills," having made a copy to keep. Maskelyne said that the rules were "well calculated to procure s the information that is wanted" and that the "dimensions of [an] extraordinary valley [Glen Tilt] deserve a more particular inquiry."[23] The committee began to draw on the treasurer,[24] and in mid-1773 the council called on Charles Mason to go to Scotland to observe its mountains and valleys.[25] At the end of July 1773 the council's instructions were made more specific, and Mason thereafter set off on horseback into the Scottish

Highlands.[26] In early 1774, a year and a half after the committee had been formed, a final decision was made on the basis of Mason's survey. The choice was a 3547 foot-high granite mountain in Pertshire, "Maiden's Pap,"[27] also known as "Schiehallien," meaning "constant storm." Schiehallien was made to Cavendish's order, a regular, detached mountain, with a narrow base in the north-south direction.[28] Losing no time, the committee selected Mason to do the experiment on the attraction of this mountain, but Mason turned down the invitation, and with it unforeseen glory. In his place, the committee hired Maskelyne's new assistant observer at the Royal Observatory, Reuben Burrow.[29] It was by then dead winter, there was no hurry; the committee had time for second thoughts. The Greenwich assistant did not seem equal to this important assignment; the committee told the council: "it would add to the lustre and authenticity of the observations to be made in Scotland, if Mr Maskelyne could be prevailed on to undertake the direction of them upon the spot." Maskelyne was prevailed upon and duly received permission from the King to absent himself from the Royal Observatory.[30] The experiment required a large number of instruments, one of which was a dipping needle in need of repair, which was refurbished under the supervision of Cavendish, who had by now succeeded his father, Lord Charles, as the society's expert on instruments in general.[31]

Loaded down with instruments in working condition, Burrow preceded Maskelyne to Schiehallien, where with William Menzies he determined the size and shape of the mountain. Maskelyne arrived at the end of June to make astronomical measurements on forty-three stars. Because of the storms, the experiment dragged on. Maskelyne returned to Greenwich only at the end of October, and Burrow and Menzies stayed on to do more surveying. When Burrow returned, he was paid off and told to give over the original papers of his survey. Cavendish and C. J. Phipps were charged by the council to compare Burrow's scarcely legible Schiehallien papers with his own fair copy, and in April 1775 Cavendish and Phipps declared the copy faithful and Burrow an excellent surveyor. Maskelyne was empowered to hire persons for the calculations.[32] This was the end of Cavendish's formal involvement in the experiment on the attraction of mountains, but it was not the end of his interest in the quantity it addressed, the mean density of the earth.

The attraction of Schiehallien was palpable, if barely. The experiment had been genuine, its success not guaranteed, as is clear from Cavendish's attempts to estimate in advance its likelihood of succeeding.[33]

True to Newton, to his own promise, and to the outcome of the experiment, Maskelyne told the Royal Society in July 1775 that "we are to conclude, that every mountain, and indeed every particle of the earth, is endued with the same property (attraction), in proportion to its quantity of matter," and further that the "law of the variation of this force, in the inverse ratio of the squares of the distances, as laid down by Sir Isaac Newton, is also confirmed." For this experiment, Maskelyne was awarded the Copley medal in 1775. In his address on the occasion, the president of the society, John Pringle, said that now the Newtonian system was "finished" and that every man must become a Newtonian.[34]

Maskelyne and the president's conclusions could have come as no surprise to Cavendish. What interested him, however, was the mean density of the earth, which had to wait for the calculations of the mathematician Charles Hutton, whom Maskelyne had employed for the purpose. It was not until early in 1778 that Hutton finished his paper. The hundred pages of "long and tedious" calculations had demanded his "close and unwearied applications for a considerable time." They came down to this number: the ratio of the mean density of the earth to the density of the mountain was 9 to 5. Hutton pointed out that the density of the mountain was not known and that only an empirical study of its internal structure could reveal it.

Hutton's calculation was not the same or as satisfying as another number, the mean density of the earth expressed in terms of the standard, the density of water. That number Hutton estimated by assuming that the mountain is "common stone," the density of which is $2\frac{1}{2}$; the density of the earth is therefore $4\frac{1}{2}$ times the density of water. Newton's best guess that the mean density of the earth is between 5 and 6 was close ("so much justness," Hutton said, "even in the surmises of this wonderful man!"). Reminding his readers that this experiment was the first of its kind, Hutton hoped that it would be repeated in other places. New methods of calculation had had to be invented, he said in explanation of why it had taken him so long. The delay was despite labor-saving methods of calculation, which he said he owed to Cavendish.[35]

Legend has it that Maskelyne threw a bacchanalian feast for the inhabitants around Schiehallien complete with a keg of Scotch whisky, and a ballad exists testifying to it.[36] It is hard to imagine Maskelyne himself or indeed Cavendish taking part in this licentious affair, but then Cavendish was not on the mountain. Cavendish had done the comprehensive planning for the experiment, but he did not go into the field to look at the mountain, nor did he make the astronomical observations on it, nor did he make the final calculations of the earth's density; others did these things. In the imagery of the experiment, Cavendish was the valley, not the mountain. As he demonstrated, the valley offered the same effect and likely greater accuracy than the mountain, but the experiment was done on a mountain, a feature of the landscape that draws the eye more than the valley does. Cavendish's work on the experiment went unseen except by others who worked on it too. Work through committees could be a haven for a man of secluded habits.

II. CAVENDISH WEIGHS THE WORLD

Some twenty-five years separated Cavendish's work with the Royal Society on the determination of the earth's density and his own, private determination of it. In the meantime Cavendish continued to study the earth. The Royal Society's experiment on the attraction of mountains coincided with what might be called scientific *mountain fever*. Throughout Europe in the 1770s mountains were being scaled not for their challenge or their sublimity but for the rarity of their air. Scientifically minded men carried their barometers up mountains in the hope of perfecting a practical method for determining heights (not a new idea but a new hope). This went on even while Maskelyne was on Schiehallien making his experiment. Among the visitors he received there was his good friend William Roy, Surveyor-General of the Coasts and Engineer of Military Surveys for Great Britain. Roy brought his own barometers in order to measure the height of the mountain, which he then compared with the geometrically determined height. In Roy's experiment, Cavendish had a part too, having assisted Roy in experiments on the expansion of mercury for just this sort of measurement.[37] In Cavendish's first publication, in 1766, in pneumatic chemistry, he referred to a rule on the density of air, given by the French astronomers "who measured the length of a degree in Peru," for "finding the height of mountains barometrically." Later Cavendish made his own experiments on the height of mountains. They were, in fact, a main objective in his singular journeys outside of London in the 1780s. These journeys carried Cavendish far from his London haunts and were as close as he came to undertaking scientific adventures of Bouguer's and Maskelyne's kind. (When these journeys are looked at closely, however, they do not seem like a great departure from Cavendish's ordinary forms of seclusion: he traveled with Blagden in a closed coach, on a predetermined course, stopping only at geologically and industrially interesting sites, and meeting there with a few persons who had been contacted by Blagden in advance.)

For the purposes of this article, the most interesting of Cavendish's journeys was his second, an eight-hundred mile tour through the eastern coal counties in 1786. He and Blagden stopped for several days in Thornhill, near Wakefield in Yorkshire, to visit the rector of St. Michael's Church there, John Michell. Cavendish had known Michell for a long time, and he had followed Michell's work ever since Michell had been Woodwardian Professor of Geology at Cambridge. Michell and Cavendish were both elected to the Royal Society in 1760. In that year, before their election, Michell's great paper on the cause of earthquakes was read in five consecutive meetings of the society. Cavendish was present at all of these meetings.[38] Michell's subject, the structure and strata of the earth's interior, would link his and Cavendish's interests thereafter.

On his visit to Thornhill in 1786, Cavendish obtained from Michell a remarkable table of strata going down 221 yards into nearby coal pits,[39] and he and Blagden looked over Michell's collection of fossils found in these strata. They all also took a geological side trip over the limestone country, where with Cavendish's barometer they hoped to measure heights; they managed to take some lower elevations, but foul weather prevented them from "ascending any mountains."[40] There was another interest: Michell had been working for years on a great reflecting telescope, $2\frac{1}{2}$ feet in aperture. It caused a stir in London, and in Blagden's opinion (but I doubt Cavendish's) it was the reason he and Cavendish were visiting Michell at all. They looked through the telescope but with disappointing results since Michell had cracked the speculum.[41]

There is no mention of any apparatus for determining the density of the earth in the letters and the journal from Cavendish and Blagden's visit with Michell in 1786. But Michell's intended experiment on the density of the earth would certainly have been discussed on this visit. Cavendish's interest and encouragement are on record in a letter he wrote to Michell three years before, in 1783. He knew that Michell was already in trouble with his telescope because of its enormous scale. He wrote: "if your health does not allow you to go on with that [the telescope] I hope it may at least permit the easier and less laborious employment of weighing the world."[42] This letter of 1783 contains the earliest mention of Michell's and ultimately Cavendish's "weighing the world."

"Experiments to Determine the Density of the Earth," Cavendish's paper in the *Philosophical Transactions* for 1798, opens with a historical paragraph that establishes his connection with Michell. There was another connection between the two of them through Francis John Hyde Wollaston:

Many years ago, the late Rev. John Michell, of this Society, contrived a method of determining the density of the earth, by rendering sensible the attraction of small quantities of matter; but, as he was engaged in other pursuits, he did not complete the apparatus till a short time before his death, and he did not live to make any experiments with it. After his death, the apparatus came to the Rev. Francis John Hyde Wollaston, Jacksonian Professor at Cambridge, who, not having conveniences for making experiments with it, in the manner he could wish, was so good as to give it to me.[43]

Michell died in 1793, and he had not finished building his apparatus until shortly before then. How the apparatus came into Wollaston's hands Cavendish does not say, nor does he say who initiated the gift of the apparatus from Wollaston to Cavendish, though from all that passed before it was almost surely Cavendish. In any case, Michell, Cavendish, and Wollaston were all on familiar terms. Wollaston belonged to a dynasty of men of science and the Church, all of whom, like all of the principals in this scientific episode – Cavendish, Maskelyne, and Michell – were Cambridge men. The educational, scientific and personal connections between the Wollastons, Michell, and Cavendish are as many as they are hard to keep in mind, given the large number of Wollastons and the family parsimony in assigned first and middle names.[44] It is – this is the point – entirely reasonable that Michell's apparatus should end up in Cambridge with one of the Wollastons, and that Cavendish knew its whereabouts, coveted it, and was given it to use.

Cavendish was nearly sixty-seven when he weighed the world. His most recent publication of experiments had been on chemistry ten years before, and it would have been his last if it had not been for Michell's work, which Cavendish finished for him. Cavendish's experiment was, in reality, several "experiments," seventeen in number, each consisting of many trials. The first experiment was done on 5 August 1797, and the first eight were done a few days apart through the rest of August and up to the last week in September. The remaining nine experiments were done the following year, from the end of April to the end of May. The paper reporting the experiments was read to the Royal Society on 21

June 1798, just three weeks after the last experiment. This lengthy paper must have been largely written before the completion of the experiment.

Cavendish began the report of his work with a promising beginning: "The apparatus is very simple." The apparatus, which Cavendish largely remade, is in truth easily described. Its moving part was a six-foot wooden rod suspended horizontally by a slender wire attached to its center, and suspended from each end of the rod was a lead ball two inches across; the whole was enclosed in a narrow wooden case to protect it from wind. Toward the ends of the case and on opposite sides of it were two massive lead balls, or "weights," each weighing about 350 pounds. The weights could be swung to either side of the case to approach the lead balls inside, and in the course of the experiment this was regularly done. The gravitational attraction between the weights and the balls was able to draw the rod sensibly aside. From the angle of twist of the rod, the density of the earth could be deduced; but for this to be done, the force needed to turn the rod against the force of the twisted wire had to be known, and for this it was necessary to set the rod moving freely as a horizontal pendulum and to observe the time of its vibrations.

To the modern reader the way Cavendish got from the mutual attraction of the lead "weights" and balls to the density of the earth seems roundabout, which is to be expected. Cavendish did not write equations, and he did not distinguish between weight and mass, so no gravitational constant appears. He introduced an artifice, a simple pendulum, the length of which was one-half the length of the wooden rod constituting the horizontal beam of his apparatus. The simple pendulum, which was not part of the experiment but only of the analysis, oscillates in a vertical plane under the action of the earth's gravity. It does not look at all like Cavendish's horizontal beam oscillating freely as a horizontal pendulum, but the two pendulums are described mathematically the same way; they are both "pendulums" performing simple harmonic motion. By combining and manipulating the formulas that relate the forces on the two pendulums, certain proportionalities result, which include the wanted expression for the density of the earth in terms of the measures of the apparatus and two things observed in the experiment, the period of the torsion balance and the displacement of the beam by the attraction of the weights. The reason why the earth enters this expression is that the "weights" have weight owing to the attraction of the earth, which is proportional to the matter of the earth. Using modern terminology and notation, this derivation can be done with a few lines of equations, but they would not correspond to Cavendish's reasoning.[45]

In the earlier experiment on the attraction of mountains, it was an open question if a mass the magnitude of a mountain was sufficient to cause a detectable effect. In Cavendish's experiment, the detectable effect was readily achieved by weights small enough to fit into an apparatus. The lead balls were what he "weighed" with his apparatus, thereby weighing, indirectly, the world. This was not an obvious weighing like the chemist's weighing with his balance (Cavendish, as chemist, was renowned for his weighings of this sort[46]). Rather, it measured the attraction of lead spheres, which led by a chain of theoretical arguments to the weight, or density, of the world.[47]

Cavendish's experiment was a precision measurement of a seemingly inaccessible magnitude. Newton had made the calculation of the attraction of two one-foot spheres of earth matter placed one quarter inch apart to show that the force was too feeble to produce a sensible motion; he thought it would take a month for the spheres to cross the quarter inch separating them.[48] The force between the spheres in Cavendish's experiment was only 1/50,000,000 part of their weight, so that the minutest disturbance could destroy the accuracy of it. To guard against any disturbance, Cavendish placed the apparatus in a small, closed "room," about ten feet high and as many feet across. From outside the room, he observed the deflection and vibration of the rod by means of telescopes installed at each end. Verniers at the end of the rod enabled him to read its position to within 1/100th of an inch. The only light admitted into the room was provided by a lamp near each telescope, which was focussed by a convex lens onto the vernier. The rod and weights were manipulated from outside the room. In doing the experiment, Cavendish brought the massive weights close to the case, setting the rod in motion. Then peering through the telescope into the semi-dark room, he took readings from the illuminated vernier at the turning points of the motion, and he timed the passing of the rod past two close-lying, predetermined divisions. The experiment was a trial of the observer's patience: depending on the stiffness of the suspension wire, a single "vibration" could take up to fifteen minutes, and a single experiment might take two and one half hours.

Much of the time Cavendish spent on the experiment was taken up with errors and corrections. He traced a minute irregular motion of the rod to a difference of temperature between the case and the weights, which gave raise to air currents. One entirely negligible correction he published as an appendix to his paper. This was the attraction on the rod and balls by the mahogany case that enclosed them, the counterpart of the attraction of ideal mountains in Cavendish's previous calculations: it amounted to an exhaustive summing of the attractions of the box on the movable part of the apparatus, only instead of the cones and other figures he had used to represent mountains, here he used rectangular planes to represent the regular boards of the wooden case. It is fitting that Cavendish's paper should read like a dissertation on errors. Errors were, after all, the point at which he had entered the subject: the first evidence of his interest in the density of the earth was his criticism of astronomical observations that ignored the attraction of mountains.

"To great exactness," Cavendish concluded, the mean density of the earth is 5.4 times the density of water.[49] That number was the object of Cavendish's last experiment, the work of ten months near the end of his life and the reward for twenty-five years of tenacity.

In addition to the precision of the technique and the knowledge of the earth's interior that it offered, there was another reason, I believe, why Cavendish did this last major experiment. He had long since completed the principal researches of his middle years; his fundamental researches in electricity, chemistry, and heat, for which he is famous. By the end of the eighteenth century, in all of these fields scientific opinion had moved away from his. But his experiment on gravity was not subject to the vagaries of scientific opinion in the same way. This is not to say that he did not expect criticism. In any case, he got it.

The challenge came in connection with continuing claims for the earlier preferred method of determining the density of the earth. Cavendish's paper brought a prompt response from Charles Hutton, who had done the calculations on Schiehallien. The paper in manuscript had been given to him by Maskelyne, and it had not given him pleasure. Just a year or so before Cavendish's paper, Hutton had called attention to his calculation of the density of the earth from the Royal Society's experiment. In the article "Earth" in his *Mathematical and Philosophical Dictionary*, Hutton wrote of the density of the earth: "This I have calculated and deduced from the observations made by Dr. Maskelyne, Astronomer Royal, at the mountain Schiehallien, in the years 1774, 5, and 6." In this work he took pride. Then came Cavendish's paper. On the same day that Hutton received a second copy of Cavendish's paper from the Royal Society, he wrote to Cavendish from the Royal

Military Academy in Woolwich where he worked. He went straight to the point: Cavendish's "ingenious" paper, which made the density of the earth 5.48 that of water, concluded with a paragraph that called attention to the earlier, much lower value of $4\frac{1}{2}$, in the "calculation of which" he, Hutton, had borne "so great a share." Anyone who has looked at Hutton's heroic paper can sympathize with the plaintive note. Hutton thought that Cavendish's wording hinted at inaccuracies in his calculations and seemed to disparage the Royal Society's experiment. That experiment, Hutton reminded Cavendish, had determined not the density of the earth but only the ratio of that density to the density of the mountain, 9 to 5. Hutton had supposed that the density of the mountain is the density of ordinary stone, $2\frac{1}{2}$ times that of water, but the actual density of the mountain was unknown, as Hutton had remarked at the time. All that was known was that Schiehallien was a "mass of stone." Hutton now believed that the mountain's density was higher, 3 or even $3\frac{1}{2}$, which would then make the density of the earth "between 5 and 6" – or exactly where Cavendish (and, inexactly, Newton) had put it – and "probably nearer the latter number." The Royal Society had not finished its experiment because it had not determined the density of the stone, Hutton said. Even now, he hoped the society would finish it, so that "an accurate conclusion, as to the density of the earth, may be thence obtained."[50]

Cavendish believed that he had just drawn that accurate conclusion and that it was 5.48. Hutton wanted the density of the earth to depend on what could never be made precise, the density of "stone." At the bottom of Hutton's letter to him, Cavendish drafted a brief response. Without referring to Hutton's guesswork or excuses, it read: "According to the experiments made by Dr Maskelyne on the attraction of the hill Schiehallien the density of the earth is $4\frac{1}{2}$ times that of water." As to which density, his or the society's, was better, Cavendish did not commit himself, since the society's determination was "affected by irregularities whose quantity I cannot measure."[51]

It would have been known to Cavendish that Hutton had not let go of the problem of determining the earth's density by the attraction of mountains. In 1780, two years after his calculation of the density of the earth, Hutton had published another paper following up "the great success of the experiment" on Schiehallien to "determine the universal attraction of matter," in which he repeated his wish that more experiments of the same kind would be made.[52] Hutton was to have his wish but not his way. In 1811 he got John Playfair to do an investigation of the structure of the rocks of Schiehallien. Playfair found the density of the rocks to be between 2.7 and 2.8. Originally, Hutton had guessed 2.5, so Playfair's result raised his calculated density of the earth, but only slightly, to 4.7. Cavendish's density, 5.48, is much closer to, within one percent of, the accepted value today, 5.52. Readers who know the history of the Royal Society in the eighteenth century will recall that the Charles Hutton of the attraction of mountains is the Charles Hutton who had last his job as foreign secretary at the Royal Society in the early 1780s, precipitating a bitter feud known as the society's "dissentions." Maskelyne who had brought Hutton into the experiment on the attraction of mountains, had earlier been a vigorous supporter of Hutton's losing side in the dissentions. By contrast, Cavendish had given decisive support to Hutton's nemesis, the society's president, Joseph Banks.[53] If this unhappy experience of Hutton's at the Royal Society and the suspected opposition of Cavendish had anything to do with Hutton's continuing efforts to keep alive his method of determining the attraction of mountains as an alternative to Michell and Cavendish's method, it is impossible to say. Hutton had a vested interest in the earlier method, after all; but for completeness, the personal circumstances are here acknowledged. Hutton lived to 1823, long enough to know of the high regard in which Cavendish's experiment came to be held, though not long enough for him to know that it was *the* Cavendish experiment.

Cavendish was the first to be asked to repeat his own experiment on the density of the earth. From Paris, Blagden wrote to Banks in 1802, telling him of a conversation what Laplace about Cavendish's experiment and suggesting that Banks pass along what Laplace had said. What Laplace said was that the attraction Cavendish measured might involve electricity as well as gravity, and Laplace expressed the wish that "Mr. Cav. would repeat it [the experiment] with another body of greater specific gravity than lead."[54] If Cavendish got the message he never repeated the experiment, but there was no need to; others would do it, and many times, ever with the desire to achieve even greater accuracy and perfection than Cavendish had. Experiments on the attraction of mountains ceased to be regarded as a precise way to determine the earth's density, although the attraction of mountains remained a consideration as a source of error in astronomical measurements of location and distance.[55]

The Cavendish experiment survived as an active research tool even after scientists had left behind them the problem of the density of the earth. That it did has to do not only with its precision but as well with its subject, a fundamental and still enigmatic force of nature, gravity, with its characteristic universal constant. It became the experiment to determine "big G," as C. V. Boys explained in 1892: "Owing to the universal character of the constant G, it seems to me to be descending from the sublime to the ridiculous to describe the object of this [Cavendish's and now Boys's] experiment as finding the mass of the earth or the mean density of the earth, or less accurately the weight of the earth."[56]

Still today, three hundred years after Newton and two hundred after Cavendish, gravity is at the center of physical research. To quote from a recent publication by contemporary researchers in gravity: "The most important advance in experiments on gravitation and other delicate measurements was the introduction of the torsion balance by Michell and its use by Cavendish ... It has been the basis of all the most significant experiments on gravitation ever since."[57] That is why Cavendish's experiment became *the* Cavendish experiment.

III. CAVENDISH IN SECLUSION

Cavendish initiated no more ambitious programs of research, and his only publication after the one on the density of the earth came some ten years later, a short paper on a typical concern, a way to improve the accuracy of astronomical instruments.[58] Except for going regularly to meetings of the Royal Society and to other meetings of scientific men, he stayed home, which is where he had done his experiment on the density of the earth.

Cavendish's main house was his country house at Clapham Common. Today Clapham Common is swallowed up by London, but in the eighteenth century it was a commuting suburb. Cavendish's house was a substantial brick villa overlooking the common, yet it was modest by comparison with some of his neighbors' houses. What set his apart was its use: by all accounts Cavendish converted his house into an eighteenth-century version of a scientific institute. The drawing room was a laboratory, the adjoining room a workroom with a forge, the upstairs an observatory, and stuck all about the house were thermometers and other gauges. Long after Cavendish's death, Clapham Common neighbors would point to the house and tell their children that that was where the world was weighed. Although Cavendish was not the first owner of that house, after his death it was known as *the Cavendish house*.[59]

Times have changed. John Henry Poynting, for his repetition of the Cavendish experiment a hundred years later, received a grant from the Royal Society, and he was given a workplace in an institute, in the laboratory at Cambridge named after Henry Cavendish. Clerk Maxwell, the first director of the Cavendish Laboratory, gave Poynting permission to do the experiment.[60] Poynting's repetition of the Cavendish experiment belongs to physics when it had become an established discipline with its principal home in places of higher learning, complete with institutes, directors and grants. Cavendish did his experiment at home.

In connection with the determination of the earth's density, Cavendish brought into his home one person from the outside, George Gilpin, not a Fellow of the Royal Society but its clerk, whom Cavendish asked to make the last two experiments. Replacing Cavendish at the telescope, Gilpin gave the world another actor and a witness and another set of observations by which to judge the experiment and the experimenter. He was no doubt cast by Cavendish as a detector of error as well as a confirmer of observations.

Mountains high on the earth and open to the sky could deflect weights too; the earth could be weighed that way, and Cavendish had worked with the astronomers who weighed it that way. But his own experiment was better suited to his temperament. With his experiment he did not need to go out into the world to know it; he could know it and know it more precisely by staying home, manipulating his apparatus and reasoning from universal principles. The world came to Cavendish. (Another way of viewing it is that Henry Cavendish was a *Cavendish*, and the Cavendishes liked to stay home and let the world come to them, but this is another discussion.) Cavendish stayed at home, inside of a building, looking inside of a room and through a slit into a case inside of which was the world – his world, on his terms.

The observer of this world is recognized by his traits: extreme caution, endless fussing over errors, tolerance only of tolerable certainty, thoroughness to the point of exhaustion, then finally lassitude. Like the cut of his coat and the style of his wig, Cavendish's experiment bore the unmistakable stamp of the man. That, I believe, is true, but it is only a half-truth. It has been noted that while there is much talk about the effect of the scientist's personality on science, there is little of the other, perhaps more profound, effect of science on the personality.[61] In Cavendish we see both effects, mutually reinforcing.

From the beginning Cavendish turned away from what he found difficult, ordinary society, and toward nature and its understanding through science, and through science he came into a society he found, if not comfortable, to his liking. Precisely those traits that in his casual contact with people gave rise to anecdotes about his eccentricities were the traits that in his scientific work made him extraordinary. To do science, Cavendish did not have to overcome his extreme diffidence, he had only to adapt it to science. That he did, and in so doing he adapted science to his personality. This most impersonal of investigators left a personal impression on science. It worked both ways. The experiment on the density of the earth is arguably not Cavendish's most important experiment, but if it is looked at for what it reveals about the experimenter, like a diary (which he did not keep) or a portrait (which he did not allow), it is the most expressive of his experiments.

No preliminary manuscripts connected with the experiments on the earth's density have survived or, anyway, surfaced.[62] That cannot be said of any other important experiment by Cavendish. The quirky history of his papers after his death enabled Cavendish this time to exclude not only his contemporaries but his biographers from his behind-the-scenes labors. With his paper of 1798, he appeared before the world finished, complete.

The man who weighed the world was a secluded figure and yet a constant companion of men of science, posing and symbolizing the historian's perennial problem of the relationship of the individual person or event to collective actions. Through the experiment on the density of the earth, Cavendish worked out his own destiny, and at the same time he was the able representative of a long development and also of a beginning in science. His experiment exemplified the drive for precision measurement, which began in Cavendish's time and which has gathered force ever since.[63] He carried out the experiment secluded in Clapham Common, but his experiment belonged to a nexus of established scientific problems, instrumental possibilities, and interested, qualified parties.[64]

The Cavendish experiment provided more than precise information about the earth, it became an ideal of scientific practice. Cavendish was not a "geophysicist" or a "physicist," he was a universal natural philosopher. But at the time of his last experiment, the discipline of physics was emerging. In Germany, for example, the early physics journal was the *Annalen der Physik und Chemie*. When after eight years of operation its founder, F. A. C. Gren, died, in 1798, the year of Cavendish's experiment, its new editor, L. W. Gilbert, wrote a foreword to the new beginning under him, and under the new, restricted title, *Annalen der Physik*. Explaining that the richest vein of material for his journal would continue to be mined from foreign sources, Gilbert hoped that in his journal work by the best physicists in Germany would stand side-by-side with the best work from abroad, such as Henry Cavendish's experiment on the density of the earth with its wonderful "exactness."[65] The ideal of the time could not be more exactly put. Cavendish's experiment, in this sense, belongs to the history of physics in the nineteenth and twentieth centuries, to which Martin Klein's writings on Gibbs, Ehrenfest, and others have given such impetus.

1204 Oak Drive
Eugene, OR 97404
U.S.A.

NOTES

* For his very helpful comments on this paper, I wish to thank Robert Deltete.
[1] Martin J. Klein, "Lives in Science: Gibbs and Duhem," presented to the History of Science Society in Philadelphia, October 29, 1982. Duhem did not originate the phrase "retiring disposition" but appropriated it, with acknowledgement, from gifted student and later colleague of Gibbs, Henry Andrews Bumstead. Bumstead wrote a biographical memoir for the *American Journal of Science* in 1903 and included it in an expanded version in his and Ralph Gibbs Van Name's edition of *The Scientific Papers of J. Willard Gibbs*, 2 vols. (New York, 1906); the phrase quoted is in vol. 1, p. xxiii.
[2] *Gentleman's Magazine*, March 1810, p. 292.
[3] Lord George Cavendish to Charles Blagden, 9 and 10 Mar. 1810, Blagden Letters, C 17 and 19, Royal Society Library.
[4] "Shut up apart" is an eighteenth-century meaning of "seclude." *Oxford Universal Dictionary*, 3rd rev. ed., 1935, p. 1825.
[5] Entries for 19 June 1766 and 12 Nov. 1767, Minutes of Council, vol. 5, 1763–1768, pp. 157, 184. Cavendish's letter to the president is dated 9 June 1766. At a council meeting on 12 Nov. 1767, Cavendish was appointed to a committee to "consider the places proper to observe the ensuing transit of Venus, – and the methods, the persons fit, – and other particulars." *Ibid.*, p. 184.
[6] Assuming a rotating fluid of uniform density, Newton calculated the flattening of the earth to be 1/230; Huygens calculated it to be 1/577; the truth lay in between. The later

understanding was that the flattening also depends on the distribution of the internal density of the earth. K. E. Bullen, *The Earth's Density* (London, 1975), pp. 8–9, 43, 62–64, 87.

[7] In his *System of the World*, Newton wrote to discourage this objection to his work but only succeeded in challenging his successors: if all bodies attract, why do we not see them do it on earth? Newton's answer was that "terrestrial bodies do not count." He calculated that "a sphere of one foot in diameter, and of a like nature to the earth, would attract a small body placed near its surface with a force 20000000 times less than the earth would do if placed near its surface; but so small a force could produce no sensible effect. If two such spheres were distant but by $\frac{1}{4}$ of an inch, they would not, even in space void of resistance, come together by the force of their mutual attraction in less than a month's time . . . Nay, whole mountains will not be sufficient to produce any sensible effect. A mountain of an hemispherical figure, three miles high, and six broad, will not, by its attraction, draw the pendulum two minutes out of the true perpendicular . . ." *Sir Isaac Newton's Mathematical Principles of Natural Philosophy and His System of the World*, trans. A. Motte, rev. F. Cajori, 2 vols. (Berkeley and Los Angeles, 1962) 2: 569–570. Derek Howse, *Nevil Maskelyne: The Seaman's Astronomer* (Cambridge, 1989), p. 129.

[8] "Precession of Equinoxes," Henry Cavendish MSS, VIII 9, pp. 14–15, Devonshire Collections, Chatsworth. I thank the Duke of Devonshire and the Trustees of the Chatsworth Settlement for their permission to quote from the papers of Henry Cavendish.

The part of the manuscript on precession I quote is written out in two versions. The one I select is different from the one printed in *The Scientific Papers of the Honourable Henry Cavendish, F.R.S.*, vol. 2: *Chemical and Dynamical*, Edward Thorpe, ed. (Cambridge, 1921), p. 436.

[9] Howse, *Maskelyne*, pp. 129–130. "Maskelyne, Nevil," *Dictionary of National Biography* 12: 1299–1301.

[10] Nevil Maskelyne, "Observations on a Clock of Mr. John Shelton, Made at St. Helena: In a Letter to the Right Honourable Lord Charles Cavendish, Vice-President of the Royal Society," *Philosophical Transactions of the Royal Society of London* 52 (1762): 434–443, on 442.

[11] Charles Hutton, "An Account of the Calculations Made from the Survey and Measures Taken at Schehallien, in Order to Ascertain the Mean Density of the Earth," *Phil. Trans.* 68 (1778): 689–788, on 784. B. E. Clotfelter, "The Cavendish Experiment as Cavendish Knew It," *American Journal of Physics* 53 (1987): 210–213, on 211.

[12] Nevil Maskelyne, "Introduction to the Following Observations, Made by Messieurs Charles Mason and Jeremiah Dixon, for Determining the Length of a Degree of Latitude, in the Provinces of Maryland and Pennsylvania in North America," *Phil. Trans.* 58 (1768): 270–273, on 273. Maskelyne said that Boscovich was the first to take notice of the effect of the attraction of mountains in his account of the measurement of the length of a degree of latitude in Italy. Maskelyne also said that Mason and Dixon's measurement could not be affected because the degree passes through level country.

[13] Nevil Maskelyne, "Postscript by the Astronomer Royal," *Phil. Trans.* 58 (1768): 325–328, on 328. Maskelyne's postscript follows the paper by Mason and Dixon on the length of a degree of latitude, which his paper, note 12, introduces; in it he took back what he had said about mountains and about Mason and Dixon. The reason was that "Cavendish has since considered this matter more minutely, . . . having mathe-

matically investigated several rules for finding the attraction of the inequalities of the Earth." Cavendish's suppositions had to do with the attraction of the Allegany and the Atlantic masses and deficits. Maskelyne mentioned that Cavendish had also found that the degrees measured in Italy and at the Cape of Good Hope were probably affected by attraction of mountains and the deficiency of attraction of seas. "Observations of the Length of a Degree of Latitude," Henry Cavendish MSS, VIII 16, Devonshire Collections, Chatsworth.

[14] "Paper Given to Maskelyne Relating to Attraction & Form of Earth," Henry Cavendish MSS, VI(b) 1, p. 18, Devonshire Collections, Chatsworth.

[15] Letter Maskelyne to Cavendish, 10 Apr. 1771, Henry Cavendish MSS, VIII 4, Devonshire Collections, Chatsworth.

[16] "Paper Given to Maskelyne," pp. 12–14, 19.

[17] "Paper Given to Maskelyne," pp. 15–16.

[18] Cavendish's reason for thinking there is irregularity is that the observations of pendulums in different places "differ more than I should think could be owing to the error of experiment." If that is so, the crust is thinner in some places than in others, and so the gravity at the top of a mountain cannot be calculated by the inverse square law reckoned from the center of the earth; "no certain conclusion could be drawn from experiments on the pendulum at the top & bottom of a mountain in such place." By contrast, with the plumb-line method it is clear that the measurements would be unaffected by an irregularity of this sort. "Paper Given to Maskelyne," pp. 19–20.

[19] Cavendish also calculated the attraction on a plumb bob by ocean tides treated as infinite slabs, and he compared his result with Boscovich's. Cavendish further calculated the errors in plumb-line measurements due to the effects of irregular refraction in viewing near-zenith stars as caused by variations in temperature on mountains. "Paper Given to Maskelyne," pp. 9–10; the three pages on refraction are unnumbered.

[20] Nevil Maskelyne, "A Proposal for Measuring the Attraction of Some Hills in this Kingdom by Astronomical Observations," Phil. Trans. 65 (1775): 495–499, on 496.

[21] Money was left over from the king's grant for observing the transit of Venus, and he approved the new use for it. Entry for 23 July 1772, Minutes of Council, vol. 6, p. 145. The members of the committee were Cavendish, Maskelyne, Samuel Horsley, Benjamin Franklin, and Daines Barrington.

[22] "On the Choice of Hills Proper for Observing Attraction Given to Dr Franklin," Cavendish MSS VI(b) 3, p. 5. Cavendish counseled Franklin on the kind of zenith sector to be used and where and how observations were to be made once a hill was selected.

[23] Letter, Maskelyne to Cavendish, 5 Jan. 1773, Henry Cavendish MSS, X(b); published in full in Cavendish, Scientific Papers 2: 402. Cavendish included his calculations on Glen Tilt in his "Rules for Computing the Attraction of Hills" (which has a covering sheet not in Cavendish's hand, "Mr Cavendish's Rules for Computing the Attraction on Mountains on Plumb Lines") and in the preliminary version of that paper, "Thoughts on the Method of Finding the Density of the Earth by Observing the Attraction of Hills," Henry Cavendish MSS, VI(b) 2 and 6, Devonshire Collections, Chatsworth.

[24] The committee recommended payment to John Greenwood, who aided the committee member Samuel Horsley in making measurements in Wales. Entry for 21 Jan. 1773, Minutes of Council, vol. 6, p. 163.

[25] Entry for 24 June 1773, Minutes of Council, vol. 6, p. 180.

[26] Entry for 29 July 1773, Minutes of Council, vol. 6, pp. 185–186.

[27] Untitled study of "Maiden's Pap" and another Scottish mountain. Henry Cavendish MSS, Misc., Devonshire Collections, Chatsworth.

[28] Nevil Maskelyne, "An Account of Observations Made on the Mountain Schehallien, for Finding Its Attraction," *Phil. Trans.* **65** (1775): 500–542, on 503.

[29] Entry for 27 Jan. 1774, Minutes of Council, vol. 6, pp. 210–211.

[30] Entry for 5 May 1774, Minutes of Council, vol. 6, p. 234.

[31] The actual repair was to be done by the instrument-maker Jeremiah Sisson. With Sisson, Cavendish was also charged with procuring a variation compass.

[32] Entries for 11 Aug., 11 Oct., 22 Dec. 1774 and 30 Mar., 6 and 27 Apr. 1775, Minutes of Council, vol. 6, pp. 242, 244, 255, 260–261, 267–269.

[33] Calculating from cones and spherical segments, Cavendish had prepared a table of deviations of the plumb-line in seconds of arc for the use of persons looking for a suitable mountain. If the observations on a steep slope could be made with the same accuracy as on level ground, Cavendish reasoned that the observer should be able to determine the difference in the zenith distances of the stars on the two sides of the mountain with "tolerable certainty" to 3", and would not be "likely to err" more than $1\frac{1}{2}$". Based on this estimate of accuracy, Cavendish further reasoned that "if the mean density of the Earth is not more than 7 times greater than that of the surface the effect of attraction must pretty certainly be sensible[,] & it is an even chance that it will come out such that we may with tolerable certainty pronounce [it] to be not owing to the error of observation[,] & even if the mean density is 14 times greater than that of the surface the effect of attraction will most likely be sensible." "Thoughts on the Method of Finding the Density of the Earth by Observing the Attraction of Hills," unnumbered sheet. There are a good many assumptions behind this cautious statement about *tolerable certainty*. To Franklin, Cavendish wrote: "It will be needless to send an account of any hill or valley if the sum of its deviations is less than 50" or 60" as I am in hopes some may found nearer home near as good as that." "On the Choice of Hills Proper for Observing Attraction Given to Dr Franklin," unnumbered sheet. Maskelyne's results fell just within Cavendish's estimated limits of tolerable certainty. The apparent difference in the position of the stars at the two sides of the mountain was 54.6", and the difference in latitude of the two stations, as determined by measuring, was 42.94"; so the difference, 11.6", was the true combined effect of the two attractions, or 5.8" was the effect of the attraction of Schiehalllien on the plumb bob of the zenith sector.

[34] John Pringle, *A Discourse on the Attraction of Mountains, Delivered at the Anniversary Meeting of the Royal Society, November 30, 1775* (London, 1775); the remark on the Newtonian system comes at the end of the discourse.

[35] Hutton, "An Account of the Calculations Made from the Survey and Measures Taken at Schehallien, in Order to Ascertain the Mean Density of the Earth," 689–690, 750, 766, 781–783, 785.

[36] Howse, *Maskelyne*, pp. 137–138.

[37] Maskelyne listed his many visitors, including Roy, in "An Account of Observations Made on the Mountain Schehallien for Finding Its Attraction," p. 525. William Roy referred to Cavendish's assistance on p. 673 and to his experiments on Schiehallien while Maskelyne was there on pp. 718–722 and 760, 775 in "Experiments and Observations Made in Britain, in Order to Obtain a Rule for Measuring Heights with the Barometer," *Phil. Trans.* **67** (1778): 653–788. Bouguer too had determined heights with

a barometer on the expedition to Peru, and Roy discussed his observations, pp. 748 ff. The practical connection of the heights of mountains and other problems of the earth is evident from Roy's comment: "the perfecting of the theory of the barometer is not the only advantage that would accrue from a combination of these observations (on mountains); for, while they were carrying on in different climates, or zones of the earth, good opportunities would offer of determining the refractions, as well as the force of gravity and figure of the globe, from the vibrations of the pendulum." Roy called for the "united labours" of philosophers in researches to perfect a solution to the barometric problem. *Ibid.*, pp. 766, 769.

[38] Henry Cavendish, "Three Papers, Containing Experiments on Factitious Air," *Phil. Trans.* **56** (1766): 141 ff.; in *Scientific Papers* **2**, 77–101, on 83. The meetings of the society were on 28 Feb., 6 and 13, 20, and 27 Mar. 1760. Journal Book of the Royal Society, vol. 23, 1757–1760, pp. 782, 795, 800, 802, and 807.

[39] "Strata Which Michell Dug Through for Coal." The table, is on p. 13 of the 14-page untitled account in Cavendish's hand of the 1786 journey. Henry Cavendish MSS, X(a) 3, Devonshire Collections, Chatsworth.

[40] At Thornhill, Blagden recorded their activities in his diary. On 2 Sep.: "At Mr Michell's took some altitudes & looked over his fossils .... At night looked thro' his telescope." On Sunday, 3 Sep.: "Mr Michell's sermon I had heard or read before. Went over the track of yellow limestone country." Blagden Diaries, Osborn Collection, Beinecke Rare Book and Manuscript Library, Yale University. I thank the Beinecke Library for permission to quote. Letter, Blagden to Joseph Banks, 1 Sep. 1786, British Museum Add MSS 33272, p. 5.

[41] The year before Blagden had written to Michell: "I endeavoured to persuade our friend Mr Cavendish to make you a visit at Thornhill in order to see it [the telescope]." Blagden too wanted to see the telescope, but so far he had failed to persuade Cavendish to go and look. Letter, Blagden to Michell, 25 Apr. 1785, Blagden Letterbook, Osborn Collection, Beinecke Rare Book and Manuscript Library, Yale University. Letter, Blagden to Banks, 19 Aug. 1786, British Museum, Add MSS 33272, p. 2.

[42] Cavendish added: "for my own part I do not know whether I had not rather hear that you had given the exper. – of weighing the world – a fair trial than that you had finished the great telescope." Letter, Cavendish to Michell, 27 May 1783, draft, Henry Cavendish MSS, Devonshire Collections, Chatsworth. The relations and the correspondence between Cavendish and Michell I discuss in my article, "John Michell and Henry Cavendish: Weighing the Stars," *British Journal for the History of Science* **4** (1968): 126–155.

[43] Henry Cavendish, "Experiments to Determine the Density of the Earth," *Phil. Trans.* **88** (1798): 469–526; in Cavendish, *Scientific Papers* **2**: 249–286, on 249.

[44] Wollaston's father, Francis, born the same year as Cavendish, took his degree in law but entered the Church instead. He had a passion for astronomy, and he had his own observatory with first-class instruments. With at least that much in common, in 1768 Cavendish brought Francis Wollaston as a guest to a meeting of the Royal Society, the usual way friends and colleagues introduced prospective members. In 1769 Francis Wollaston's certificate of membership was put up in the society's public meeting room, signed by Cavendish along with Maskelyne and several other prominent members; Wollaston was elected that year. Cavendish brought Francis Wollaston as a guest on 8 Dec. 1768; Wollaston's certificate is dated 12 Jan. 1769. Journal Book of

the Royal Society, vol. 26, 1767–1770. "Wollaston, Francis," *Dictionary of Scientific Biography* **21**, 778–779. One of Francis Wollaston's sons, William Hyde Wollaston, was an eminent chemist. Cavendish proposed him, as he had his father, as a member of the Royal Society; he too was elected, in 1793. "Wollaston, William Hyde," *Dictionary of National Biography* **21**: 782–787, on 782. Another of Francis's sons, George Hyde Wollaston, was one of Cavendish's neighbors at Clapham Common, where Cavendish performed his experiment on the density of the earth. "Wollaston of Shenton," *Burke's Geneological and Heraldic History of the Landed Gentry* (London, 1939), p. 2479. George Hyde Wollaston's house as well as Cavendish's are in the map of Clapham Common, "Perambulation of Clapham Common 1800," from C. Smith, *Actual Survey of the Road from London to Brighthelmston*. Yet another of Francis's sons was Francis John Hyde Wollaston, Jacksonian Professor of Chemistry, from whom Cavendish received Michell's apparatus. Michell's association with the Wollastons went back as far as Cavendish's. To give but one indication: as a recently elected Fellow of the Royal Society, Michell's first recommendation for a new member, in 1762, was for Francis's youngest brother, George Wollaston, Fellow and Mathematical Lecturer of Sidney-Sussex College, Cambridge. "Wollaston, Francis," p. 779.

[45] As is to be expected, the modern analysis of Cavendish's experiment is simpler than Cavendish's. But what modern accounts usually say that Cavendish did, he did not do. He did not derive the universal gravitational constant, though it can be readily got from the results of his experiment, which is the point of B. E. Clotfelter, "The Cavendish Experiment as Cavendish Knew It," *American Journal of Physics* **55** (1987): 210–213. Cavendish wanted the density of the earth, and there is nothing in his analysis to require the gravitational constant nor any reason why, at that time, he should have regarded such a formulation as desirable. Although the unit of force is not necessary to derive the gravitational constant, the unit suggests it, and the unit did not yet exist for expressing $F = GM_1M_2/r^2$, the attraction between two masses, M_1 and M_2, separated by a distance r.

[46] Maurice Daumas, *Scientific Instruments of the Seventeenth and Eighteenth Centuries*, trans. M. Holbrook (New York and Washington, 1972), p. 134.

[47] The chemist's balance *was* used to determine the earth's density, but only later, in attempts to improve upon Cavendish's experiment; notably by P. J. G. von Jolly in 1878–80, J. H. Poynting in 1890, and F. Richarz and O. Krigar-Menzel in 1898. Edward Thorpe, "Introduction," Cavendish *Scientific Papers* **2**: 1–74, on 72–73.

[48] For the quotation from Newton, see note 7 above.

[49] "Experiments to Determine the Density of the Earth," p. 284.

[50] Quotation from "Earth," in Charles Hutton, *Mathematical and Philosophical Dictionary*, vol. 2 (London, 1796): 407. Letter, Charles Hutton to Cavendish, 17 Nov. 1798, Henry Cavendish MSS, Devonshire Collections, Chatsworth.

[51] *Ibid.* "Experiments to Determine the Density of the Earth," p. 284.

[52] Charles Hutton, "Calculations to Determine at What Point in the Side of a Hill Its Attraction Will Be the Greatest, etc.," *Phil. Trans.* **70** (1780): 1–14, on 3.

[53] The full part played by Cavendish was hidden because of his form of seclusion. It is recoverable from the correspondence of the time, which provides documentation for my article, "Henry Cavendish on the Proper Method of Correcting Abuses," in *Beyond History of Science: Essays in Honor of Robert E. Schofield*, E. Garbor, ed. (Bethlehem, 1990), pp. 35–51.

[54] Letter, Blagden to Banks, 1 Apr. 1802, British Museum, Add MSS 33272, pp. 172–173. Notable repetitions include R. Reich, *Versuch über die Mittlere Dichtigkeit der Erde mittelst der Drehwage* (Freiburg, 1838); Francis Baily, *Memoires of the (Royal) Astronomical Society of London* **14** (1843): 1–120; C. V. Boys, "On the Newtonian Constant of Gravitation," *Phil. Trans.* **186** (1895): 1–72.

[55] For example, John Henry Pratt's criticism of the observations taken in the Great Indian Survey in the middle of the nineteenth century: his criticism was based on the neglect of the attraction of the Himalayas and his own calculation of their attraction: Mott T. Greene, *Geology in the Nineteenth Century* (Ithaca and London, 1982), pp. 238–243.

[56] Boys is quoted by Clotfelder to make the point about the shift in interest in Cavendish's experiment. "The Cavendish Experiment as Cavendish Knew It," p. 211. Boys first calculated G from the Cavendish experiment, and then from it he calculated the mean density of the earth. Conversely, to obtain G from the density of the earth, Boys said, he could have recalculated the attraction of the earth by viewing it as an ellipsoid of similar shells of equal density, which is the way J. H. Poynting had calculated it in 1892. Boys recommended using a room with a more uniform temperature than Oxford's, a detail that will be appreciated by anyone who has experienced the chill of Oxford's rooms. His accuracy was very great, despite his room; he believed that his G had an accuracy of 1 in 10,000.

[57] A. H. Cook, "Experiments on Gravitation," in *Three Hundred Years of Gravitation*, S. W. Hawking and W. Israel, eds. (Cambridge, 1987), p. 52. Significantly, Cook talks of the Cavendish experiment only in connection with G and not with the density of the earth. Only recently, he says, has the accuracy of G been improved upon over what can be obtained from Cavendish's own experiment, and although in the study of materials we can achieve an accuracy of 1 part in 10^{12}, we still know G only to about 1 part in 10^3. Cook refers the torsion balance to electrostatics as well as to gravitation. In a footnote in his paper of 1798, on p. 250, Cavendish too referred to Coulomb, who had used an apparatus of the same kind for measuring small electric and magnetic attractions. Cavendish said: "Mr. Michell informed me of his intention of making this experiment, and of the method he intended to use, before the publication of any of Mr. Coulomb's experiments." As far as Cavendish knew, the torsion balance was independently invented by Michell and by Coulomb. Coulomb's biographer discusses the question of priority: Stewart Gillmor, *Coulomb and the Evolution of Physics and Engineering in Eighteenth-Century France* (Princeton, 1971), pp. 613–665.

[58] Henry Cavendish, "On an Improvement in the Manner of Dividing Astronomical Instruments," *Phil. Trans.* **99** (1809): 221–231.

[59] Cavendish weighed the world not in his house proper but in an outbuilding in his garden. No picture or plan or description of the house as it was in Cavendish's day has survived, but among Cavendish's scientific manuscripts is a sketch in Cavendish's hand of the drains at his Clapham property. Included in the sketch is a building separate from the house, and in the accompanying commentary on the drains and cess pools, there is mention of a "greenhouse." Although a greenhouse seems unlikely, it is possible that the building Cavendish sketched is the separate building in which he weighed the world. It does not matter; Cavendish weighed it at home. "Plan of Drains at Clapham and Measures Relating to Bason," Henry Cavendish MSS, Misc., Devonshire Collections, Chatsworth. A good case for an outbuilding shown in a map of 1827 is made by P. F. Titchmarsh, "The Michell–Cavendish Experiment," *The School Review*, No. 162 (Mar.

1966), pp. 320–330, on p. 322.

[60] J. H. Poynting, "On a Determination of the Mean Density of the Earth and the Gravitation Constant by Means of the Common Balance," *Phil. Trans.* **182** (1892): 565–656, on 565–566. It is a noteworthy coincidence that Poynting should do this experiment in Cavendish's spirit, to improve upon Cavendish's accuracy, in the Cavendish Laboratory directed by Maxwell; for Maxwell was the editor of Henry Cavendish's electrical papers, and his edition was reprinted as the first volume of Cavendish's *Scientific Papers*.

[61] Philip J. Hilts, *Scientific Temperaments: Three Lives in Contemporary Science* (New York, 1982), p. 11.

[62] In this denial, one manuscript should be mentioned. Cavendish experimentally determined what we would call the moduli of bend and twist for wires and glass tubes. He compared the vibrations of his twisting apparatus with the vibrations of a simple seconds pendulum. He tried wires of different materials, iron, copper, silver, and brass, suspending from them rods of wood, brass, and zinc. His undated experiments on twist show Cavendish's interest in torsion, but they are not necessary for his experiment with Michell's torsion balance. "Exper. on Twisting of Wire Tried by the Time of Vibration," Henry Cavendish MSS, VI(b) 22, Devonshire Collections, Chatsworth.

[63] It was only at the end of the eighteenth century that precision measurement "becomes a really essential factor in scientific progress." Maurice Daumas, "Precision of Measurement and Physical and Chemical Research in the Eighteenth Century," in A. C. Crombie, ed., *Scientific Change* (New York, 1963), pp. 418–430, on p. 429.

[64] L. W. Gilbert, "Vorrede," *Annalen der Physik* **1** (1799): unnumbered page in the three-page foreword. This quotation connects Henry Cavendish with the starting point of Christa Jungnickel and Russell McCormmach, *Intellectual Mastery of Nature*, 2 vols. (Chicago, 1986) 1: 35.

ELIZABETH GARBER

READING MATHEMATICS, CONSTRUCTING PHYSICS: FOURIER AND HIS READERS, 1822–1850

Historians consider that early nineteenth-century French mathematical physics is one of the roots of modern theoretical physics. This is problematic: French mathematical physics is mathematics, not physics. What evolved from this mathematics was the first logically defensible form of the calculus.

Throughout the eighteenth century, the calculus had developed through the mathematical exploration of the equations of mechanics.[1] Physical imagery may appear at the beginning of those same papers, but on searching the papers, the physics content is absent.[2] Nevertheless, seen from the late twentieth century, and given the power of mathematics in the development of contemporary physics, mathematics seems a natural language for physics. We expect physics, and we then read it into the papers, whether its demonstrably there or not.

In the eighteenth century, physics was experiment and used a rich store of imagery expressed in the vernacular.[3] Mathematics was not a part of its explanatory arsenal. Even after the development of quantitative experiments in France, physics was experimental physics. Mathematical physics was a branch of mathematics.

The process of transformation from explanations developed in the vernacular to those developed in the languages of mathematics fused with physical imagery, to create theoretical physics, is usually seen as a linear and inevitable appropriation of a mathematized physics already developed in France by 1820.[4] Modern physics is characterized by theory and experiment integrated into one discipline unified by general physical laws. If French mathematical physics is mathematics rather than physics, however, then the process of its transformation – together with its sister discipline of quantitative experimental physics – into modern physics needs careful reexamination. This reexamination requires us to reread early nineteenth-century mathematical physics as mathematics, removing much of the later interpolation of readers determined to see this work as physics.

31

A.J. Kox and D.M. Siegel (eds.), No Truth Except in the Details, 31–54.
© 1995 *Kluwer Academic Publishers.*

One problem we have to face is that historians tend to treat mathematics as merely a tool that has been used, then replaced on the shelf. Georg Simon Ohm's work on galvanic currents has been classified as physics, not mathematics, because Ohm did not treat series and differential equations as mathematicians do. Yet mathematicians in the 1820s preferred functional equivalents of infinite Fourier series, and most assumed that solutions existed for all differential equations that emerged from physical processes. Thus, Ohm was acting as a mathematician of the 1820s.[5] Mathematics can no longer be regarded as a tool with no history of its own. Rather, it must be discussed as having developed and interacted with the other sciences in complex ways that need deciphering if we are to understand how mathematics became the language of physics in the mid-nineteenth century.

It seems particularly useful to look at Joseph Fourier's work on the conduction of heat, and how it was read in the mid-nineteenth century, in order to trace some of the twists and turns in the development of modern physics. Fourier has been seen as seminal by many historians of physics.[6] The most extensive examinations of the importance of Fourier for physics, however, treat Fourier in terms of twentieth-century measures of the content of Fourier's theory of heat, reading back into it matter that is frankly not present. For example, while seeing Fourier as a theoretical physicist, John Herivel notes that Fourier's most significant physical concept was the notion of flux. Yet Fourier never treated this notion physically, nor did he name it.[7] What we need to do is examine Fourier's work in light of the standards of early nineteenth-century mathematics, then see how Fourier was read by his contemporaries, both in France and beyond, in the crucial middle decades of the nineteenth century.[8]

"Physique-mathématique" (mathematical physics) was mathematics, not physics.[9] French mathematics began in experimental results and with physical imagery, but the end-product was mathematics. Sometimes particular physical models were used to formulate the actual differential equations to be solved; but whether the model was actually connected to the structure of the mathematical expression of the problems, or used to interpret the meaning of that problem's solution in physical terms, is another matter. In fact, the mathematics was never used to interpret the physical world, nor to develop the physical model with which the mathematician began his analysis. The model never changed, nor did the model limit the generality of mathematical solutions sought.[10] Experiment guaranteed that a mathematical solution existed. It also guided the mathematician to the expression for the partial differential equations whose solution the mathematicians then sought. Once experimental results were translated into mathematical form, all connections with the original physical problem were lost. The solutions to the partial differential equations were mathematical, interpreted and argued over as mathematical results, not physical ones. If a particular solution was found, rather than the mathematically general one, it was through mathematical devices, such as chopping off series, changing variables, expressing functions as constants, and ignoring factors as small with respect to others, without any reference to the physical situation in which the problem originated, or whether, under the physical conditions of the problem, these processes made sense. Sometimes the mathematician would reach a numerical result, especially if previous experiments offered points of comparison. Whether the conditions of the mathematics matched those of experiment was fortuitous, not planned. If the results coincided with experiment, it was seen as validating the mathematics.

Important results in the development of analysis emerged from the mathematical consideration of a series of particular physical problems. For example, Jean le Rond d'Alembert's expression, then solution of the wave equation, led to a dispute with Leonard Euler on what constituted a function. For d'Alembert, a function had to be continuous in the same manner as a geometrical figure. For Euler, continuity was an algebraic notion and could be defined within limits of values for the function. Daniel Bernoulli's suggestion for a physically guided solution to the wave equation was simply ignored.[11] It was Jean-Louis Lagrange who offered the definitive eighteenth-century, functional solution to the wave equation.[12]

Standard approaches to the solutions of such problems were well defined by the end of the eighteenth century. The mathematician sought the most general solution to the partial differential equation, then turned to specific particular forms. Solutions were in functional form. Trigonometric series, unless used as particular solutions, were problematic, as there was no way to prove in general that they converged. From 1800 onwards, quantitative experimental results in electrostatics, as well as the conduction of heat and light offered the mathematicians new starting points for their explorations of mathematics.[13]

Much more was at stake in French mathematical physics than the priority of reaching the general solutions to new partial differential equations. In the early nineteenth century, physics problems and the equations they generated were the battleground over the very foundations of the calculus. This began with Lagrange in the 1780s, who was faced with teaching the calculus. He realized that there was no systematic way of proceeding from axioms, definitions, and theorems to known results. Lagrange based his form of the calculus on Taylor's theorem.[14] The shortcomings of Lagrange's version of the calculus were perceived as more and more French mathematicians turned to the same issues. André-Marie Ampère, Poisson, Fourier and others were participants in the acrimonious debates that led in the 1820s to the form of the calculus established by Augustin Cauchy. These internal rifts within mathematical physics were deep and critical for the direction of the development of mathematics. Issues of the legitimacy of initial physical models, hence quarrels over the validity of the basic partial differential equations to be solved, as well as questions of the rigor of proofs and the permissibility of mathematical methods and results, divided individuals from one another and decided careers in the fiercely competitive world of Napoleonic French science. This led to long, discursive papers, with acrimonious discussions of other's results, which were often rederived, or generalized one step further, or developed along different mathematical paths.[15]

Into this developing Parisian, professional, and disciplinary quagmire waded the outsider Joseph Fourier. His initial paper, submitted to the Institut in 1807, was less complete than the prize essay of 1811, but was a stunning achievement: It was also subversive of the basic standards of mathematics. Fourier considered a series of physical problems that led to particular solutions of the "equations of motion for heat," which were themselves of limited mathematical generality. In some cases, only particular solutions were open to him, but in finding his solutions, Fourier used the separation of variables. Some of his solutions were in the form of trigonometric series, not in their equivalent, functional forms. This was legitimate, if the limits on the series allowed for convergence. However, Fourier proceeded to an extended discourse on the expression of arbitrary functions in terms of these finite series and also in their integral form – that is, Fourier analysis. Yet he presented no proof that in general these infinite series were convergent. The power of his mathematics was clear, but was it correct?

The mathematical lacunae were quickly pointed out by Denis Poisson, even before any of Fourier's work appeared in print.[16] Poisson only commented on Fourier's physics in the most general terms, with no mention of the idea of flux. Poisson's detailed remarks were confined to Fourier's mathematics, its lack of generality and rigor.[17] Although the contents of Fourier's paper of 1811 were generally known among mathematicians in Paris by 1815, the paper itself, crowned in 1811, was only published, in two parts, in 1824 and 1826, by the Académie des Sciences, of which he was by then the permanent secretary.[18] Between 1811 and 1822, Fourier published three articles on particular mathematical points of his work and a separate vernacular discussion of the physics of radiant heat.[19]

Fourier's 1822 text was both a defense and his first chance to present his work fully to the mathematical public. And it was this version to which most of his readers, especially outside of France, referred.[20] In his "Introduction" Fourier placed his work firmly in the tradition of mathematical physics. The source of the problems might be physical, but the goal of the enterprise was mathematics. Physics guaranteed the existence of a solution and that it be meaningful:

Profound study of nature is the most fertile source of mathematical discoveries. Not only has this study, in offering a determinate object to investigation, the advantage of excluding vague questions and calculations without issue; it is besides a sure method of forming analysis itself, and of discovering the element which it concerns us to know, and which natural science ought always to preserve: these are the fundamental elements which are reproduced in all natural effects.[20b]

The conduction of heat was the occasion for Fourier to explore new areas of analysis that lay undiscovered using conventional methods. This publication was his first chance to persuade his readers that his new methods were legitimate. The consideration of particular examples could lead to the mathematically general solution, and trigonometric series were the key to that general solution. He approached this issue first through informal appeals to simplicity:

We might form the general equations which represent the movement of heat in solid bodies of any form whatever, and apply them to particular cases. But this method would often involve very complicated calculations which may easily be avoided. There are several problems which it is preferable to treat in a special manner by expressing the conditions which are appropriate to them.[20b]

He also demonstrated that solutions to very particular forms of the equations of motion for heat became the most general ones if one generalized the problem as far as possible, for example by considering a ring of masses generalized to a continuum. However, at some point, these demonstrations depended upon the reader accepting Fourier's expressions for arbitrary functions in terms of trigonometric series. Proliferation of examples substituted for more formal proof. This made his extended discussion of the representation of arbitrary functions, Fourier's theorem, plausible, but he did not prove the necessary theorems in general.[21] The problem of generality and the convergence properties of this series were still not satisfactorily solved, and functional alternatives were already available in the literature.[22]

Even on the technical level Fourier broke with his peers in his use of the method of separation of variables. If we consider the equation for the motion of heat in two dimensions at stationary temperatures T,

$$\frac{\partial^2 T}{\partial x^2} + \frac{\partial^2 T}{\partial y^2} = 0,$$

usually the solution would be written as $T = F(x + iy) + f(x - iy)$. Fourier chose the solution $T = F(x)f(y)$, demonstrating that this was indeed a legitimate solution to the equation. He then constructed the general solution

$$T = \sum_{r=1}^{\infty} a_r e^{-n_1 x} \cos n_2 y$$

with the boundary conditions $T = 0$, when $y = \pm 1$, and $T = 1$ when $x = 0$.[23] Here, as elsewhere, the boundary conditions are defined mathematically, not physically.[24]

The problems that led to the restricted forms for the equations of motion for heat were disparate. What these examples allowed Fourier to do was to generalize his mathematics in ordered steps.[25] Initially, Fourier used physical cases to establish particular forms for the equations of motion for heat, but the solution of these was delayed until he had stated the most general equation of motion for an infinite three-dimensional solid. He then solved the particular cases when the temperature at a point in the solid is a constant, and when it is a function of time. In this case, he had to bring in more physical observations – namely, he used Newton's law of cooling at the surfaces of the various solids. Within these particular cases, he introduced trigonometric series solutions to his equation of motion, with additional analytical conditions.[26] In his discussion of the examples, he did not develop the basic notion of "the linear motion of heat" physically in any more depth. The physical observations upon which Fourier based his mathematics were qualitative, but clearly and precisely stated. He used no specific imagery to transform these physical images into mathematical form. The linear flow of heat was transformed into mathematical form without discussion of any process on the microscopic level within the solid.

Fourier had tried a microscopic model in the one-dimensional case, but it had not worked mathematically, so he abandoned the attempt.[27] This was the point at which Biot and Poisson questioned Fourier's methods and their disputes with him began. Fourier explicitly stated that the equations of motion for heat were not reducible to those of mechanics; this was one of Poisson's points of difference with him. In one important aspect of his work, Fourier did try to go beyond his peers in discussing the physical implications of his results. However, all he could show was that his analytical deductions were compatible with the principles with which he had initiated the argument.[28]

Experiment is marginal in the text. Experiments delineate the behavior of heat and hence set the principles upon which the equations of motion must be based. They are mentioned as confirming certain results at some points in the argument, i.e., at a point where mathematical deductions are worked out and the form of the function representing the temperature can be compared with experiment. Yet no details of the experiments are given, nor is the data.[29] Fourier describes, briefly, an experiment on heating a metal ring at different points and measuring the temperature at various points when the ring reached equilibrium. He claims that these experiments "fully confirm" the results of theory.[30] The only difference between Fourier's citation of experiment and those of Poisson or Laplace is that they are his own experiments. Fourier was also more careful to reduce the mathematics to a case that matched experiment. In either case, experiment confirmed the rightness of the mathematics; it was not meant to probe the nature of heat or conductivity.[31]

For the reader of this text – and this is the Fourier known to most of his readers – physical problems are transformed into an elegant, new form of analysis, with explorations of some of its mathematical possibilities. This is not, however, an example of theoretical physics. While particular physical cases were necessary for Fourier to establish his mathematics, the physical cases were chosen for their mathematical possibilities. Experiment and observation enter at the beginning and end of the process, to help set up the mathematical equations to be solved and to assure the reader that the analysis is correct. Fourier's *Theory* is mathematics recognizable as such to his readers, but with many of the usual elements of that tradition reworked to redirect that tradition into new paths. While we see it as a triumph, some of Fourier's contemporaries were less sanguine. Exception was taken to his assumption of Newton's law of cooling, seriously undermined by 1820 by the work of Dulong and Petit and Delaroche and Bérard. There were also questions about his assumption of the linear motion of heat. This was sniping at the very foundations of his mathematics – its experimental base. These and his mathematical solutions were all suspect.

However much his contemporaries might criticize his work, the younger generation of French mathematicians extended and developed it along new mathematical paths. Joseph Liouville expanded Fourier's work by treating non-homogeneous bodies and the case of an unequally polished bar, as well as the two-dimensional problem of a plate. This was mathematical as was the contemporary work of Joseph Bertrand, Michel Chasles, J. M. C. Duhamel and Gabriel Lamé.[32] Thus, the tradition of mathematical physics continued in France through the nineteenth century, as did the separate discipline of experimental physics. However, this has been lost to view in the twentieth century, as we have seen the beginnings of contemporary mathematics and physics in Germany and Britain. French science has even been seen as in "decline." Yet for three decades from 1820, French mathematical physics and experimental physics were the models for mathematics and physics in both Germany and Britain.

The earliest reactions to Fourier's mathematics outside of France were from the Germans. Peter Lejeune Dirichlet had spent the years 1822 to 1826 in Paris, accepted into the circle around the now powerful Fourier. In many of his later papers Dirichlet drew upon the work of French mathematics and hence began with mathematical problems suggested by physics. However, Dirichlet's research was directed to mathematics: His work on definite integrals and infinite series grew out of Fourier's work, but addressed a new mathematical rigor. He helped to establish a tradition with German mathematics which was firmly oriented towards mathematics, not physics. Thus Dirichlet's work on the stability of the solar system was directed to criticizing the mathematical methods of Poisson and Laplace, not to the astronomical problem that gave rise to the mathematical issues. In his examination of boundary values problems, his attention was on extending the analysis of the potential function to any number of dimensions. While boundary values problems were important in physics, Dirichlet did not connect his mathematical work to physics.[33]

Much the same can be said of some of the work of Bernard Riemann. His text is on definite integral solutions to the partial differential equations of Fourier's theory of heat. First, Riemann discusses all the necessary mathematical techniques required; then he launches into a section on ordinary differential equations, before turning to partial differential equations of the second order. Fourier is introduced by setting up the most general form for the equation for the flow of heat in three dimensions. Riemann then takes a series of mathematically special cases that lead to definite integral solutions. Restrictions are mathematical boundary conditions. None of the mathematical development is accomplished with any hint of physical explanation.[34] The mathematical goals are also manifest in Karl Friedrich Gauss's work on the potential, Carl Gustav Jacob Jacobi's lectures in mechanics, and Alfred Clebsch's work on elasticity and other "physical" subjects.[35] These were the direct descendants of the mathematical tradition of French mathematical physics. Of what possible use these treatises could have been to their contemporaries in physics is problematic. And they have been rather problematic for historians.[36]

For the physicists in the German States, consciously breaking away from the hyperbole and hypotheses of Naturphilosophie, the French offered a highly successful alternative. Physics was quantitative experiment, whose results were taken over into the separate discipline of mathematical physics. The most complete, successful, and available example of the latter was Fourier. There are few indications that the Germans saw mathematical physics as either exclusively mathematical or physical. Even if the German physicists read Fourier as physics, we still have to examine how, given that the text makes its mathematical ends manifest and does not return to the original physical problem, physicists related this text to their goal of understanding certain phenomena and processes of nature. As research, mathematical physics usually appeared in Crelle's journal, rather than Poggendorff's.[37] In the case of handbooks of physics that appeared in the 1830s, mathematical physics was in its own separate section.[38]

All of French mathematical physics was absorbed and used as it suited the problem in hand. Fourier had followed the general pattern of French mathematical physics, using experiment to establish the equations to be solved. However, only Georg Simon Ohm had tried to use those equations in his work on galvanic currents. Yet in the 1820s he chose the less controversial functional solutions to those equations.[39] Despite Franz Neumann's detailed reading of Fourier's *Theory*, its influence on him was programmatic rather than specific. He, too, separated experimental and mathematical explorations of phenomena, but he did not choose to follow Fourier's mathematical example.[40] This and Ohm's behavior reflects available mathematical methods in Berlin and the state of German mathematics, rather than disciplinary differences. Physics as a discipline was at this time still experiment.[41]

The only specific use Neumann made of Fourier was in his experimental work on the specific heats of crystals. This was after his appointment at Königsberg in mineralogy, and what he did there was to follow French mathematical work also in mineralogy. However, Neumann was investigating nature, not mathematics, and this led him later to reject Poisson's model of matter as nonsense and to choose particular mathematical paths in optics as choices based in physics. While apologizing for using the wave theory of light, Neumann argued physically about the conditions at the boundary of two media and chose the mechanically physical one to develop mathematically. However, his solutions omitted any discussion of physical processes because he had rejected physical models. Similarly, his published lectures on mathematical physics follow the French model.

Other physicists were also turning mathematical physics to their own needs. In his examination of elasticity, Gustav Theodor Fechner discussed various theories of elasticity and their shortcomings. This was in the section of his text on experimental physics and was expressed in the vernacular. In his discussion of mathematical physics, however, Fechner particularly criticized general solutions to the partial differential equations of elasticity by Poisson as useless. More important, Fechner tried to extract physical meaning from the mathematics, hence his need for specific solutions. In this mathematical section Fechner tried to compare experimental results with deductions from the mathematics.[42]

As Neumann's and Fechner's initial steps indicate, German physicists needed to go beyond the limitations of French methodologies to create a new kind of physics. The first step was to make mathematical physics accountable to experiment, then to make physical choices to guide the initial direction of the mathematics. The final break was accomplished by Wilhelm Weber, who integrated hypotheses into the exploration of the mathematics itself. Fourier was not enough, either in terms of his general method or his mathematics.

The British travelled the same general route, but with more intimacy with French mathematical physics than the German physicists. By the time Fourier's *Theory* was published, the British had already embraced French mathematical physics and were busy embedding it into the mathematical Tripos. Their model was, however, Lagrange, whose calculus was geometrical, elegant and expressed in the problems of mechanics – the core of the educational system of Cambridge. The characterization of "geometrical" was given to Lagrange's work by Poisson as a form of derision. Lagrange translated the variables of mechanics into points in space, removing their connection to the physical world that guaranteed the validity of the mathematics deduced from them. However, Cambridge mathematicians were not the first to examine French mathematics and see in it possibilities that Newton's fluxions could not give them.[43] Institutions as well as individual mathematicians responded to French mathematics. Driven by the Tripos, the Cambridge curriculum changed. Systematic teaching in the new mathematical methods was necessary.[44] Fourier published a new kind of mathematics just as the first generation of young men were in the thick of committing Cambridge to a rival mathematics.

A decade later this generation was split over the place for such technically demanding mathematics within that same curriculum. While William Whewell might have admired the new French mathematics, including Fourier's, he questioned the need for it within the Tripos. John Herschel placed mathematics in a decidedly inferior role in the sciences, as a tool for observation and experiment. To understand those reversals by Whewell and Herschel, we need to examine their respective careers. Whewell became responsible for the education of all the students at Trinity, and his perspective on the role of mathematics in education had to change from that of a very bright young undergraduate out to modernize his University. Mathematics as "training of the mind" is not the same as the training needed for mathematical physics. The latter developed only the narrow, technical expert. Whewell could hardly sell that as the result of a "liberal education." In the 1820s, Herschel, along with Charles Babbage, had published on aspects of the new mathematics that the French developed further. Herschel stayed a very restless young man until he followed into his father's profession, observational astronomy. Here, mathematics became a tool for the reduction of results and for corrections of measurements, and that was the only place he allowed for it. George Biddell Airy travelled the same path, from Tripos scholar to observational Astronomer Royal.[45]

While Whewell and Herschel, both firebrands of reform and active in placing French mathematics into the Cambridge curriculum, modified their positions on mathematical physics, it was too late. Other members of this generation embraced the French mathematics, and with it Fourier. George Green was the first to mention Fourier's mathematics, but Fourier did not influence him; Green's mathematics remained Lagrangian.[46] It was George Peacock who introduced Fourier to the general scientific public, in his report to the British Association in 1833. Much of this report placed French mathematics within the domain of Peacock's own interest, symbolic algebra. Fourier was only one of many French mathematicians whose work intersected with Peacock's own. Fourier entered in the section on the representation of discontinuous functions, and Peacock noted the unsatisfactory nature of Fourier's proof with respect to his use of series.[47]

In the following year, before the same group, Whewell examined the mathematical theories of heat, light, magnetism, and electrostatics. His charge was to compare the mathematical theories with the facts that were thought to confirm them, not to discuss their details. Whewell treated all mathematical theories here as theories about physical phenomena, not explorations into analysis. Indeed, in his section on heat, Whewell noted that mathematicians were led beyond the needs of "physical science" into "that deep and charmed labyrinth", leaving most of the rest of us behind. No doubt this could have been otherwise, had they chosen a mathematics such as Newton would have used. And although some rigor might be lost, important though that might be, "such solutions would have been just in all the material points." Whewell was already very distant from French mathematical physics and its goals.[48]

Just how many in his audience or at Cambridge would have agreed with him is unknown at this point. There are some indications that he was speaking for very few. While some, like Philip Kelland, would try to develop a physical model for heat, the mathematics remained firmly that of Fourier. Kelland rejected the caloric theory as incapable of accounting for radiation and turned to Poisson's microscopic model for a physical process. This was qualitative and in the vernacular. When it came to the mathematical theory of heat, Kelland switched completely to Fourier, and his ideas remained as separate halves captured between the same book covers. Four years later, however, Kelland tried to compare the various physical theories of heat with the available experimental evidence, to judge their validity. In doing so, he was careful to "examine for ourselves such of the formulae as appear parallel to the experiments we possess." And he worked out specific examples that might be tried experimentally, although not for all four theories he identified. Unfortunately, "theoretical writers on this branch of physics" – and he pointed directly to Poisson – "have not presented their results in a form sufficiently tangible to direct or suggest the application of experiment to them." His own choice was then in limbo.[49] In his textbook, Kelland worked out a limited number of mathematical cases, all of which suggest experimental ones. He clearly considered the exercise as one within physics; discussed the theories as physical ones reflected in the mathematics and went on to construct a meeting point between theory and experiment. There is much more here than a misunderstanding of Fourier.[50]

It would take longer for others trained like Kelland in the Tripos to reach this point. By the time Whewell addressed the British Association, there was a coterie of mathematicians, in Cambridge, London, Dublin, and Edinburgh, interested in all French mathematical methods and eager for more. And in 1839 they got their own outlet, the *Cambridge Mathematical Journal*. In the opinion of its first editor, David Gregory, mathematical problems have arisen that were not encompassed within the mathematical methods of mechanics. The mathematical theories of heat and light had introduced problems of discontinuities, whose representation was intimately connected with the theory of definite integrals. Given the journal's contents in the following decade, we can assume that Gregory's purpose was to provide a forum for this new analysis at Cambridge. In addition, the journal would discuss problems likely to appear on the Tripos examination and would publish these, as well as research papers. Initially, the journal published both kinds of articles. The research subjects began as small and timid but became bolder. It was not a journal for run-of-the-mill students, but for would-be wranglers, faculty, and interested mathematicians across Britain.

As if to emphasize the journal's commitment to the new analysis, Gregory himself wrote the first article to appear, on Fourier. Ironically, he tried to reexpress some of Fourier's results in functional form, partly to address the problems of Fourier's proofs.[51] In these early papers, as in William Thomson's paper of 1841, Fourier's proofs were discussed and improved, rather than considering the broader implications of his work.[52]

In general, in its first decade, this journal did publish a small number of articles on the physical principles behind some of the mathematics, separately and in the vernacular. For the majority of papers, if any began with physical problems, the actual point of the paper, original or not, was mathematical not physical. Yet as the decade of the 1840s rolled on, a distinctly British style of mathematics emerged in its pages. The operational algebra instituted by Peacock became so popular that the editor – by this time William Thomson, who succeeded Robert Ellis – begged papers from Stokes, so that something else might appear on its pages. And for Thomson, it was getting to be too much of a research journal. By this date it was also a recognizable modern journal of mathematics. Cayley and Boole, for example, were treating the problems of mechanics or light as particular cases of a type of partial differential equation, as well as publishing their work in "pure" mathematics.

William Thomson became editor of this journal in 1845 and continued until 1850. Looking through those pages, I do not see that he changed the direction of the journal, other than including the Irish within its title and pages. Thomson expanded coverage by including electrostatics and hydrodynamics, largely written by himself and George Gabriel Stokes. In 1841, the connection that Thomson saw between the motion of heat and electrostatics lay in the mathematical forms they shared. This I think everyone agrees with. But rather than seeing these as a path to physical analogy (as Maxwell used this paper), Thomson used those mathematical samenesses to mediate solutions to mathematically tricky problems that occur in both subjects. Then, by inverting the process, he converted theorems in the theory of the attraction of ellipsoids into statements about the flow of heat. He did so by demonstrating how one can replace a series of sources of heat, electrical charge, gravitational attraction by an "Isothermal" surface.

None of this is new in the history of physics. However, historians do not note that Thomson in these first papers included problems in gravitational attraction in his mathematical net. Attraction was the term used, with electrostatics only as one example of attraction. In one place he stated (as did Gauss) that by replacing particular constants in his basic equations one could do the problem in all three areas – heat, electrostatics, and gravitation.

Thomson's examination of Isothermal surfaces followed that of the French mathematicians Duhamel and Liouville. Isothermal refers to a surface with certain mathematical properties; it has no physical significance or existence here. Thomson finished up by using the theory of heat to play with ideas about orthogonal surfaces, defining a point using curvilinear coordinates and then following what kinds of surfaces are generated by the equations of motion of heat in as general a case as he could possibly handle.[53] Some aspects of these papers are confusing, as Thomson moves back and forth between the physical cases that carry with them the mathematics he wants to connect together. For a reader trained in French mathematical physics, the physical names label a type of mathematics and do not necessarily refer to anything beyond the mathematics. To further confirm the rightness of this mathematical approach, Thomson verified a fundamental proposition in Gauss's work on attraction by replacing the material points with his mathematical surface. Thomson also discussed the differences of the electrostatic and gravitational cases. Gauss's proposition was put in the mathematical terms in which Gauss expressed it, as were his deductions, either analytical or geometrical. Thomson was dealing with the properties of functions and surfaces, although he concluded that the experimental result that there is no electric charge within a hollow conductor "is confirmed," by the mathematics.[54]

Physics was the instrument for generating mathematics, not the reverse. What Thomson was moving towards was a generalized mathematical method of treating these disparate physical cases through one mathematical approach, as opened up by Fourier. The picture of Thomson that I am drawing is markedly at odds with the usual one. Thomson is normally seen as a physicist first and always, building a new physics based upon the mathematics of "geometric," i.e., macroscopic, entities. This approach is traced back to the unique philosophical framework given natural philosophy in the Scottish higher education system by the Common Sense philosophy.[55] Yet at Glasgow University mathematical and experimental natural philosophy, although taught by the natural philosophy professor, were separate courses, and there is no evidence as to whether mathematical natural philosophy grew directly out of the theoretical ideas presented in the other course, or whether, as with Kelland, they lived alongside each other yet unconnected.

Most historians of physics who analyze Thomson's first paper on Fourier forget that gravitation was also involved, and that Thomson was doing a mathematics built on a different foundation than we expect. That foundation is physical problems, not other mathematical ones, although that is what it develops into. The usual interpretation also reflects Maxwell's characterization of Thomson's work. We forget that Maxwell used Thomson for his own purposes, which may have had little to do with what Thomson was groping towards in the 1840s.

For the same reasons, Thomson's article in Liouville's journal is French mathematical physics. It is more than reconciling different approaches to electrostatics. Thomson is building a mathematical theory that transcends all physical theories and uses Fourier to accomplish it. The experiments of Coulomb and Faraday cannot be at odds. Both are true. And what's more, a mathematical theory will bring them together. Which does not mean that we have a physical theory of what is going on. In three places at least he points out that his is a mathematical theory, "independent of physical hypothesis." Fourier's laws for the motion of heat

constitute a mathematical theory, properly so-called; and when we find the corresponding laws to be true for phenomena presented by electrified bodies we may make them the foundation of the mathematical theory of electricity: and this may be done if we consider them merely as actual truths, without adopting any physical hypothesis.[56]

This was written in Paris, in that summer of 1846 in which his whole life changed. Thomson spent 8–10 hours a day in Regnault's laboratory, then rushed over to Liouville for mathematical company. The dual nature of his existence and of the separation of mathematics and physics was duplicated in the geography of his summer. The dual nature of this existence continued for at least five more years with his appointment as Professor of Natural Philosophy at Glasgow. The demands on him in Glasgow began to push him in entirely different directions. He had to deal with students becoming engineers, not liberally educated gentlemen, and to continue to show that he was not merely a mathematician. He turned to experiment, and to the volunteer labor of students.

Also that summer he was introduced to Clapeyron's mathematization of Carnot's theory of heat. And here Thomson performed again as a conforming mathematician, extending the mathematics of Clapeyron. But this time he drew out of the mathematics implications about the measurement of heat – the Absolute Scale of Temperature. Physics came directly from the mathematics. This was reinforced in his second paper on Carnot's theory, when his brother James deduced the result of the decrease in freezing temperature with pressure that William Thomson then confirmed by experiment.

Joule's results – equally experimental posed a very real problem to the foundations of all Thomson had done with Carnot's theory. In this case, Thomson needed to go beyond the bounds of mathematics and experiment to embrace the mechanical theory of heat. He had to accept hypotheses as a necessary aspect of the construction of physics, and as an explanation of the processes of nature.

Some measure of the change by 1850 lies in treatment of irreversibility. He had always dealt with irreversibility in the mathematics of Fourier. Yet in the 1840s the problem of negative time-values was simply a mathematical, not a physical one. From the middle of 1850s, the mathematical expression of a system at zero time carried physical and even cosmic meaning. Negative time was meaningless.[57]

From 1845 there was also the influence of George Gabriel Stokes. Stokes had already begun to divert mathematical physics from an exclusively mathematical path into one useful for physics. Although he wrote on some of the mathematical aspects of Fourier's work, Fourier was only one of many French mathematical sources that informed his work. Many papers with seemingly physical titles turn out to be mathematical exercises. Yet in others Stokes does draw strictly physical conclusions from his mathematical derivations.[58] And, he does not necessarily pursue those mathematical implications if they have no important mathematical or physical point to them. However, in this decade he discussed physical hypotheses in separate articles, usually in the *Philosophical Magazine.*[59]

He understood when he wanted to address a physics, not a mathematics problem. When he tackled hydrodynamics, he only used mathematics that made physical sense. And he taught that to Thomson. Yet he was also a careful mathematician. In his paper to the Cambridge Philosophical Society in 1847 on Fourier, he is all mathematical business. While noting that the mathematics under discussion was useful for solving physical problems (heat, electrostatics), Stokes did not include these in the pages of this paper. They are his concern elsewhere.[60]

Stokes's gift, freely given to Thomson and Maxwell, was his ability to separate the physically necessary from the mathematically interesting. This was absolutely crucial to both, and they acknowledged it. And in Thomson's case, Stokes forced Thomson to specify his physical ideas in ways that could be mathematized. Stokes also understood the

amount of mathematics needed to solve the physics problems mean-
ingfully. Conversely, he clothed the mathematics in physical meaning
while remaining aware of his own assumptions. He also attributed to
other mathematicians the same physical goals as he had himself, an
interpretation he passed to Thomson and Maxwell.

French mathematical physics was then influential, especially in
Britain, but not in quite the ways depicted by historians. Whether we
can point to any one French mathematician with singular influence is
debatable. French mathematical physics was absorbed whole, and it is
difficult to isolate Fourier and his influence as unique even on Thomson, despite his later testimony. Both the British and the Germans found
French mathematical physics insufficient to explain the physical processes of nature. To fully explore nature, hypotheses were necessary,
and these had ostensibly been rejected by the French. Fourier and others
were the means both of opening up the possibilities of the mathematical
explorations of physical phenomena, and of demonstrating the limitations of this strictly mathematical approach.

Fourier and French mathematical physics also spawned a number of
disciplines, all of which claimed mathematics and physics within their
domain. By the end of the nineteenth century these disciplines merged at
their boundaries. Both mathematicians and physicists worked under the
rubric of mathematical physics and argued over how much mathematics
or physics is necessary within that discipline. Felix Klein discerned the
differences in the approaches of physicists and mathematicians who carried the same label but practiced different disciplines.[61] Within physics,
theoretical and mathematical physicists argued over the same issues.
With this work reread as physics, not mathematics, and with the kind
of mathematics he practiced superceded in mathematics itself, Fourier
became a model for a seemingly powerful approach to the solution of
physical problems. This was reinforced by the increasing importance of
his mathematical techniques within physics. Yet, experimental physics,
reaffirmed as the core of the discipline in the nineteenth century, could
still claim in the 1930s that physics was experiment, the rest was only
mathematics.[62]

Department of History
State University of New York at Stony Brook
U.S.A.

NOTES

[1] Judith Grabiner, *The Origins of Cauchy's Rigorous Calculus* (Cambridge MA: MIT Press, 1981) and Grattan-Guinness, *The Development of the Foundations of Mathematical Analysis from Euler to Riemann* (Cambridge: Cambridge University Press, 1970).

[2] The papers may have a title we take to be physics but are like the problem that "a man has four apples, then picks up another three then eats one, how many apples has he left?" This is hardly a problem in pomology or agriculture.

[3] See John Heilbron, *Electricity in the Seventeenth and Eighteenth Centuries* (Berkeley CA: University of California Press, 1979). Here we are dealing with Physics, narrowly defined.

[4] For the most recent example see Kathryn M. Olesko, *Physics as a Calling: Discipline and Practice in the Königsberg Seminar for Physics* (Ithaca NY: Cornell University Press, 1991) who simply assumes that French mathematical physics is the same as modern physics.

[5] See Christa Jungknickel and Russell McCormmach, *Intellectual Mastery of Nature*, 2 vols. (Chicago IL: University of Chicago Press, 1986), vol. 1, p. 55.

[6] For example see Robert Friedman, "The Creation of a New Science: Joseph Fourier's Analytical Theory of Heat," *Historical Studies in the Physical Sciences* 8 (1977): 73–100. In this account mathematics is supplementary to the essential physical problem of heat conduction. Friedman focusses on the concepts Fourier used but does not look at the problem of how, or if, Fourier used these ideas in his mathematical theory of heat.

[7] John Herivel, *Joseph Fourier: The Man and the Physicist* (Oxford: Clarendon Press, 1975).

[8] Many of the assertions made here are explored more fully in a forthcoming study on the relationships between mathematics and physics from 1750 to 1850 that is almost complete and will appear as Garber, *Mathematics as Language*.

[9] For an extended account see Ivor Grattan-Guinness, *Convolutions in French Mathematics, 1800–1840: From the Calculus and Mechanics to Mathematical Analysis and Mathematical Physics*, 3 vols. (Boston MA: Birkhäuser, 1990).

[10] On Poisson and Laplace, for example, see Robert Fox, "The Rise and Fall of Laplacian Physics," *Historical Studies in the Physical Sciences* 4 (1974): 89–136, and Garber, "Siméon-Denis Poisson: Mathematics versus Physics in Early Nineteenth Century France," in *Beyond History of Science: Essays in Honor of Robert E. Schofield*, Garber, ed. (Bethlehem PA: Lehigh University Press, 1990), pp. 156–176. Physical limitation on the development and depth of mathematics is one of the hallmarks of theoretical physics in the mid-nineteenth century.

[11] Jean le Rond d'Alembert, "Recherches sur la courbe que forme une corde tendue mise en vibration," *Histoire et Mémoire de l'Académie Royale des Sciences, Berlin* 3 (1747): 214–219 and "Suites des Recherches sur la courbe que forme une corde tendue, mise en vibration," *ibid.* 220–249. Leonard Euler, "Sur les vibrations des cordes," *ibid.* 4 (1748): 69–85. Daniel Bernoulli, "Réflexions et éclairissements sur les nouvelles vibrations des corps exposées dans les Mémoires de l'Académie de 1747 et 1748," *ibid.* 9 (1753): 146–172.

[12] Lagrange "Recherche sur la nature, et la propagation du son," *Miscell. Taurenis* 1 (1759): 1–112 and 2 (1760–1761): 11–172, reprinted in *Oeuvres de Lagrange*, 14 vols. (Paris: Gauthier-Villars, 1867), vol. 1, 39–1488, 151–316. See also *Oeuvres*, vol. 14,

50   ELIZABETH GARBER

"Correspondence between Lagrange and Euler," August 1758, p. 157, July 1759, October 1759, and November 1759, pp. 170–172.

[13] While explicitly dismissing hypotheses as having no place in their science, Fresnel's wave theory of light was used, with apologies into the 1840s, by mathematicians unable to resist the mathematical explorations that it offered them.

[14] See Grabiner, *Origins* and Grattan-Guinness, *Development of the Foundations*.

[15] Perhaps the longest standing dispute of this kind is in Poisson's reworking of Fourier's theory of heat that appeared in its final form in 1835. Among other points Poisson objected to Fourier's derivation of the equations of motion of heat, which he then rederived. But he obtained the same partial differential equations. He then solved them functionally again duplicating many of Fourier's results. Poisson, *Théorie mathématique de la chaleur* (Paris: Bachelier, 1835).

[16] Poisson, "Mémoire sur la propagation de la chaleur dans les corps solides," *Bulletin des sciences par la Société Philomathique de Paris* (1807): 112–116.

[17] For detailed discussions of this critique see Ivor Grattan-Guinness and Jerome Ravetz, *Joseph Fourier (1768–1830)* (Cambridge MA: MIT Press, 1972), pp. 442–443, and John Herivel and Pierre Costabel, *Joseph Fourier face aux objections contre sa théorie de la chaleur, lettres inédits, 1808–1816* (Paris: Bibliothèque Nationale, 1980).

[18] Fourier, "Théorie du mouvement de la chaleur dans les corps solides," *Mémoires de l'Académie Royale des Sciences* 4 (1819–1820), 185–555, and "Suite," *ibid.* 5 (1821–1822)): 153–246. Dated for the years before the appearance of his text publication was, as usual, delayed. The content of these papers is almost identical with Fourier's text.

[19] Fourier, "Questions sur la théorie-physique de la chaleur rayonnante," *Annales de chimie et physique* 6 (1817): 259–303. For those on the mathematics of particular cases see Fourier, "Théorie de la chaleur," *ibid.* 3 (1816): 350–375, "Sur la température des habilitations et sur le mouvement varié de la chaleur dans les prisms rectangulaires," *Bulletin des sciences par la Société Philomathique de Paris* (1818): 1–11, "Extrait d'un mémoire sur la refroidissement séculaire du globe terrestre," *Ann. Chim. Phys.* 13 (1820): 418–438, also in *Bull. Soc. Philomathique* (1820): 58–70, abstracted in *Journal de Physique* 90 (1820): 234–236.

[20] The version used here is Fourier, *The Analytical Theory of Heat* (New York: Dover, reprint of translation of 1878, 1955), translated by Alexander Freeman. In comparing this with the reprint in Fourier's collected papers the Freeman version stays closer to the text. In the collected papers there are unmarked editorial additions and changes in mathematical notation. Freeman seems scrupulous in noting his own additions and in keeping the original notation.

[20a] Fourier, *Theory*, p. 7, see also p. 131.

[20b] Fourier, *Theory*, p. 85.

[21] Grattan-Guinness, *Convolutions*, notes that in the articles in the *Mémoires* of the Académie Fourier introduced geometrical figures to illustrate how certain functions could be constructed from his series. He also notes that Fourier's attempts at a general demonstration in his *Theory*, (1822) paragraphs 280–281 are "rather woolly."

[22] Poisson, "Sur la distribution de la chaleur dans les corps solides," *Mémoires de l'Académie des Sciences, Paris* 1 (1816): 1–70.

[23] Fourier, *Theory*, pp. 131–159. This derivation is long because Fourier needs to demonstrate how trigonometric series can be used in this case, p. 135 onwards.

[24] Fourier, *Theory*, pp. 90–94 where Fourier establishes the motion of heat in a sphere,

the initial conditions and the conditions at the surface of the sphere, p. 94.

[25] The physical examples are respectively a row of disparate bodies developed into a one-dimensional line, a ring for the two-dimensional case, then various forms of bars (prismatic, cylindrical) where the temperature can be taken as constant across the bar that lead into the three-dimensional case of the cube and infinite solid.

[26] Fourier, *Theory*, p. 137.

[27] See Grattan-Guinness and Ravetz, *Fourier*, pp. 39–81, commentary, pp. 36–38. He returned to this mathematical approach later.

[28] Fourier, *Theory*, p. 316.

[29] Fourier, *Theory*, p. 90.

[30] Herivel, *Fourier*, sees Fourier using experiment only as a confirmation of theory.

[31] Details of some of his experiments were published in the 1820s. Fourier, "Recherche expérimentales sur la facilité conductrice des corps mince soumis à l'action de la chaleur; et description d'une nouvelle thermomètre de contact," *Annales de Chimie et Physique* 37 (1827): 291–315. He also published two articles, with no mathematics on the physics of a cooling earth, Fourier, "Mémoire sur la température du globe terrestre et des espaces planétaires," *Mémoires de l'Académie des Sciences, Paris* 7 (1827): 570–604, abstract in *Annales de Chimie et Physique* 27 (1824): 136–167.

[32] On Liouville see Jesper Lutzen, *Joseph Liouville, 1809–1882: Master of Pure and Applied Mathematics* (New York: Springer-Verlag, 1990), pp. 589–595. For a discussion of Duhamel, Lamé and Liouville in the 1830s see Grattan-Guinness, *Convolutions*, vol. 2, pp. 1168–1176.

[33] See his lectures on the inverse square forces, Dirichlet, *Vorlesungen über die im umgekehrten Verhältnisse des Quadrats der Entfernung wirkenden Kräfte*, F. Grube, ed. (Leipzig: B. G. Teubner, 1876). The Electrostatics chapter is an exercise in the theory of functions not in electrostatics. See Nikolai Stuloff, "Die mathematische Methoden in 19. Jahrhundert und ihre Wechselbeziehungen zu einigen Fragen der Physik," *NTM* 33 (1966): 52–71, for a survey of developments in "pure" mathematics through the consideration of problems originating in physics.

[34] Riemann, *Partielle Differentialgleichungen und deren Anwendung auf physikalische Fragen. Vorlesungen*, K. Hattendorf, ed. (Brunswich, 1896).

[35] Gauss, "Allgemeine Lehrsätze in Beziehung auf bis im verkehrten Verhältnisse des Quadrats der Entfernung wirkenden Anziehungs- und Abstrossungskräfte", in Gauss *Werke*, vol. 5, 196–242, p. 200. Trans. in *Taylor's Scientific Memoirs* 3 (1843): 153–196. That Gauss was doing mathematics not physics was noted by Kenneth O. May, "Gauss, Karl Friedrich," *Dictionary of Scientific Biography*, vol. 5, pp. 298–314. See also Ch.-J. de la Vallee Poussin, "Gauss et la théorie du potential," *Revue des Questions Scientifiques* 23 (1962): 314–330. Jacobi, *Vorlesungen über Dynamik*, A. Clebsch, ed., vol. 8 of Jacobi, *Gesammelte Werke*, K. Weierstrass, ed. (New York: Chelsea, reprint of second edition, 1969), Alfred Clebsch, *Theorie der Elasticität fester Korper* (Leipzig: Teubner, 1862).

[36] While Dirichlet and Jacobi's work has been noted repeatedly as important for physics the notes have been short on specifics. See "Gustav Peter Lejeune Dirichlet," *Dictionary of Scientific Biography*, vol. 4: p. 127, and Jungknickel and McCormmach, *Intellectual Mastery*, vol. 1, p. 174. These remarks were made by mathematicians trained in the tradition of nineteenth century German mathematics.

[37] Auguste Leopold Crelle established the *Journal für die reine und angewandte Math-*

ematik in 1826. Johann Christian Poggendorff became editor of the *Annalen der Physik und Chemie* in 1824. For Crelle's notion of mathematics see Gert Schubring, "The Conception of Pure Mathematics as an Instrument in the Professionalization of Mathematics," in , *Social History of Nineteenth Century Mathematics*, Herbert Mertens, Henk Bos, Ivo Schneider, eds. (Boston MA: Birkhäuser, 1981), pp. 111–134. Poggendorff only expected mathematics to appear in his journal linked to the needs of experiment. See Poggendorff, "Vorwort," *Annalen der Physik* **1** (1824): v–viii.

[38] For example, Gustav Theodor Fechner, *Repertorium der Experimental Physik, enthaltend eine vollständige Zusammenstellung der neuern Fortschritte dieser Wissenschaft*, 3 vols. (Leipzig: Leopold Bok, 1832). Most of the text is experimental, with treatment of theory in the vernacular. However, Section IV of Part I, pp. 35–77 is a discussion of the mathematical treatment of elasticity, motion and equilibrium. See also Dirichlet, "Ueber die Darstellung ganz willkührlicher Functionen durch Sinus- und Cosinusreihen," in *Repertorium der Physik*, 3 vols., Heinrich Dove and Ludwig Moeser, eds. (Berlin, 1837), vol. 1, pp. 152–174.

[39] Ohm's experiments and the law he deducted from them were presented separate from his mathematical theory. See Ohm, "Vorläufige Anzeige des Gesetzes, nach welche Metalle die Contact-Electricität leitern," *Journal für Chemie und Physik* **44** (1825): 79–86, 87–88. These initial papers are followed by two more on the conductivity of metals. The fourth, Ohm, "Bestimmung des Gesetzes, nach welchem Metalle die Contactelectricität leitern, nebst einem Entwerf zu einer Theorie des Voltaischen Apparates und des Schweiggerischen Multiplicators," *ibid.* **46** (1826): 137–166 includes Ohm's law. The mathematical theory appeared as Ohm, *Die galvanische Kette, mathematisch bearbeitet* (Berlin, 1827), trans. *Taylors Scientific Memoirs* **2** (1841): 401–506, which lacks Ohm's introductory remarks. The equation Ohm used is in Fourier, *Theory*, p. 88.

[40] See Olesko, *Physics as a Calling*, p. 35 for Neumann's notes on Fourier and his request to the Kultus Ministerium for a post as a mathematical physicist.

[41] For the level of mathematics in Berlin in the 1820s see Joseph Dauben, "Mathematics in Germany and France in the Early Nineteenth Century: Transmission and Transformations," in *Epistemological and Social Problems of the Sciences in the Early Nineteenth Century*, Hans Niels Jahnke and Michael Otte, eds. (Dordrecht: Reidel, 1981), pp. 371–399.

[42] Fechner, *Repertorium*, vol. 1, p. 41.

[43] See Philip Enros, "Cambridge University and the Adoption of Analytics in Early Nineteenth-Century England," in *Social History of Nineteenth-Century Mathematics*, Herbert Mertens, Henk Bos, Ivo Schneider, eds., pp. 135–148, Niccolò Guicciardini, *The Development of Newtonian Calculus in Britain, 1700–1800* (Cambridge: Cambridge University Press, 1989), ch. 9, Ivor Grattan-Guinness, "Before Bowditch: Henry Harte's Translation of Books 1 and 2 of Laplace's "Mécanique Céleste"," *NTM* **24** (1987): 53–55 and "French "Calcul" and English "Fluxions" around 1800; Some Comparisons and Contrasts," in *Jahrbuch Überblicke Mathematik* (1986): pp. 167–178.

[44] On this score the Analytical Society is seen as less important than the role of its members within Cambridge as examiners. See Philip Enros, "Cambridge University."

[45] See Martha Mackin Garland, *Cambridge before Darwin: The Idea of a Liberal Education* (Cambridge: Cambridge University Press, 1980) for both Airy and Whewell on mathematics and liberal education. For Whewell's impact on the Tripos see Harvey Becher, "William Whewell and Cambridge Mathematics," *Historical Studies in*

the Physical Sciences **11** (1980): 1–48. Herschel's opinion are in Herschel, "Mathematics," in *Edinburgh Encyclopedia* (Edinburgh: Blackwood, 1830), vol. 13, part 1, pp. 359–383, in *Treatise on Physical Astronomy, Light and Sound,* contributed to the *Encyclopedia Metropolitana* (London: Richard Griffin and Co., nd). This is a reprint of the articles with their original pagination in the *Encyclopedia.* His message repeatedly is that mathematics must be useful.

[46] Green, "An Essay on the Application of Mathematical Analysis to the Theories of Electricity and Magnetism," in *Mathematical Papers,* N. M. Ferrers, ed. (New York: Chelsea Pub. Co., reprint of 1871 edition, 1970), p. 1–115. His remarks on Fourier appear pp. 7–8. Despite its title this text is mathematical. I had wondered how on earth Green, a miller's son, no matter how bright, would get his hands on French mathematical treatises in the 1820s when it was supposedly so avant garde and he was in Nottingham. Green's major patron, Sir Edward Ffrench Bromhead, is one of the original members of the Analytical Society, at whose estate the Analysts used to plot how to take over the Cambridge examination system. See Garland, *Liberal Education.*

[47] George Peacock, "Report on the Recent Progress and Present State of Certain Branches of Analysis," *British Association Report* (1833): 185–352. Fourier appears pp. 254–259, p. 254 on Fourier's proofs. For Peacock's mathematics see Elaine Koppelmann, "The Calculus of Operations and the Rise of Abstract Algebra," *Archive for History of Exact Sciences* **8** (1972): 155–242.

[48] Whewell, "Report on the Recent Progress and Present Condition of the Mathematical Theories of Electricity, Magnetism, and Heat," *British Association Report* (1835): 1–34, 29.

[49] Philip Kelland, *Theory of Heat* (Cambridge: J. J. Deighton, 1837) and "On the Present State of Our Theoretical and Experimental Knowledge of the Laws of Conduction of Heat," *British Association Report* (1841): 1–25.

[50] George Chrystal and Peter Guthrie Tait, "Philip Kelland," *Proceedings of the Royal Society of Edinburgh* **10** (1879): 321–329, saw Kelland's work in heat as showing great mathematical ingenuity but was based "upon an unsound physical assumption." Kelland's real work lay in pure mathematics.

[51] David Gregory, "Preface," *Cambridge Journal of Mathematics* **1** (1837–1839): 1–2. Gregory, "Notes on Fourier's Heat," *ibid.,* 115–118.

[52] Thomson, "On Fourier's expansion of functions in trigonometric series," *Cambridge Journal of Mathematics* **2** (1839–1841): 258–262. This paper is on a purely mathematical point in Fourier and Kelland. Thomson, "Notes on a Passage in Fourier's Heat," *ibid.* **3** (1841): 25–27, which is an exercise in Fourier analysis.

[53] Thomson, "Uniform Motion of Heat in Homogeneous Solid Bodies and Its Connection with the Mathematical Theory of Electricity," *Cambridge Journal of Mathematics* **3** (1841–1843): 71–84. Gravitation is introduced on p. 83. Thomson, "On the Equations of the Motion of Heat Referred to Curvilinear Coordinates," *ibid.* **4** (1843–1845): 33–42.

[54] Thomson, "Demonstrates of a Fundamental Proposition in the Mechanical Theory of Electricity,," *ibid.* 223–226. The essay Thomson refers to is Gauss, "Allgemeine Lehrsätze." Gauss states that gravitation, electrostatics and magnetism are "special cases of the particular mathematical solution being sought," p. 241.

[55] See Crosbie Smith and Norton Wise, *Energy and Empire: A Biographical Study of Lord Kelvin* (Cambridge: Cambridge University Press, 1989). While Norton Wise has argued that Thomson was lead to conservation through mathematics, he does not see

Thomson as anything but a physicist. Wise, "William Thomson's Mathematical Route to Energy Conservation: A Case Study in the Role of Mathematics in Concept Formation," *Historical Studies in the Physical Sciences* 10 (1979): 49–83.

[56] Thomson, "On the Mathematical Theory of Electricity in Equilibrium," *Cambridge Journal of Mathematics* 5 (1846): 75–93, on 86 a translation with additions to his paper for Liouville.

[57] Thomson, "Note on Some Points in the Theory of Heat," *Cambridge Journal of Mathematics* 4 (1843–1845): 67–72 for Thomson's mathematical discussion of negative time. For the implications of his physical understanding of time see Joe D. Burchfield, *Lord Kelvin and the Age of the Earth* (New York: Science History Pub., 1975).

[58] Stokes, "Supplement to a Memoir on Some Cases in Fluid Motion," *Cambridge Philosophical Society, Transactions* 8 (1847): 409–416. He compared mathematical results with his own experiments, concluding that his experiments were too crude to decide the issue.

[59] Stokes, "On Fresnel's Theory of the Aberration of Light," *Philosophical Magazine* 28 (1846): 76–81. The only mathematics here is geometry and algebra, "On the Constitution of the Luminiferous Ether, Viewed with Reference to the Phenomenon of the Aberration of Light," *ibid.* 29 (1846): 6–10. In the first paper of this series Stokes gave a simplified explanation of the aberration of light consistent with wave theory and minimum calculus. "On the Aberration of Light," *ibid.* 27 (1845): 9–15.

[60] Stokes, "On the Critical Value of the Sums of Periodic Series," *Cambridge Philosophical Society, Transactions* 8 (1847) in Stokes, *Mathematical and Physical Papers*, 6 vols. (Cambridge: Cambridge University Press, 1880), vol. 1, pp. 236–313. The applications in this paper are mathematical, not physical.

[61] David E. Rowe, "Klein, Hilbert and the Göttingen Mathematical Tradition," *Osiris* 5 (1989): 186–213.

[62] For the affirmation of physics as experiment see Jungknickel and McCormmach, *Intellectual Mastery*, vol. 1, and Olesko, *Physics as a Calling*. See Laurie Brown and Helmut Rechenberg, "The Development of the Vector Meson Theory in Britain and Japan (1937–38)," *British Journal of the History of Science* 24 (1991): 401–433, on 417 for attitudes in Britain in the 1930s.

OLE KNUDSEN

ELECTROMAGNETIC ENERGY AND THE EARLY HISTORY OF THE ENERGY PRINCIPLE

I. INTRODUCTION

The story of the concept of electromagnetic energy began, one might claim, in 1845, when Franz Neumann realized that electromagnetic forces and torques could be derived from a potential function, and that the law of electromagnetic induction could be formulated by means of the same potential function.

Neumann's paper appeared in the memoirs of the Berlin Academy for 1845 (Neumann 1889), and a substantial extract was published in Poggendorff's *Annalen* (Neumann 1846). Two years later Neumann published yet another paper, in which the electrodynamic potential figured as the basic quantity in electrodynamics (Neumann 1892).

The 1845 extract caught the attention of the young army surgeon Hermann Helmholtz, who was just then working on his famous essay on energy conservation. (Stationed in Potsdam, Helmholtz did not have access to the Berlin memoirs; fortunately, however, the local *Gymnasium* subscribed to the *Annalen*.)

When Helmholtz's essay appeared in 1847 (Helmholtz 1907) it contained a final section entitled "Kraftäquivalent des Magnetismus und Electromagnetismus." In this section Helmholtz made use of Neumann's potential to work out the consequences of the energy principle as applied to the interaction of electrical circuits and magnets. One consequence seemed to be that the existence of electromagnetic induction, as well as the mathematical law for induced electromotive forces, could be derived from the energy principle (cf. Bevilacqua 1993, pp. 330–332).

As I show below, Helmholtz's reasoning was erroneous. It is in fact not possible to derive the law of induction from the energy principle, and modern textbooks invariably reverse the procedure, i.e., they first postulate the law of induction and then derive from this an expression for the energy of an electromagnetic system. Nevertheless, Helmholtz's way

55

A.J. Kox and D.M. Siegel (eds.), No Truth Except in the Details, 55–78.

of reasoning is very seductive. A year later, William Thomson published an argument very similar to Helmholtz's (Thomson 1882, 91–92), and James Clerk Maxwell published a similarly erroneous treatment in his first article on electricity and magnetism, written in 1855–56. Even in his 1873 *Treatise*, Maxwell repeated Helmholtz's argument almost verbatim, remarking without any reservation that

they [i.e. Helmholtz and Thomson] showed that the induction of electric currents discovered by Faraday could be mathematically deduced from the electromagnetic actions discovered by Örsted and Ampère by the application of the principle of the Conservation of Energy (Maxwell 1891, II, art. 543).

The earliest correct treatment of the energy of two electric circuits is found in an unpublished memorandum, written in 1851 by Thomson. A report of the result of this investigation was published by Thomson in 1860, while the memorandum itself was published only in 1872 (Thomson 1872, 441–443). It is interesting that Maxwell followed his account in the *Treatise* of Helmholtz's argument by quoting Thomson's correct result, namely

... that when work is done by the mutual action of two constant currents, their mechanical action is *increased* by the same amount, so that the battery has to supply *double* that amount of work, in addition to that required to maintain the currents against the resistance of the circuits (Maxwell 1891, II, art. 544, his emphasis).

Apparently, Maxwell did not see that the derivation of this statement was in conflict with the earlier treatments by Helmholtz and Thomson. In the third edition of Maxwell's *Treatise*, J.J. Thomson inserted a footnote showing that the law of induction cannot be derived from the energy principle. This argument was, however, based on Maxwell's theory of electric circuits, in which the expression for electromagnetic energy was based on the postulate that electromagnetic energy is of the nature of kinetic energy, so that a system of circuits may be treated by analytical mechanics. For this reason J.J. Thomson's argument does not give a clear indication of the precise nature of Helmholtz's error, and his reasoning is difficult to follow for a modern reader.

Evidently, a historian interested in the early history of the concept of electromagnetic energy meets with a certain amount of error and confusion in the primary sources. Secondary accounts are often of little help, witness the above-mentioned discussion by J.J. Thomson, or E.T. Whittaker's few scattered remarks, which are purely verbal and so brief as to be unintelligible to the uninitiated reader (Whittaker 1951, 218–219, 221–222).

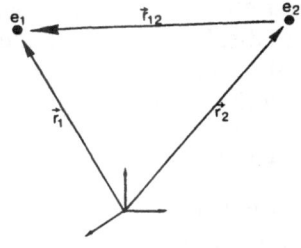

Fig. 1.

In this paper I analyze in detail the writings of Helmholtz, William Thomson, and Maxwell on this point. As my aim is to clarify the logical structure of the various arguments, I use a standardized modern notation (with SI units) almost everywhere, but I shall of course take care not to invest the symbols with a physical content foreign to the writer in question.

In Sections II and III, I present a modern textbook account of the concepts of electrostatic and electromagnetic energy, with special emphasis on the question of the relation between energy conservation and the law of induction. This account is fairly trivial and serves merely as a background for the historical discussion in the following sections.

II. TWO POINT CHARGES

As a preliminary to our discussion of electromagnetic energy, let us consider a simple electrostatic system consisting of two point charges (Figure 1) e_1 and e_2, with position vectors r_1 and r_2, and with relative position vector

$$\mathbf{r}_{12} = \mathbf{r}_1 - \mathbf{r}_2.$$

This system has a potential energy U_p given by

(2.1) $U_p = \dfrac{e_1 e_2}{4\pi\varepsilon_0 r_{12}}$

with the property that the electrostatic force $\mathbf{F}_1$ on e_1 is given by

(2.2) $\mathbf{F}_1 = -\nabla_1 U_p$

where the gradient operator refers to r_1, the position of e_1.

If e_1 undergoes a small displacement δr_1 under the action of the force F_1, e_2 being held fixed, the kinetic energy T of the system increases by an amount equal to the work performed by the force F_1:

$$(2.3) \quad \delta T = \delta W = F_1 \cdot \delta r_1.$$

Note that (2.3) is a consequence solely of Newton's second law of motion and the usual definition of kinetic energy in terms of masses and velocities of the two particles.

Now, the existence of the potential energy function U_p with the property (2.2) means that we may rewrite Equation (2.3) as

$$(2.4) \quad \delta T = -\delta U_p$$

or as

$$(2.5) \quad \delta T + \delta U_p = 0$$

and this leads to

$$(2.6) \quad T + U_p = E$$

where E is a constant representing the total energy of the system.

These three equations are just slightly different expressions of the principle of conservation of mechanical energy, defined as the sum of kinetic and potential energy. Helmholtz stated this principle as follows

... der Verlust an Quantität der Spannkraft [ist] stets gleich dem Gewinn an lebendiger Kraft, und der Gewinn der ersteren dem Verlust der letzteren. *Es ist also stets die Summe der vorhandenen lebendigen und Spannkräfte constant* (Helmholtz 1907, 14, his emphasis).

From Helmholtz's definitions of "lebendiger Kraft" and "Spannkraft" it is unproblematic to identify these concepts with kinetic and potential energy, respectively, and to see the whole passage as a verbal explanation of Equations (2.4) and (2.6). The word "Spannkraft" or "tension force," with its connotations of springs and elastic materials in general, suggests in this context something like "force stored in the system," so that there is no large step to a modern conception of Equation (2.6) as expressing a balance between kinetic energy and internal electrostatic energy stored in the system.

A word must be added on Helmholtz's use of the term "potential." It is introduced in the beginning of the section on electrostatics (with a reference to Gauss) as the negative of our U_p, so that the increase in kinetic energy is equal to the *increase* in the potential of the system. This was of no great conceptual significance to Helmholtz, since the potential was a mathematical function devoid of physical meaning; but the difference in algebraic sign is an additional source of confusion to a modern reader.

Finally, it is worth emphasizing a fairly trivial point, namely that Equations (2.4) and (2.5) hold for any system which possesses a potential energy function. For arbitrary small displacements within such a system, the differential work performed by the internal forces and torques will always be given by

(2.7) $\delta W = -\delta U_p.$

The replacement of δW by δT follows, as already noted, from Newton's second law without any further assumption (or, in the case of angular displacements, from the angular momentum principle).

III. TWO CIRCUITS

After these elementary observations, we are ready to consider a system of two electric circuits ℓ_1 and ℓ_2, carrying currents I_1 and I_2, produced by galvanic batteries supplying electromotive forces V_1 and V_2. With the symbols shown in Figure 2, the total electromagnetic force $\mathbf{F}_1$ exerted by circuit ℓ_2 on circuit ℓ_1 is given by the following formula, which combines the Lorentz force expression with the Biot–Savart law for the magnetic field produced by ℓ_2:

(3.1) $\mathbf{F}_1 = \dfrac{\mu_0}{4\pi} I_1 I_2 \displaystyle\oint_{l_1} \oint_{l_2} \dfrac{dl_1 \times (dl_2 \times \mathbf{r}_{12})}{r_{12}^3}$

where dl_1 and dl_2 are line elements of circuits 1 and 2 respectively, and the integrations are over the closed circuits. The total force $\mathbf{F}_2$ on circuit ℓ_2 is given by a similar expression, and one can show that these forces satisfy the third law of motion:

(3.2) $\mathbf{F}_1 = -\mathbf{F}_2.$

It can also be shown that these forces may be derived from a potential energy function of the form

(3.3) $U_p = -I_1 \Phi_{12} = -I_2 \Phi_{21}$

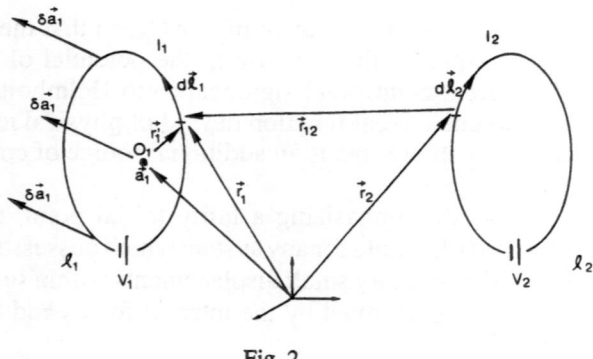

Fig. 2.

where Φ_{12} is the magnetic flux through ℓ_1 due to the current in ℓ_2, and Φ_{21} is the flux through ℓ_2 due to ℓ_1.

The fluxes in Equation (3.3) are given by the formulas

(3.4) $\Phi_{12} = MI_2$

(3.5) $\Phi_{21} = MI_1$

where M is the mutual inductance, given by an expression known today as Franz Neumann's formula:

(3.6) $M = \dfrac{\mu_0}{4\pi} \displaystyle\oint_{l_1} \oint_{l_2} \dfrac{dl_1 \cdot dl_2}{r_{12}}$.

The result is that the potential energy function may be written as

(3.7) $U_p = -MI_1I_2 = -\dfrac{\mu_0}{4\pi} I_1 I_2 \displaystyle\oint_{l_1} \oint_{l_2} \dfrac{dl_1 \cdot dl_2}{r_{12}}$.

Proofs of these formulas may be found in for instance (Grant and Phillips 1975, 154–159).

The function U_p is essentially the electrodynamic potential that Neumann introduced in 1845. His major contribution was the theorem that the induced electromotive force in the circuit ℓ_1, produced by ℓ_2, can be expressed mathematically by putting I_1 equal to unity and then taking the time derivative of U_p (Neumann 1846, 38–40). From Equation (3.3) it is easily seen that this theorem is equivalent to the modern expression of induced electromotive force as minus the time derivative of the magnetic flux (cf. Equation (3.16) below).

As a simple illustration, analogous to the one in Section II above, of the relation of U_p to kinetic energy and work, we consider a small, rigid translation, $\delta \mathbf{a}_1$, of the circuit ℓ_1. Let $\mathbf{a}_1$ be the position vector of an arbitrary point O_1, fixed with respect to ℓ_1, so that the position vector $\mathbf{r}_1$ of any circuit element $d\ell_1$ is

$$(3.8) \quad \mathbf{r}_1 = \mathbf{a}_1 + \mathbf{r}_1'$$

where $\mathbf{r}_1'$ is constant during the translation. Then

$$(3.9) \quad \mathbf{F}_1 = -\nabla_1 U_p$$

where ∇_1 means the gradient operator referring to $\mathbf{a}_1$. From this it follows that the work performed by the electromagnetic force on ℓ_1 is given by

$$(3.10) \quad \delta W = \mathbf{F}_1 \cdot \delta \mathbf{a}_1 = (-\nabla_1 U_p) \cdot \delta \mathbf{a}_1 = -\delta U_p.$$

If ℓ_1 is moving freely we may, just as in the electrostatic case, use the second law of motion to replace the work performed by the increase in kinetic energy, and write

$$(3.11) \quad \delta T + \delta U_p = 0.$$

So far, we have a complete formal analogy with the electrostatic case. However, as textbook writers sometimes take care to warn their readers, Equation (3.11) cannot be interpreted as a correct expression for the energy balance in the system. In particular, U_p does not represent the energy stored in the system (cf. Grant and Phillips 1975, 154n).

To see why this is so, we must involve Ohm's law and Faraday's law of electromagnetic induction. If our two circuits are at rest, we have by Ohm's law

$$(3.12) \quad V_1 = R_1 I_1$$

and

$$(3.13) \quad V_2 = R_2 I_2$$

where R_1 and R_2 denote resistances. Multiplying by, respectively, $I_1 \delta t$ and $I_2 \delta t$, we get the following expressions for the energy balance in the two circuits during a short time interval δt:

$$(3.14) \quad V_1 I_1 \delta t = R_1 I_1^2 \delta t$$

$$(3.15) \quad V_2 I_2 \delta t = R_2 I_2^2 \delta t.$$

Here the left-hand sides are the amounts of energy supplied during δt by the two batteries, while the right-hand sides are the amounts of Joulean heat developed in the two circuits.

Let us now go back to the case where ℓ_1 undergoes a rigid translation $\delta \mathbf{a}_1$ during the time interval δt, while ℓ_2 is kept in a fixed position. By Faraday's law the induced electromotive force in a circuit is given by

$$(3.16) \quad V = -\frac{d\Phi}{dt}$$

where Φ is the magnetic flux through the circuit. Hence Equations (3.12–3.13) must now be replaced by

$$(3.17) \quad V_1 - \frac{d\Phi_1}{dt} = R_1 I_1$$

$$(3.18) \quad V_2 - \frac{d\Phi_2}{dt} = R_2 I_2.$$

The flux through ℓ_1 is given by

$$(3.19) \quad \Phi_1 = L_1 I_1 + M I_2$$

where L_1 is the self-inductance of ℓ_1 and M is the mutual inductance given by Equation (3.6). As we have assumed ℓ_1 to be rigid, L_1 is a constant, while M varies with the changing relative position of the two circuits during the translation of ℓ_1. As a further simplification we assume the batteries to be regulated in such a way that the two currents keep constant values during the translation. Then the only non-constant quantity in Equation (3.19) is M, and we may rewrite Equations (3.17–3.18) as

$$(3.20) \quad V_1 - I_2 \frac{dM}{dt} = R_1 I_1$$

$$(3.21) \quad V_2 - I_1 \frac{dM}{dt} = R_2 I_2.$$

Before we use these equations to analyze the energy balance in the system, let us take stock of the energy contributions we have considered so far. These are the following:

- The energy δW_b supplied by the batteries. This is given by

$$(3.22) \quad \delta W_b = (V_1 I_1 + V_2 I_2)\delta t.$$

- The Joulean heat δQ produced by the currents in the two circuits, given by

$$(3.23) \quad \delta Q = (R_1 I_1^2 + R_2 I_2^2)\delta t.$$

- The work δW performed by the electromagnetic forces. By the discussion leading up to Equation (3.10) we may write

$$(3.24) \quad \delta W = -\delta U_p = I_1 I_2 \delta M.$$

- The change δU in the internal energy U of the system.

Since the energy supplied by the batteries must equal the changes in the various forms of energy appearing in the system, we can write the following equation for the energy balance during the translation:

$$(3.25) \quad \delta W_b = \delta Q + \delta W + \delta U.$$

Here the only unknown quantity is δU. We might replace the other terms by the expressions (3.22–3.24), and in a certain sense δU would then be determined from given quantities. These quantities would, however, include the electromotive forces of the batteries and the resistances of the circuits, and so we would not be able to express U solely as a function of the currents and of the form and relative position of the two circuits.

To reach this goal, we multiply Equations (3.20–3.21) by $I_1 \delta t$ and $I_2 \delta t$, respectively, and add the results. This gives

$$(3.26) \quad (V_1 I_1 + V_2 I_2)\delta t = (R_1 I_1^2 + R_2 I_2^2)\delta t + 2I_1 I_2 \frac{dM}{dt} \delta t$$

or, replacing $(dM/dt)\delta t$ by δM and using Equations (3.22–3.24),

(3.27) $\delta W_b = \delta Q + 2\delta W.$

Combined with Equation (3.25) this leads to

(3.28) $\delta U = \delta W = -\delta U_p = I_1 I_2 \delta M.$

Although our analysis has been confined to the very special case of a rigid translation with constant currents, Equations (3.27–3.28) show some features which are true in general. One is that for every motion with constant currents, the increase in the energy stored as internal energy is precisely equal to the mechanical work performed by the electromagnetic forces, so that the batteries have to supply an amount of energy equal to twice the work performed (in addition to the amount spent in Joulean heat). This means that the potential energy function U_p does not play the role it does in electrostatics or gravitational theory, that of energy stored in the system. This role is played instead by the function U, given by

(3.29) $U = -U_p.$

In fact, a more complete analysis including variable currents gives the result that Equation (3.29) is true in general, and that the internal energy of a system of two circuits is given by

(3.30) $U = \dfrac{1}{2} I_1 \Phi_1 + \dfrac{1}{2} I_2 \Phi_2 = \dfrac{1}{2} L_1 I_1^2 + M I_1 I_2 + \dfrac{1}{2} L_2 I_2^2$

(cf. Grant and Phillips 1975, 236–239; or Reitz and Milford 1967, 231–236). Clearly, the last term in Equation (3.28) follows from Equation (3.30) in the special case of constant currents and constant self-inductances.

The above analysis gives a clear view of the possibility of deriving the law of electromagnetic induction from the energy principle, Equation (3.25), for we have derived three of the four energies involved without using the law of induction or even presupposing the existence of the phenomenon of induction. However, in Equation (3.25) the internal energy appears as an unknown quantity, so before one could derive anything from this equation one would have to make some kind of conjecture about this quantity. Given the analogy with the electrostatic case, the most obvious procedure would be to identify the internal energy with the potential energy U_p. By Equation (3.10) the two last terms in Equation (3.25) would then cancel out and the latter would reduce to

(3.31) $\delta W_b = \delta Q.$

This is simply the sum of the two equations (3.14–3.15) which have been derived from Ohm's law applied to *stationary* circuits. One would thus obtain two conservation equations, one for mechanical energy, namely Equation (3.10), and one for electrical energy, namely Equation (3.31). The latter would imply that Equations (3.12–3.13) would hold also for *moving* circuits, in other words that induced electromotive forces do not exist. It is thus possible to frame a theory, even a very plausible one, which conforms to the principle of energy conservation but denies the existence of electromagnetic induction. This proves conclusively that the law of induction cannot be derived from the energy principle. Only by using the law of induction, Equation (3.16), can one obtain a correct expression for the internal energy stored in a system of two circuits.

IV. HELMHOLTZ

I shall now take a closer look at the arguments by which Helmholtz, Thomson, and Maxwell claimed to derive the law of induction from the energy principle. One difficulty about these arguments is that Helmholtz and Thomson both began by considering a system of a circuit and a permanent magnet in relative motion and then went on to extend the argument to a system of two circuits. They no doubt regarded the former system as the simpler one; however from a modern point of view a permanent magnet is a very complicated system, so modern textbooks tend to avoid discussing it. I shall return to this problem in Section V below.

Helmholtz's argument, as it is found in his 1847 essay (Helmholtz 1907, 47–50), runs somewhat as follows: The amount of *vis viva* gained by a magnet which is moving under the influence of a current, must be supplied from the tension forces ("Spannkräfte") consumed in the current. According to (Neumann 1846), this amount of *vis viva* gained in a short time interval δt is $I(dE/dt)\delta t$, where I is the intensity of the current and E is the potential of the magnet relative to the conductor when the current in the latter is unity. Thus,

$$(4.1) \quad VI\delta t = RI^2\delta t + I\frac{dE}{dt}\,\delta t$$

where V is the electromotive force driving the current and R is the resistance of the conductor. Hence

$$(4.2) \quad I = \frac{V - (dE/dt)}{R}$$

and we have thus deduced, from the energy principle, the existence of a new kind of electromotive force, induced electromotive force, represented by $-dE/dt$.

In the case of two conductors in relative motion, the same reasoning leads, according to Helmholtz, immediately to the equation

$$(4.3) \quad (V_1I_1 + V_2I_2)\delta t = (R_1I_1^2 + R_2I_2^2)\delta t + I_1I_2\frac{dE}{dt}\,\delta t$$

where E now denotes the potential of the two conductors when both are carrying currents of unit intensity. If $I_2 \ll I_1$ we may set $V_1 \approx R_1I_1$, and we then have

$$(4.4) \quad I_2 \simeq \frac{V_2 - I_1(dE/dt)}{R_2}\,.$$

Again we have derived from the energy principle the existence of an induced electromotive force, given by the expression $-I_1(dE/dt)$.

Apart from insignificant changes in notation, this is an almost verbatim rendering of Helmholtz's argument. In the case of two currents, this argument is clearly at variance with our analysis in Section III. To pinpoint the difference, we may compare Equation (4.3) with our Equation (3.25). The left-hand side of Equation (4.3) and the first term on the right-hand side can immediately be identified with our δW_b and δQ, respectively, so that the difference has to do with the last term in Equation (4.3). In view of what Helmholtz says about the corresponding term in Equation (4.1), the most natural interpretation of his reasoning seems to be that it represents the "amount of *vis viva* gained" which Helmholtz then expresses in terms of Neumann's potential. This leads us to identify this term with our δT or δW, expressed by means of Equation (3.24); the quantity E in Equation (4.3) would then be the same as our M, the mutual inductance. What is missing in Helmholtz's equation, then, is a term corresponding to our δU, the change in the energy stored in the system.

At first sight, the defect in Helmholtz's argument is not terribly surprising. After all, the missing term is, in modern theory, the change in the magnetic field energy, and this is a concept which was completely foreign to Helmholtz in 1847. On the other hand, one may argue that Helmholtz's treatment was inconsistent from his own point of view as well, for the following reason.

As we have seen in Section II, Helmholtz set up an equation for a system of electrical particles which expressed a balance between the *vis viva* of the particles' motion and the "tension forces" of the system, the latter being expressed mathematically by the electrostatic potential. He went on to show that a similar balance existed in a purely magnetic system, consisting of permanent magnets and soft iron, and that the tension forces could be expressed by a magnetic potential of the same mathematical form as the electrostatic one. He then proceeded to quote Ampère's result that an electric current may be replaced by a certain distribution of magnetism on a surface bounded by the circuit, and said that in this way Neumann had transferred the notion of potential to closed currents, by equating the electrodynamic potential with the potential of the magnetic surfaces representing the currents (Helmholtz 1907, 47). Thus, for Helmholtz, Neumann's electrodynamic potential was a quantity of the same nature as the electrostatic potential, furnishing a convenient mathematical expression for the tension forces in the system. Since he identified the three terms in Equation (4.1) with "the tension forces consumed in the current," "the *vis viva* produced in the circuit," and "the *vis viva* gained by the magnet," respectively, there is some justification for accusing him of omitting a term corresponding to the magnetic tension forces in the system.

My conclusion is, then, that Helmholtz's argument was not only defective from a modern point of view, but also inconsistent with his own treatments of electro- and magnetostatics. This statement is, however, rather too blunt as a historical conclusion and needs to be qualified by a closer look at the historical context. First of all, if Helmholtz had reasoned "consistently," the analogy between the electrodynamic and the magnetostatic case would undoubtedly have led him to identify the tension forces in the system with Neumann's potential. As we have seen at the end of Section III, this would have entailed the non-existence of electromagnetic induction, which would have been as absurd to Helmholtz as it is to us. Secondly, it must be remembered that before Helmholtz, "potential" denoted an almost purely mathematical entity, which Helmholtz was just beginning to give a physical meaning in terms of "tension force." This, in turn, was a new concept that Helmholtz himself had introduced in his general formulation of the conservation principle: it had not yet acquired the full range of physical significance associated with the modern concept of potential energy. In his applications of the general principle to electricity and magnetism, Helmholtz preferred using the more familiar concepts of *vis viva* and potential. In his section on electrostatics Helmholtz used "tension force" only once, and in the section on magnetism not at all; in his section on electromagnetism he used it only to denote the energy supplied by a battery. Thus, although in order to arrive at Equations (4.1) and (4.3) he must have used something like Equation (3.11) to equate the gain in *vis viva* with the change in potential, he seems to have reverted to an older understanding of Equation (3.11) as a mathematical formality, forgetting its new role as an expression of the conservation principle, entailing a notion of potential as tension force stored in the system.

One further remark is needed. Helmholtz's Equation (4.3) conflicts with our Equation (3.26), which we derived from Faraday's law of induction. So how could Helmholtz derive from it Equation (4.4), which is a correct statement of the Faraday law? The answer is that he did so by an erroneous argument, as we can easily see by applying his argument to our Equation (3.26); the result will be an incorrect factor of two in front of the induced electromotive force. To show exactly where his argument is at fault, we divide Equation (4.3) by $I_1 \delta t$ and obtain the following equation:

$$(4.5) \quad V_1 - R_1 I_1 + \frac{I_2}{I_1} \left(V_2 - I_1 \frac{dE}{dt} - R_2 I_2 \right) = 0.$$

Now it is true that if I_2/I_1 is small then $V_1 \approx R_1 I_1$, but this does not imply that the bracket multiplying I_2/I_1 is small as well. So the argument is simply a mathematical blunder, involving the mishandling of an approximation. (The manifestation of this in the straightforward route from Equation (4.3) to Equation (4.4) is in dividing by I_2, which is of the order of zero in this approximation.)

V. A CIRCUIT AND A PERMANENT MAGNET

Both Helmholtz and Thomson based their discussion of a circuit and a magnet in relative motion on the tacit assumption that a permanent magnet is totally uninfluenced by the changing magnetic field from the circuit. Modern physics would probably not accept this as even an idealisation of the behaviour of a ferromagnetic material. Since I am concerned here with the consistency of 19th-century theories rather than with the properties of real materials, I shall follow Helmholtz and Thomson and consider a magnet at rest to produce a magnetic field which is constant in time although of course variable in space.

We may now consider a system like that shown in Figure 2, except that the circuit ℓ_2 is replaced by a permanent magnet producing a time-independent magnetic field. If Φ_m denotes the magnetic flux from the magnet through the circuit ℓ (as there is now only one circuit we drop the indices), the total force on the circuit can still be derived from a potential energy function, given by

$$(5.1) \quad U_p = -I\Phi_m$$

where I is the current in ℓ. For a small rigid displacement of ℓ, with I kept constant, we again have an energy balance equation like Equation (3.25), and inserting known quantities we get

$$(5.2) \quad VI\delta t = RI^2\delta t + I\delta\Phi_m + \delta U.$$

Faraday's law of induction gives for this case

$$(5.3) \quad V - \frac{d\Phi_m}{dt} = RI$$

from which it follows that

$$(5.4) \quad VI\delta t = RI^2\delta t + I\frac{d\Phi_m}{dt}\delta t.$$

Since Φ_m has no explicit time-dependence the Φ_m-terms in Equations (5.2) and (5.4) are identical. We may thus write Equation (5.4) as

$$(5.5) \quad \delta W_b = \delta Q + \delta W$$

and combining this with Equation (5.2) we get the surprising result

$$(5.6) \quad \delta U = 0.$$

This means that the magnetic energy stored in the system does not depend on the relative position of the circuit and the magnet. It also means that Helmholtz's Equation (4.1) is in fact correct, in the sense that it is consistent with his tacit assumption on the nature of a permanent magnet. This does not mean that Helmholtz's reasoning is valid, for it was only by invoking the Faraday law that we were able to derive Equation (5.5); the energy principle alone leads to Equation (5.2) in which δU is still an unknown quantity.

There is thus an essential difference between the two cases we are considering, even though Helmholtz's reasoning in both cases was flawed in the same way by his neglect of the internal energy term. In the case of two circuits this led to an equation (4.3) that was in manifest conflict with the law of induction (but from which Helmholtz nevertheless derived the law of induction by a mathematical error), whereas in the case of a circuit and a magnet the resulting equation (4.1) agreed perfectly with the law of induction.

Finally a few remarks are in order about Equation (5.6) which has some implications that seem almost absurd at first sight. As an example, consider a cylindrical bar magnet homogeneously magnetized throughout its length and with a solenoid wound tightly around it. Let the current in the solenoid be adjusted so that its magnetic field everywhere cancels that of the magnet. Now, with the current kept constant, let the magnet be pulled out of the solenoid and removed to a great distance. Equation (5.6) then implies that the magnetic field energy of the system undergoes no change during this process. This is surprising because in the initial state there is no magnetic field anywhere, while in the final state we have a solenoid and a magnet, each surrounded by its own magnetic field. There would thus seem to be a positive increase in magnetic field energy during the process, in violation of Equation (5.6). However, inside the substance of the magnet the B-field is by assumption zero in the initial state, while in the final state the field vectors B and H are both non-zero but in opposite directions (assuming the magnetization to be constant). The magnetic energy inside the magnet therefore decreases during the process – energy is being extracted from the interior of the magnet – and this decrease compensates for the increase in the field energy outside the magnet.

VI. THOMSON AND MAXWELL

William Thomson first discussed these matters in a short note published in the British Association Report for 1848 (Thomson 1882, 91–92). Although Thomson almost certainly had not read Helmholtz's essay, which had been privately printed less than a year earlier, his argument was essentially the same. He considered a closed conductor subject to induction by a moving magnet, and equated "the amount of work expended in producing the relative motion" with "the mechanical effects lost by the current induced in the wire." Having no battery in his circuit, he arrived at two equations which correspond to Equations (4.1–4.2) with V equal to zero, and claimed that he had given an *a priori* demonstration of Neumann's theorem on induction, founded on the axiom of the equality of the work spent in sustaining the motion and the mechanical effect lost in the wire. Like Helmholtz, Thomson in this note made no mention of the possibility that energy might be stored in the system, and, again like Helmholtz, he therefore arrived at a correct result by a defective argument.

Three years later Thomson reconsidered the question in a memorandum dated "Oct. 13, 1851" but only published much later as a footnote in his *Reprint of Papers on Electrostatics and Magnetism*, which appeared in 1872 (Thomson 1872, 441–442n). The memorandum shows that Thomson had now mastered the problem completely. Instead of attempting to derive the law of induction from the energy principle, he now used Neumann's formulation of the law of induction as a basis for the derivation of energy balance equations. His procedure was essentially similar to the one I have used above in deriving Equations (5.5) and (3.27); his results correspond exactly to Equations (5.6) and (3.28), namely that "the mechanical values of the current and the magnet together are not altered," while in the case of two circuits "the mechanical value of two currents is diminished by $(1/J)wdt$." (The latter quantity denotes the work spent by external forces and is therefore equal to $-\delta W$ in my notation.)

In 1860 Thomson found occasion to publish, in an article for Nichol's *Cyclopædia*, a verbal statement of the insight he had reached in 1851. The following passage is a very clear explanation of the content of our Equations (3.27–3.28), and since it is the earliest published correct amount of the energy balance of two circuits, it is worth quoting in full:

If two conductors, with a current sustained in each by a constant electro-motive force, be slowly moved towards one another, and there be a certain *gain of work* on the whole, by electro-dynamic force, operating during the motion, there will be twice as much as this of work spent by the electro-motive forces (for instance, twice the equivalent of chemical action in the batteries, should the electro-motive forces be chemical) over and above that which they would have had to spend in the same time, merely to keep up the currents, if the conductors had been at rest, because the electro-dynamic induction produced by the motion will augment the currents; while on the other hand, if the motion be such as to require the *expenditure* of work against electro-dynamic forces to produce it, there will be twice as much work saved off the action of the electro-motive forces by the currents being diminished during the motion. Hence the aggregate mechanical value of the currents in the two conductors, when brought to rest, will be increased in the one case by an amount equal to the work done by mutual electro-dynamic forces in the motion, and will be diminished by the corresponding amount in the other case (Thomson 1872, 442–443n, his emphasis).

In addition this article contained a result not found in the 1851 memorandum. Thomson explained that by applying similar considerations to the relative motion of portions of the same circuit he had obtained a mathematical expression for the "mechanical value" of a single current. From his verbal description it is clear that he had found the expression

$$(6.1) \quad U = \frac{1}{2} LI^2$$

which is a special case of Equation (3.30).

He went on to state another important result:

The mechanical value of a current in a closed circuit, determined on these principles, may be calculated by means of the following simple formula, not hitherto published:

$$(6.2) \quad \frac{1}{8\pi} \int\int\int R^2 dx dy dz$$

where R denotes the resultant electro-magnetic force at any point (x, y, z).

Since the expression (6.2) is completely equivalent to the modern expression for magnetic field energy,

$$(6.3) \quad U = \frac{1}{2\mu_0} \int \mathbf{B}^2 d\tau$$

– where $\mathbf{B}$ is the magnetic induction, μ_0 is the magnetic permeability of free space, and the integration is over all spatial elements $d\tau$ – we may conclude that the concept of magnetic field energy was born sometime between 1851 and 1860. It is at least certain that the formula (6.2) would have allowed Thomson to think of the "mechanical value" of a current as residing, not in the current itself, but in the magnetic field associated with it. As I have argued elsewhere, the mathematical analogy between (6.2) and the expression for the kinetic energy of a moving fluid would have supported this interpretation (Knudsen 1985, 158–164; cf. Smith and Wise 1989, 255).

VII. MAXWELL

In his first electromagnetic paper, "On Faraday's Lines of Force," written in 1855, Maxwell presented a generalized version of Helmholtz's argument (Maxwell 1890, I, 203–205). With modernized notation and disregarding missing factors as well as inconsistencies in algebraic signs, Maxwell's treatment may be rendered as follows:

He considered first a distribution of magnetism and introduced a quantity called the total potential of the system. As this quantity plays the same role as our potential energy function, I denote it by U_p in the following. Maxwell first expressed his potential in terms of "real magnetic density" and "magnetic tension" and then transformed it into the expression

$$(7.1) \quad U_p = -\frac{1}{2} \int \mathbf{H} \cdot \mathbf{B} d\tau$$

where $\mathbf{H}$ is the magnetic field. Although Equation (7.1) had been derived for a system of permanent magnets, Maxwell assumed it to be true also for an electromagnetic system. He then transformed it once again into

$$(7.2) \quad U_p = -\frac{1}{2} \int \mathbf{A} \cdot \mathbf{j} d\tau$$

where $\mathbf{j}$ is the current density and $\mathbf{A}$ is the "Electro-tonic intensity," related to the "quantity of magnetic induction" $\mathbf{B}$ by

$$(7.3) \quad \mathbf{B} = \nabla \times \mathbf{A}.$$

Maxwell accompanied the transformation from Equation (7.1) to Equation (7.2) by the following comment:

We have now obtained in the functions [A] the means of avoiding the consideration of magnetic induction which *passes through* the circuit. Instead of this artificial method we have the natural one of considering the current with reference to quantities existing in the same space with the current itself (Maxwell 1890, I, 203, his emphasis).

In view of the strong commitment to the primacy of the field concept which characterizes Maxwell's later work (cf. Siegel 1991, 98), it is interesting that he at this stage should have regarded a field expression as artificial and an expression in terms of sources as the more natural one. In an abstract of his paper Maxwell elaborated this point by showing that $\mathbf{A}$ is indeed a natural representation of Faraday's concept of an electrotonic state (Maxwell 1991, 373–375); this made its use especially attractive to Maxwell.

Maxwell's next step was to use Helmholtz's method to derive the law of induction in terms of the "electro-tonic intensity." He first wrote an expression for the work "in the form of resistance overcome":

$$(7.4) \quad \delta Q = \delta t \int \mathbf{E} \cdot \mathbf{j} d\tau$$

where $\mathbf{E}$ is the "electromotive force." He then expressed the "work done mechanically by the electro-magnetic action of these currents" in terms of the "total potential," as

$$(7.5) \quad \delta W = -\delta U_p = \delta t \frac{1}{2} \frac{d}{dt} \int \mathbf{A} \cdot \mathbf{j} d\tau.$$

The sum of δQ and δW represented, so Maxwell argued, the whole work done by the external causes producing the currents. If no such causes operated this quantity must vanish, and so, for an isolated system,

$$(7.6) \quad \delta Q + \delta W = \delta t \int \mathbf{E} \cdot \mathbf{j} d\tau + \delta t \frac{1}{2} \frac{d}{dt} \int \mathbf{A} \cdot \mathbf{j} d\tau = 0$$

"where the integrals are taken through any arbitrary space." Assuming the currents to be constant, Maxwell concluded from Equation (7.6) that the equation

$$(7.7) \quad \mathbf{E} = -\frac{1}{2} \frac{d\mathbf{A}}{dt} = -\frac{1}{2} \left[(\mathbf{v} \cdot \nabla)\mathbf{A} + \frac{\partial \mathbf{A}}{\partial t} \right]$$

must hold for every point in the system. (Here $\mathbf{v}$ is the velocity of a moving part of the material system.)

Maxwell's treatment was historically important because he succeeded in formulating the law of induction in the form of a partial differential equation linking the induced electric field with changes in the vector potential $\mathbf{A}$. However, comparing Equation (7.6) with Equation (3.25) we see that Maxwell, like Helmholtz, lacked a term corresponding to our δU, the energy stored in the system (cf. the discussion in Section IV above). And, by translating Maxwell's equations into a consistent modern notation, we have seen that Maxwell's reasoning leads to the wrong result, since the resulting Equation (7.7) has an erroneous factor of 1/2 on the right-hand side. (If, instead of Equation (7.6), we use the correct Equation (3.27), the result comes out right.) It should be added that all of Maxwell's expressions for the "total potential" were too large by a factor of two, so that his final result corresponding to Equation (7.7) in fact did not contain this erroneous factor.

VIII. CONCLUSION

In this paper I have been concerned with an episode in the early history of the energy principle. It took place in a period when the energy concept was being extended to embrace more and more new areas of science, and when the concept itself was receiving an increasingly well-defined physical meaning and mathematical expression. This episode shows one of the problems involved in this process, the problem of arriving at a clear understanding of energy stored in the form of internal energy in an electromagnetic system. A central aspect of this problem had to do with the transformation of the concept of potential from a purely mathematical auxiliary function, by means of which electrostatic problems could be given a convenient mathematical formulation in terms of the differential equations of Laplace and Poisson, to a physical quantity, potential energy, having an important role to play in the principle of energy conservation. For its mathematical uses, potential could be defined as a function with the property that a certain vector field of interest – such as an electric, magnetic, or gravitational force, or a velocity – could be derived as either plus or minus the gradient of the potential, the choice of algebraic sign being of no consequence whatever. Thus Gauss chose the plus sign, while Green chose the minus. Of the authors we have considered here, Helmholtz and Maxwell followed Gauss, Thomson followed Green, and Franz Neumann changed from plus in 1845 to minus in 1847. For potential energy, on the other hand, it is essential to choose the minus sign, as in Equation (2.2), if the concept is to function properly as an expression for stored energy; thus, the vacillations in choice of sign indicate the extent to which potential was still primarily a mathematical concept, with little physical significance.

There were two ways in which potential began to acquire shades of physical meaning. One was through analogy. Thus Thomson explored the analogy between potential/force and temperature/heat flux, while Neumann and Maxwell used pressure in a fluid as an analogy for potential. The second way consisted in relating changes in potential to work performed, and thereby to *vis viva* gained or lost by the system. This resulted in equations like our Equation (2.4), with a plus or minus sign on the right-hand side. From this there was only a short mathematical step to a conservation equation like our Equation (2.6). Conceptually, however, this step was not so small, as is shown by the fact that Helmholtz had to invent a new concept, *Spannkraft*, in order to state his conservation principle. It is also characteristic that when Helmholtz came to apply his principle to electromagnetism, he made no mention of *Spannkraft* (except in connection with the batteries); thus, although he expressed the *vis viva* gained in terms of changes in potential, he did not see changes in potential as associated directly with *Spannkraft* lost or gained.

It is not, I think, an accident that William Thomson was the first to reach a full understanding of the energy balance in an electromagnetic system, nor that he achieved this in the years between 1848 and 1851. During this period he was hard at work on the development of mathematical field theories of electricity and magnetism, at the same time as he was striving to grasp Joule's views on the equivalence of heat and work and extend them into a dynamical theory of heat. As shown by Smith and Wise, the concept of "mechanical value" emerged as a central, unifying concept in Thomson's thinking in this period. This concept, with its financial connotation of accumulated capital, served Thomson as a measure of the ability of a system to perform mechanical work, and could be equated with the "total potential" of a system – a concept that played an important role in his field theories. This means that Thomson, unlike everybody else, had his attention firmly directed to a notion corresponding very closely to the internal energy of a physical system, be it thermodynamic or electromagnetic (Smith and Wise 1989, chs. 8–10, particularly pp. 255–256). This is also illustrated by the fact that while Thomson in 1851 published a whole paper entitled "On the Quantities of Mechanical Energy Contained in a Fluid ... " (Thomson 1882, 222–232), in Clausius's first paper on thermodynamics the internal energy appeared only once, as an unnamed "arbitrary function of v

and t," which accounted for part of the heat given to a body, but had no physical significance in itself (Clausius 1850, 384).

If Clausius in 1850 did not have a working concept of internal energy, it is no wonder that Helmholtz did not have one in 1847. It is more surprising that Maxwell in 1855 had not learned about this from his mentor, Thomson, with whom he was in frequent correspondence while working on his paper. It seems that what Maxwell acquired from Thomson was primarily the mathematics of field theory and the use of mathematical analogies; there is no evidence that he studied Thomson's thermodynamical work or that he knew of Thomson's unpublished memorandum.

My conclusion is, then, that Thomson in 1850 was the only one who possessed a fully developed conception of the internal energy of a physical system, and that he, therefore, was in a unique position to clear up the difficulties associated with the application of the energy conservation principle to electromagnetic systems.

Department of History of Science
University of Aarhus
Denmark

REFERENCES

Bevilacqua, F. (1993). "Helmholtz's *Ueber die Erhaltung der Kraft*: The Emergence of a Theoretical Physicist," in *Hermann von Helmholtz and the Foundations of Nineteenth-Century Science*, D. Cahan, ed. (Berkeley: University of California Press), pp. 291–333.

Clausius, R. (1850). "Ueber die bewegende Kraft der Wärme und die Gesetze, welche sich daraus für die Wärmelehre selbst ableiten lassen," *Annalen der Physik und Chemie* **79**: 368–397, 500–524.

Grant, I.S. and W.R. Phillips (1975). *Electromagnetism* (Chichester: John Wiley & Sons).

Helmholtz, H. (1907). *Über die Erhaltung der Kraft* (Berlin: G. Reimer, 1847), reprinted as *Ostwald's Klassiker der exakten Wissenschaften, Nr. 1* (Leipzig: W. Engelmann).

Knudsen, O. (1985). "Mathematics and Physical Reality in William Thomson's Electromagnetic Theory," in *Wranglers and Physicists: Studies on Cambridge Physics in the Nineteenth Century*, P.M. Harman, ed. (Manchester: Manchester University Press), pp. 149–179.

Maxwell, J.C. (1890). *The Scientific Papers of James Clerk Maxwell*, 2 vols., W.D. Niven, ed. (Cambridge: Cambridge University Press).

Maxwell, J.C. (1891). *A Treatise on Electricity and Magnetism*, 3rd ed., 2 vols. (Oxford: Clarendon Press).

Maxwell, J.C. (1991). *The Scientific Letters and Papers of James Clerk Maxwell*, vol. I, P.M. Harman, ed. (Cambridge: Cambridge University Press).

Neumann, F.E. (1846). "Allgemeine Gesetze der inducirten elektrischen Ströme," *Annalen der Physik und Chemie* **67**: 31–44.

Neumann, F.E. (1889). "Die mathematischen Gesetze der inducirten elektrischen Ströme" (1845), reprinted as *Ostwald's Klassiker der exakten Wissenschaften, Nr. 10* (Leipzig: W. Engelmann).

Neumann, F.E. (1892). "Ueber ein allgemeines Princip der mathematischen Theorie inducirter elektrischer Ströme" (1847), reprinted as *Ostwald's Klassiker der exakten Wissenschaften, Nr. 36* (Leipzig: W. Engelmann).

Reitz, J.R. and F.J. Milford (1967). *Foundations of Electromagnetic Theory*, 2d ed. (Reading: Addison-Wesley).

Siegel, D. M. (1991). *Innovation in Maxwell's Electromagnetic Theory: Molecular Vortices, Displacement Current, and Light* (Cambridge: Cambridge University Press).

Smith, C. and M.N. Wise (1989). *Energy and Empire; A Biographical Study of Lord Kelvin* (Cambridge: Cambridge University Press).

Thomson, Sir W. (1872). *Reprint of Papers on Electrostatics and Magnetism* (London: Macmillan & Co.).

Thomson, Sir W. (1882). *Mathematical and Physical Papers*, vol. 1 (Cambridge: Cambridge University Press).

Whittaker, Sir E. (1951). *A History of the Theories of Aether and Electricity*, vol. 1 (London: Thomas Nelson and Sons Ltd.).

PETER M. HARMAN

THROUGH THE LOOKING-GLASS, AND WHAT MAXWELL FOUND THERE*

MAXWELL AND OPTICS

In his essay on "Mechanical Explanation at the End of the Nineteenth Century," Martin Klein remarked on "the complexity and variety of the ideas that were current then": this was "a time of probing and testing."[1] These judgements are aptly descriptive of the physics of James Clerk Maxwell, and especially of his most famous innovation, the electro-magnetic theory of light. His statement in 1862, that *"light consists in the transverse undulations of the same medium which is the cause of electric and magnetic phenomena,"*[2] implied the unification of optics and electromagnetism in terms of a mechanical theory of the ether that had both optical and electromagnetic correlates.[3] When he wrote his seminal *Treatise on Electricity and Magnetism* (1873) it might have been anticipated that Maxwell would broaden the scope of his electro-magnetic theory of light to encompass an electromagnetic theory of the reflection and refraction of light. But he did not do so; and though he gave a detailed treatment of the Faraday magneto-optic rotation (where he appealed to the rotation of molecular vortices in the ether), the range of his optical theory remained essentially similar in its physical content to that first advanced in 1862 and subsequently amplified in a major paper published in 1865.

An explanation of reflection and refraction in terms of the electro-magnetic theory would have strengthened the argument of the *Treatise*. Maxwell explained this lacuna in his February 1879 referee report for the Royal Society on a paper submitted by George Francis FitzGerald "On the Electromagnetic Theory of the Reflection and Refraction of Light."[4] FitzGerald outlined the situation in the introductory paragraph of his paper as subsequently published in the *Philosophical Transactions*:

79

A.J. Kox and D.M. Siegel (eds.), No Truth Except in the Details, 79–93.
© 1995 Kluwer Academic Publishers.

In the second volume of his 'Electricity and Magnetism' Professor J. Clerk Maxwell has proposed a very remarkable electromagnetic theory of light, and has worked out the results as far as the transmission of light through uniform crystalline and magnetic media are concerned, leaving the questions of reflection and refraction untouched. These, however, may be very conveniently studied from his point of view.[5]

The problem involved formulation of appropriate boundary conditions at the surface of separation of two media, and interpretation of the boundary conditions in terms of electromagnetic variables. Reviewing FitzGerald's argument, and referring to a related discussion of the derivation of the optical laws from electromagnetic principles by H.A. Lorentz in his Leiden dissertation of 1875, Maxwell observed that "in my book I did not attempt to discuss reflexion at all. I found that the propagation of light in a magnetized medium was a hard enough subject."

In 1864, in the course of writing his Royal Society paper on "A Dynamical Theory of the Electromagnetic Field" (1865), Maxwell had attempted to derive the laws of reflection and refraction from his electromagnetic theory of light, proposing electromagnetic analogues for the elastic variables employed in theories of the luminiferous ether. He reported on his efforts to George Gabriel Stokes – both in his capacity as Secretary of the Royal Society, and as an authority on optics and theories of the luminiferous ether – in a letter of 15 October 1864.[6] He declared that "I am not yet able to satisfy myself about the conditions to be fulfilled at the surface [of separation of two media]," finding the subject to be "a stiff one." The inherent complexity of the problem led Maxwell to exclude discussion of optical reflection and refraction from his paper.

In his report (for Stokes) on FitzGerald's paper, Maxwell described his own theory of the Faraday magneto-optic effect in the *Treatise* as a "hybrid theory, in which bodily motion of the [electromagnetic] medium is made to cooperate with the electric current"; it was not a "purely electromagnetic hypothesis." While he recognized that an explanation of the Faraday effect by a purely electromagnetic theory, as proposed by FitzGerald (whose ultimate aim it was "to emancipate our minds from the thraldom of a material ether"),[7] would be "a very important step in science," and thus envisaged a weakening of the link between electromagnetism and its mechanical representation, he nevertheless remarked that the value of FitzGerald's mathematical theory would have been greatly increased by an interpretation in terms of a "dynamical hypothesis." These comments show Maxwell to be committed to the dynamical programme of the *Treatise*.

Though cautious about the complexities of the wave theory of light, by the 1860s Maxwell had established himself as an authority in two fields within the science of optics: the study of colour vision, including the composition of colours and the problem of colour blindness; and the theory of geometrical optics. Yet this work was not central to the main thrust of optical theory and experiment in the first half of the nineteenth century, which was concerned with the wave theory of light and the dynamical theory of the luminiferous ether. The rather marginal status of work on colour vision in the 1850s can be deduced from correspondence between William Thomson (Maxwell's most intimate scientific correspondent in this decade) and George Gabriel Stokes in January and February 1856, on Maxwell's paper "Experiments on Colour, as Perceived by the Eye" (1855).[8] Thomson queried,

Have you seen Clerk Maxwell's paper in the Trans. R.S.E. on colour as seen by the eye? Are you satisfied with the perfect accuracy of Newton's centre of gravity principle on w[h] all theories & nomenclatures on the subject are founded? That is to say do you believe that the whites produced by various combinations, such as two homogeneous colours, three homogeneous colours &[c] are absolutely indistinguishable from one another & from solar white by the best eye? . . . [9]

In response, Stokes, a leading authority on optics, stated that "I have not made any experiments on the mixture of colours, nor attended particularly to the subject."[10] But by November 1857 his acquaintance with Maxwell's work was such as to lead him to compliment the younger physicist that his "results afford most remarkable and important evidence in favour of the theory of 3 primary colour perceptions, a theory which you and you alone so far as I know have established on an exact numerical basis."[11] This enthusiastic endorsement led to concrete public approbation. At a meeting of the Council of the Royal Society in June 1859, Stokes and William Whewell nominated Maxwell for a Royal Medal "for his Mathematical Theory of the Composition of Colours, verified by quantitative experiments, and for his Memoirs on Mathematical and Physical subjects," and another (unsuccessful) nomination was made the following year.[12] But a nomination in May 1860 by Stokes and the Cambridge Professor of Mineralogy W.H. Miller for the Rumford Medal (which was awarded especially for studies of light and heat), for Maxwell's "Researches on the Composition of Colours, and other optical papers," met with success;[13] this followed the submission of his paper "On the Theory of Compound Colours" to the *Philosophical Transactions*,[14] and his appointment to read the paper as the Society's Bakerian Lecture for 1860. (As he was not, however, at the time a Fellow of the Royal Society, he was found to be ineligible for appointment as Bakerian Lecturer.[15])

Following his election as a Fellow of the Royal Society in May 1861,[16] Maxwell was soon pressed into service as a referee on papers submitted for publication in the Society's *Transactions*. Stokes, a Secretary of the Society, had every reason to consider Maxwell a suitable referee for papers on optics; Maxwell was soon asked to report on a paper by Samuel Haughton, "On the Reflexion of Polarized Light from Polished Surfaces, Transparent and Metallic."[17] Writing to Stokes on 16 July 1862, Maxwell found that the paper contained "many valuable observations" and recommended publication, giving a résumé of Haughton's argument and conclusions.[18] As an authority on optics, Stokes had himself already written a report on Haughton's paper;[19] this he forwarded to Maxwell in response to Maxwell's letter. Stokes was severely critical of Haughton's mode of experimentation and his failure to fully cite previous experimental work on the subject, notably by Jules Jamin;[20] Stokes also strongly questioned Haughton's claim that the ratio of the amplitudes of the components of the reflected waves polarized in and perpendicular to the plane of incidence varied with the azimuth of the polarizer. Nevertheless, Stokes declared himself "prepared, in case the other referee be decidedly favourable to the publication, to recommend that the paper be printed subject to slight modification . . . ."

In response to Stokes's communication of this report, Maxwell immediately acknowledged Stokes's greater expertise. In a letter of 21 July 1862 he concurred with Stokes's recommendations: "I therefore agree with you that the author should be requested to point out the claims of his paper to publication . . . ."[21] This Haughton was able to do,[22] and the paper proceeded to publication, with the relation to Jamin's earlier work more clearly specified.[23]

This episode indicates that Maxwell was not familiar with the breadth of contemporary work in optics, admitting (in his second letter to Stokes) that he had not read Jamin's work, which developed the classic study by Brewster earlier in the century on the reflection of polarized light. Until Stokes raised the issue, Maxwell did not question Haughton's discussion of the ratio of amplitudes. It would seem therefore that Maxwell did not have command over some of the intricacies of the wave theory of light.

PROBLEMS OF ETHER THEORY

Despite Maxwell's interest in topics of optical science, his lack of expertise in what had, up to 1850, been seen as the optical mainstream is not wholly surprising. The wave theory had certainly been a major focus of debate during the second quarter of the century, following the assimilation of the work of Young and especially Fresnel. Its importance for mathematical physics was certainly well recognized in Cambridge, where it was the only field of contemporary mathematical physics to retain its place in the Mathematical Tripos after the exclusion of the mathematical theories of electricity, magnetism, and heat in the reformed Tripos of 1849.[24] Maxwell's undergraduate notes on optics are restricted to topics in geometrical optics; but his coach William Hopkins did consider physical optics to be important, and he attended Stokes's lectures, which included discussion of the wave theory of light, in 1853.[25] Maxwell was certainly familiar with Airy's *Mathematical Tracts*, a major Tripos text, where the undulatory theory of light was given prominence. But Airy did not venture into the recondite technicalities of ether dynamics. He discussed Fresnel's first (and simpler) theory of optical reflection and refraction, based on the conservation of *vis viva* and of momentum, but not the subsequent and more labyrinthine developments by Fresnel and others.[26] But after 1850 the subject lost its status as a major field of research in mathematical physics, capable of attracting the attention of the most creative younger physicists. Yet while Maxwell's admission to Stokes in October 1864 that ether theory was "stiff," even "to the best skilled in undulations," indicates his lack of expertise in this field, his creation of the electromagnetic theory of light in the early 1860s led him to investigate two central areas of ether theory.

In April 1864 Maxwell set up an "Experiment to Determine Whether the Motion of the Earth Influences the Refraction of Light."[27] Having read Fizeau's paper[28] which established that the velocity of light in a tube carrying a stream of water "takes place with greater velocity in the direction in which the water moves than in the opposite direction," as Maxwell summarized its result, he wished to investigate the matter using a different experimental arrangement (in a manner in part suggested by subsequent work of Fizeau's).[29] Fizeau had explained his result in terms of Fresnel's theory of partial ether drag; but in calculating the effect of the Fresnel drag on the refraction of light by a glass prism,

Maxwell ignored the compensating change in the density of the medium. According to Fresnel's theory, the ether and the transparent medium satisfy a continuity equation at their boundary; this has the consequence that the retardation due to the refractive medium is not affected by the motion of the earth.[30] Stokes drew Maxwell's attention to his error when his paper on the subject was sent to the Royal Society.[31] Indeed, in a paper "On Fresnel's Theory of the Aberration of Light" (1846) Stokes himself had considered the effect of the motion of the ether on the refraction of light, concluding that the motion of the ether would have no effect on refraction.[32]

But in the paper sent to Stokes, Maxwell calculated the deflection that would result from an arrangement of prisms, as in the spectroscopic apparatus constructed by the leading London instrument-maker Carl Becker, which involved an observing telescope, three prisms, and a second telescope with a plane mirror at its focus, so that after refraction through the prisms light rays would be returned along their path. Thus "if the effect due to motion takes place . . . the ray will no longer return to its starting point but will be displaced to an extent double of its original displacement." He mounted the apparatus on a turntable so that the effect could be reversed, and he predicted a total deflection of $2\frac{1}{2}$ arc-minutes; but he found that "no displacement could be observed." Hence "the result of the experiment is decidedly negative to the hypothesis about the motion of the ether in the form stated here."

Maxwell withdrew the paper in response to Stokes's criticism of his argument, but he did give an account of the experiment in a letter to the astronomer William Huggins dated 10 June 1867. Here he declared that the earth's motion would not have an effect on the refraction of light, now pointing out that Stokes had proved this conclusion, which was also supported by an earlier experiment by Arago. Maxwell's own experiments – "tried . . . at various times of the year since the year 1864" – had "never detected the slightest effect due to the earth's motion." William Huggins included Maxwell's letter in a paper of his own (of 1868), where it appears as a separate section of the paper;[33] and Maxwell later described the experiment in his *Encyclopaedia Britannica* article on "Ether."[34] He again discussed the possible measurement of ether drag, suggesting that the ether could perhaps be detected by measuring the variation in the velocity of light when light was propagated in opposite directions, in a letter of 19 March 1879 to the American astronomer David Peck Todd. After Maxwell's death later that year, Todd sent his letter to Stokes; the letter was published in the Royal Society's *Proceedings* and in *Nature* in 1880.[35] Maxwell's discussion of a double track arrangement led Michelson to undertake his famous experiments on ether drag in the 1880s, so Maxwell's work on the problem did have wider influence.

As far as Maxwell was concerned, the episode made him aware of the difficulties of incorporating a full theory of the luminiferous ether within his electromagnetic theory. Responding in a letter of 6 May 1864 to Stokes's critique of his attempt to detect ether drag (this letter being the source for reconstructing Stokes's comments on the paper),[36] Maxwell declared that

I am not inclined and I do not think I am able to do the dynamical theory of reflexion and refraction on different hypotheses & unless I see some good in getting it up, I would rather gather the result from men who have gone into the subject.

He did not, in the event, let the matter rest there. He was here alluding to the incorporation of the laws of reflection and refraction within the terms of his paper "A Dynamical Theory of the Electromagnetic Field" (1865), and in writing the paper for submission to the Royal Society he decided to confront the issue. The evidence consists in a letter to Stokes of 15 October 1864 and a related manuscript fragment.

In his letter to Stokes, Maxwell explained that he had been reading a paper on the reflection and refraction of light published by Jules Jamin in 1860.[37] His sketchily outlined argument in the draft is based on Jamin's presentation.[38] Jamin had discussed the boundary conditions that determine the oscillation of the ether at the interface between two media, following, rather than Fresnel,[39] James MacCullagh and Franz Neumann, who had supposed that "the vibrations in two contiguous media are equivalent," as MacCullagh expressed it.[40] From this condition, Jamin proceeded to derive the optical laws, developing equations connecting the oscillations with the angles of incidence and refraction.

In his draft, Maxwell was attempting to establish an electromagnetic theory of optical reflection and refraction, seeking to interpolate results drawn from the electromagnetic theory of light into Jamin's and Fresnel's expressions for the oscillations of the ether at the interface between two media. Thus, he equated the displacement in the ether with the "electric displacement" in the electromagnetic medium, one of the cardinal concepts which Maxwell had deployed in obtaining his electromagnetic theory of light.[41]

The conservation of *vis viva* at the interface is stated by Fresnel as well as by MacCullagh and Neumann. This is the one feature of the dynamical ether models that Maxwell could accept. He informed Stokes that "I am not yet able to satisfy myself about the conditions to be fulfilled at the surface except of course the condition of conservation of energy." His terminology of course reflects the establishment of the principle of the conservation of "energy" in the late 1840s and 1850s. Yet Fresnel, and MacCullagh and Neumann, had derived the energy equation on different assumptions. Fresnel obtained the energy equation on the supposition that the densities of the ether in the two media were different; while MacCullagh and Neumann had supposed the equality of the density of the ether in the two media. Maxwell questions Jamin's solution, based on the MacCullagh–Neumann boundary condition of the "equality of the motion both horizontal & vertical in the two media" and the assumption that "no such vibrations could exist in the media unless they were of equal density." Maxwell criticized the selectivity of the conditions and assumptions, an endemic feature of dynamical ether theories.

Therefore the general theory, which ought to be able to explain the case of media of unequal density (even if there were none such) must not assume equality of displacements, of contiguous particles on each side of the surface.[42]

Maxwell did not pursue this attempt to derive the laws of reflection and refraction from an electromagnetic theory of light. In the draft fragment he failed to apply Fresnel's theory consistently, as the result of a trivial slip. The fragmentary nature of this endeavour; his admission to Stokes that "I think you once told me that the subject was a stiff one to the best skilled in undulations"; and his earlier remark to Stokes of his disinclination to attack the subject and preference to "gather the result from men who have gone into the subject," all indicate his lack of easy familiarity with ether dynamics, a judgement that is confirmed by his error on the Fresnel drag. As a result he informed Stokes, in the letter of 15 October 1864, that "I have written out so much of the theory as does not involve the conditions at bounding surfaces and will send it to the R.S. in a week." Thus "A Dynamical Theory of the Electromagnetic Field," received by the Royal Society on 27 October 1864,[43] contains no treatment of "the conditions at a surface for reflexion and refraction." Nor, as Maxwell explained to Stokes in February 1879, did the *Treatise*: the subject was too "hard." In both the Royal Society paper and the *Treatise* he confined his discussion of optical problems to the propagation of light through magnetic fields and crystalline media; to the relation between electric conductivity and opacity; and to the relation between the dielectric constant and the index of refraction of transparent media. All these topics are of interest, and gave rise to much theoretical and experimental investigation in the years following Maxwell's triumphant announcement, in letters of October and December 1861 to Michael Faraday and William Thomson, of the first version of his most famous work, the electromagnetic theory of light. Of these problems, Maxwell's treatment of the Faraday magneto-optic effect is of special interest and significance, for this result has a central bearing on the development and articulation of his field theory. The present account of the role of optics in Maxwell's field theory will therefore conclude with a review of this topic.

THE FARADAY EFFECT AND TOPOLOGY

The basis of Maxwell's physical theory of the electromagnetic field presented in his paper "On Physical Lines of Force" (1861–62) was a model of "molecular vortices" oriented along magnetic field lines. In writing to Faraday in November 1857, Maxwell had looked to the further development of his theory of the field – originally presented in his paper "On Faraday's Lines of Force" (1856) – in relation to Thomson's explanation of the Faraday effect (that is, the rotation of the plane of polarization of linearly polarized light by a magnetic field).[44] Thomson had supposed that this phenomenon was caused by the rotation of molecular vortices in the ether, the axes of revolution of the vortices being aligned along the direction of the lines of force. Thomson's paper, published in the Royal Society's *Proceedings* in 1856 and reprinted in the *Philosophical Magazine* the following March,[45] soon excited Maxwell's interest. Writing to his Cambridge friend C.J. Monro in May 1857, he remarked that he was working at "a Vortical theory of magnetism & electricity which is very crude but has some merits." The problem of the rotation of molecular vortices in a fluid, of special interest to Thomson at the time, is discussed in a letter to Thomson of November 1857; and in early 1858 he outlined an experiment on a freely rotating magnet which could establish the effect of revolving vortices within the magnet.[46] He referred to this experiment again in letters to Faraday and Thomson of October and December 1861, having had the apparatus constructed and having tried the experiment, though without success.[47] The theory of molecular vortices of "On Physical Lines of Force" was therefore of long gestation.

It seems likely that Maxwell had originally envisaged the paper "On Physical Lines of Force" as being in two parts, on the theory of molecular vortices as applied first to magnetism and second to electric currents. But during the summer of 1861 "in the country" at his home Glenlair in Galloway (as he informed Faraday and Thomson, his two mentors in field theory), he developed his mechanical ether theory along new lines. He calculated the velocity of transverse elastic waves in a cellular ether, supposing the elastic properties of the ether to have electromagnetic correlates. As he informed Thomson in December 1861, he established "the nearness between two values of the velocity of propagation of magnetic effects and that of light." This discovery was apparently unexpected, leading him to triumphantly announce his "electromagnetic theory of light," as he later termed it,[48] in the third part of the paper. To complete his theory of physical lines of force he wished to give a quantitative treatment of the Faraday effect in terms of the rotation of molecular vortices. This had provided the starting-point of the whole investigation. In October 1861 he asked Faraday for information about experiments on the rotation of polarized light by magnets; he gave a preliminary account of his theory of the Faraday effect, which formed the substance of the fourth part of "On Physical Lines of Force," in writing to Thomson two months later:

I have also examined the propagation of light through a medium containing vortices and I find that the *only* effect is the rotation of the plane of polarization in the *same* direction as the angular momentum of all the vortices being proportional to
   A the thickness of the medium
   B the magnetic intensity along the axis
   C the index of refraction in the medium
   D inversely as the square of the wave length in air
   F directly as the radius of the vortices
   G ....the magnetic capacity ....[49]

With the alphabetization corrected, this is essentially how the argument is presented in the published paper, where – as in the letter to Thomson – he is able to appeal to various experimental results, notably work by Émile Verdet, in support of his theory.[50]

In the *Treatise* Maxwell presented an account of magneto-optics in terms of the mathematical style that pervades the work. This style emphasised the mathematical expression of physical quantities freed from their direct representation by a mechanical model. Thus, while he continued to affirm his commitment to the idea that the Faraday effect is caused by the rotation of vortices, he disclaimed the precise model advanced in "On Physical Lines of Force." The relation between electricity as a rotational phenomenon was now expressed geometrically rather than in terms of the mechanical model, in vectorial (quaternion) terms: thus the relation between the electric current C and magnetic force H is expressed by the equation

$$4\pi C = V \nabla H,$$

this last term representing the rotational character of magnetism, and (using Maxwell's own terminology) may be written as curl H.[51]

In considering the magneto-optic effect in terms of some rotational motion in space, Maxwell introduced arguments drawn from contemporary discussions in topology, including Johann Benedict Listing's "Vorstudien zur Topologie" (1847) and his "Der Census räumlicher Complexe" (1861),[52] in an attempt to classify the relations between curves and surfaces; these were of central importance in the physics of the *Treatise*, where the relation between "forces" acting along lines and "fluxes" acting across surfaces was fundamental. Maxwell employed ideas drawn from Listing's "Topologie" in discussing the problem of defining the directionality of linear and rotational motions. The feature of his discussion of this problem that relates directly to his treatment of the magneto-optic effect concerns the operation that Listing termed *Perversion*, "an effect similar to that of the reflexion in a mirror" as Maxwell explained it to Tait in May 1871.[53]

Maxwell pointed out that a plane-polarized ray of light can be represented by two circularly-polarized rays, one right-handed, the other left-handed (as regards the observer).

Any undulation, the motion of which at each point is circular, may be represented by a helix or screw . . . [and] the propagation of the undulation will be represented by the apparent longitudinal motion of the similarly situated parts of the thread of the screw.

A plane-polarized ray can thus be represented by a left-handed and a right-handed helix. This geometrical representation of the two circularly-polarized rays shows that the rays of the same wave-length are "geometrically alike in all respects, except that one is the *perversion* of the other, like its image in a looking-glass." He maintains that the Faraday effect cannot be explained simply on the supposition that one of these rays has a shorter period of rotation than the other. The Faraday effect is a directional phenomenon which does "not depend solely on the configuration of the ray, but also on the direction of the motion of its individual parts." The rotation of light in the Faraday effect is "affected by the relation of the direction of rotation of the light to the direction of the magnetic force." This leads to the conclusion that "in a medium under the action of magnetic force something belonging to the same mathematical class as an angular velocity, whose axis is in the direction of the magnetic force, forms a part of the phenomenon." Thus "some rotatory motion is going on," and the angular velocity must be conceived as "the rotation ... of very small portions of the medium, each rotating on its own axis:" "This is the hypothesis of molecular vortices."[54]

The geometrical, dynamical, and optical arguments thus coalesce in this explanation of the Faraday effect. Geometry and optics are entwined in the "looking-glass" analogy. Writing to Tait in a "perverted" script in March 1873, following the publication of Lewis Carroll's *Through the Looking-Glass, and What Alice Found There*, Maxwell asked:

Why have *you* forgotten to send Alice. We remain in Wonderland till she appears. Till then no more from yours truly dp/dt.[55]

As Martin Klein has explained, it was Tait who put Maxwell's initials, JCM [$= dp/dt$], into the equations of thermodynamics.[56]

Department of History
Lancaster University
U.K.

NOTES

* This essay draws on work carried out in the preparation of my edition of *The Scientific Letters and Papers of James Clerk Maxwell*, which is in course of publication by Cambridge University Press. I am grateful to the Syndics of the Cambridge University Library and to the Council of the Royal Society for kind permission to reproduce documents. I am grateful to the Council of the Royal Society and the National Science Foundation for generous financial support of my work on the edition; and to the Department of the History of Science, Harvard University for providing facilities for this work. I thank Jed Buchwald and Alan Shapiro for discussion of the theme of the paper at a conference sponsored by the Dibner Institute and organized by A.I. Sabra, to whom I am grateful for the invitation.

[1] Martin J. Klein, "Mechanical Explanation at the End of the Nineteenth Century," *Centaurus* 17 (1972): 58–82, on 58–59.

[2] In "On Physical Lines of Force. Part III" (1862); see *The Scientific Papers of James Clerk Maxwell*, W.D. Niven, ed., 2 vols. (Cambridge: Cambridge University Press, 1890), vol. 1, p. 500; henceforth cited as Maxwell, *Scientific Papers*.

[3] See, especially, Daniel M. Siegel, *Innovation in Maxwell's Electromagnetic Theory. Molecular Vortices. Displacement Current, and Light* (Cambridge: Cambridge University Press, 1991), pp. 120–143.

[4] See *Memoir and Scientific Correspondence of the Late Sir George Gabriel Stokes, Bart*, J. Larmor, ed., 2 vols. (Cambridge: Cambridge University Press, 1907), vol. 2, pp. 40–43.

[5] G.F. FitzGerald, "On the Electromagnetic Theory of the Reflection and Refraction of Light," *Philosophical Transactions of the Royal Society* 171 (1880): 691–711, on 691.

[6] *The Scientific Letters and Papers of James Clerk Maxwell. Volume II: 1862–1873*, P. M. Harman, ed. (Cambridge: Cambridge University Press, 1995), pp. 186–188.

[7] FitzGerald, "On the Electromagnetic Theory of the Reflection and Refraction of Light," 711; and see Bruce Hunt, *The Maxwellians* (Ithaca: Cornell University Press, 1991), pp. 5–23.

[8] Maxwell, *Scientific Papers*, vol. 1, pp. 126–154.

[9] Thomson to Stokes, 28 January 1856, in *The Correspondence between Sir George Gabriel Stokes and Sir William Thomson, Baron Kelvin of Largs*, David B. Wilson, ed., 2 vols. (Cambridge: Cambridge University Press, 1990), vol. 1, p. 209.

[10] Stokes to Thomson, 4 February 1856, in *Correspondence*, Wilson, ed., vol. 1, p. 210.

[11] Stokes to Maxwell, 7 November 1857, in *The Scientific Letters and Papers of James Clerk Maxwell. Volume I: 1846–1862*, P.M. Harman, ed. (Cambridge: Cambridge University Press, 1990), p. 568n.

[12] *Minutes of Council of the Royal Society from December 16th 1858 to December 16th 1869*, vol. 3 (London, 1870), pp. 27, 62.

[13] *Minutes of Council of the Royal Society*, vol. 3, pp. 62–63, 72.

[14] Maxwell, *Scientific Papers*, vol. 1, pp. 410–444.

[15] *Minutes of Council of the Royal Society*, vol. 3, pp. 52, 59.

[16] *Minutes of Council of the Royal Society*, vol. 3, p. 85.

[17] The manuscript of Haughton's paper is preserved in Royal Society, PT. 68.4.

[18] Harman, ed., *Letters and Papers of Maxwell*, vol. 2, pp. 46–47.

[19] Harman, ed., *Letters and Papers of Maxwell*, vol. 2, pp. 50–42n.

[20] See especially Jules Jamin, "Mémoire sur la réflexion à la surface des corps transparentes," *Annales de Chimie et de Physique* ser. 3, **29** (1850): 263–304.

[21] Harman, ed., *Letters and Papers of Maxwell*, vol. 2, pp. 52–53.

[22] Harman, ed., *Letters and Papers of Maxwell*, vol. 2, p. 52n.

[23] See *Philosophical Transactions* **153** (1863): 81–125.

[24] See David B. Wilson, "The Educational Matrix: Physics Education at Early-Victorian Cambridge, Edinburgh and Glasgow Universities," in *Wranglers and Physicists: Studies on Cambridge Physics in the Nineteenth Century*, P.M. Harman, ed. (Manchester: Manchester University Press, 1985), pp. 14–19.

[25] See Harman, ed., *Letters and Papers of Maxwell*, vol. 1, pp. 9–11, 219 and Wilson, "The Educational Matrix," p. 16.

[26] G.B. Airy, *Mathematical Tracts on the Lunar and Planetary Theories ... and the Undulatory Theory of Optics*, 3rd ed. (Cambridge, 1842), pp. 342–343. For analysis of Fresnel's theories of optical reflection and refraction see Jed Z. Buchwald, *The Rise of the Wave Theory of Light. Optical Theory and Experiment in the Early Nineteenth Century* (Chicago: The University of Chicago Press, 1989), pp. 387–394.

[27] Harman, ed., *Letters and Papers of Maxwell*, vol. 2, pp. 148–153. For discussion see also C. W. F. Everitt, *James Clerk Maxwell: Physicist and Natural Philosopher* (New York: Charles Scribner's Sons, 1975), pp. 118–123.

[28] Hippolyte Fizeau, "Sur les hypothèses relatives à l'éther lumineux," *Annales de Chimie et de Physique* ser. 3, **57** (1859): 384–404; trans. in *Philosophical Magazine* ser. 4, **19** (1860): 245–258.

[29] A summary of this work was appended to the translation of Fizeau's paper cited in note 28 in *Philosophical Magazine* ser. 4, **19** (1860): 258–260.

[30] See the "Lettre de M. Fresnel à M. Arago," *Annales de Chimie et de Physique* **9** (1818): 57–66.

[31] Stokes's response to Maxwell's paper may be judged from Maxwell's letter of 6 May 1864: see below and note 36.

[32] G.G. Stokes, "On Fresnel's Theory of the Aberration of Light," *Philosophical Magazine* ser. 3, **28** (1846): 76–81.

[33] See Maxwell, "On the Influence of the Motions of the Heavenly Bodies on the Index of Refraction of Light," *Philosophical Transactions* **158** (1868): 532–535. See Harman, ed., *Letters and Papers of Maxwell*, vol. 2, pp. 306–311.

[34] Maxwell, *Scientific Papers*, vol. 2, pp. 769–770.

[35] See Maxwell, "On a Possible Mode of Detecting a Motion of the Solar System through the Luminiferous Aether," *Proceedings of the Royal Society* **30** (1880): 108–110 and *Nature* **21** (1880): 314–315.

[36] Harman, ed., *Letters and Papers of Maxwell*, vol. 2, pp. 154–156.

[37] Jules Jamin, "Note sur la théorie de la réflection et de la réfraction," *Annales de Chimie et de Physique* ser. 3, **59** (1860): 413–426.

[38] Harman, ed., *Letters and Papers of Maxwell*, vol. 2, pp. 182–185. This manuscript is discussed by Paul Theerman in his 1980 University of Chicago Ph.D. dissertation "James Clerk Maxwell: Physicist and Intellectual in Victorian Britain."

[39] A.J. Fresnel, "Mémoire sur la loi des modifications que la réflexion imprime à la lumière polarisée," *Mémoires de l'Académie Royale des Sciences* **11** (1832): 393–433.

[40] James MacCullagh, "On the Laws of Crystalline Reflexion and Refraction," *Transactions of the Royal Irish Academy* **18** (1837): 31–74; and Franz Neumann, "Theoretische

Untersuchung der Gesetze, nach welchen das Licht an der Grenze zweier vollkommen durchsichtigen Medien reflectirt und gebrochen wird," *Mathematische Abhandlungen der Königlichen Akademie der Wissenschaften zu Berlin aus dem Jahre 1835* (Berlin, 1837), pp. 1–160.

[41] Maxwell, *Scientific Papers*, vol. 1, pp. 489–502; and see Siegel, *Innovation in Maxwell's Electromagnetic Theory*, pp. 85–143.

[42] Maxwell to Stokes, 15 October 1864, cited in note 6.

[43] *Proceedings of the Royal Society* 13 (1864): 531–536.

[44] Maxwell to Faraday, 9 November 1857, in Harman, ed., *Letters and Papers of Maxwell*, vol. 1, pp. 548–552.

[45] William Thomson, "Dynamical Illustrations of the Magnetic and the Heliocoidal Rotatory Effects of Transparent Bodies on Polarized Light," *Proceedings of the Royal Society* 8 (1856): 150–158 (= *Philosophical Magazine* ser. 4, 13 (1857): 198–204).

[46] Harman, ed., *Letters and Papers of Maxwell*, vol. 1, pp. 507, 560–562, 579–580.

[47] Maxwell to Faraday, 19 October 1861; Maxwell to Thomson, 10 December 1861, in Harman, ed., *Letters and Papers of Maxwell*, vol. 1, pp. 683–688, 692–698.

[48] In "A Dynamical Theory of the Electromagnetic Field" (1865); see Maxwell, *Scientific Papers*, vol. 1, p. 577.

[49] Harman, ed., *Letters and Papers of Maxwell*, vol. 1, p. 697.

[50] Maxwell, *Scientific Papers*, vol. 1, pp. 502–513.

[51] James Clerk Maxwell, *A Treatise on Electricity and Magnetism*, 2 vols (Oxford: Clarendon Press), vol. 1, p. 28, vol. 2, p. 238.

[52] Johann Benedict Listing, "Vorstudien zur Topologie," in *Göttinger Studien. 1847. Erste Abteilung* (Göttingen, 1847), pp. 811–875; "Der Census räumlicher Complexe," *Abhandlungen der Mathematische Classe der Königlichen Gesellschaft der Wissenschaften zu Göttingen* 10 (1861): 97–182.

[53] Maxwell to Tait, 12 May 1871, Harman, ed., *Letters and Papers of Maxwell*, vol. 2, p. 644.

[54] Maxwell, *Treatise*, vol. 2, pp. 403–408.

[55] Maxwell to Tait, 5 March 1873, Harman, ed., *Letters and Papers of Maxwell*, vol. 2, p. 832.

[56] Martin J. Klein, "Maxwell, His Demon, and the Second Law of Thermodynamics," *American Scientist* 58 (1970): 84–97, esp. 94–95.

ERWIN N. HIEBERT

ELECTRIC DISCHARGE IN RAREFIED GASES: THE DOMINION OF EXPERIMENT. FARADAY. PLÜCKER. HITTORF.

The phenomena connected with the discharge of electricity through rarefied gases were observed shortly after the invention, in mid-17th century, of the air pump and the static electrical machine.[1] The colourful displays usually were attributed to chemical changes occurring in the gas, but on occasion they were thought to represent the metallic spectra caused by volatilization and ignition of the electrodes.[2] These phenomena often were regarded mainly as curiosities to be demonstrated in public by amateur investigators, of which there were many in England. In the late 1830s, Michael Faraday's experiments on electrical discharge in attenuated gases gave a new prominence to the subject. Over several decades these discoveries served to stimulate critical investigations at home and on the Continent, notably in Germany.

I. FARADAY: PATRON SAINT OF ELECTRICITY

Michael Faraday (1791–1867) was a natural philosopher whose formal training was almost nil. He identified himself primarily with the field of chemistry, which at the beginning of the 19th century included the study of heat, electricity, magnetism, and radiation. All of these domains, in due time, were absorbed – at least in part – into the discipline of physics. Faraday spoke of "physics" but had an intense dislike for the term "physicist." He was actively engaged in experimental research in his laboratory for over three decades at the Royal Institution on Albermarle Street; although he lived there, with his wife, Sarah, beyond his working days, a total of 46 years. Faraday, as John Tyndall remarked, "swerved incessantly from chemistry to physics."[3] While under the influence of Sir Humphrey Davy, he had devoted himself to chemistry. In 1820 he threw himself wholeheartedly and with phenomenal success into electromagnetic studies, shortly after he became aware of the chain

95

A.J. Kox and D.M. Siegel (eds.), *No Truth Except in the Details*, 95–134.
© 1995 *Kluwer Academic Publishers*.

of exciting experimental discoveries that had been made at Copenhagen and Paris: Hans Christian Oersted's observation of an electromagnetic effect, André-Marie Ampère's demonstration that the electrification of a wire gave rise to temporary magnetization and the alignment of iron filings in the vicinity, and François Arago's magnetization of steel needles by means of an electric current.[4]

With his discovery in 1821 of electromagnetic rotation, Faraday set in motion an ambitious research programme that led to the discovery in 1831 of electromagnetic induction. By the time he had formulated general views on electricity, around 1838, he had become convinced that a comprehensive theory of the flow of electricity necessarily would have to include phenomena associated with the conduction of electricity in gases under conditions of varying attenuation.

The discovery that some gases decomposed when subjected to electric tension led to a temporary shift in Faraday's focus from electromagnetism to chemistry. Before long he essentially had exhausted what could be achieved in the area of gas discharge phenomena with the instruments available to him. Those findings occupy but a small niche in Faraday's overall chemical and physical contributions. However, as we shall see, most of the investigators who pioneered in subsequent electrical discharge studies – persons such as Julius Plücker, Wilhelm Hittorf and Heinrich Hertz, all of whom greatly admired Faraday as an experimentalist – came to the subject indirectly.

As recorded in the *Diary*, in June of 1836, Faraday began a two-year study on this subject by having "an apparatus [glass globe] made for passage of sparks, brushes, glow, etc. between wire ends in different gases ... The glass of the Globe was thin but good and it bore exhaustion well and was very tight." Faraday discovered that "the phenomena vary with: Size of the ends. Distance of ends apart. P[ositive] of N[egative] end primarily electrified. Nearness of ends or wires to glass. Size therefore of the vessel. Nature of atmosphere within. Degree of rarefaction. Temperature of atmosphere. Quantity of Electricity. Substance of ends? Mixture of atmospheres?" The ramifications, seemingly endless, produced appearances "of great beauty."[5]

The results of these experiments were brought together in 1838 in Faraday's *Experimental Researches in Electricity*. There he spoke of the "very remarkable circumstances in the luminous discharge accompanied by negative glow" when electricity passes through rarefied air or other gases. On separating the two rods (electrodes), "a continuous glow came over the end of the negative rod, the positive termination remaining quite dark. As the distance was increased, a purple stream or haze appeared at the end of the positive rod, and proceeded directly outwards toward the negative rod; elongating as the interval was enlarged but never joining the negative glow, there being always a short dark space between them." Faraday explained the dark space by suggesting that the "discharge is taking place across the dark part of the dielectric ... [such that] the two electric forces are brought into equilibrium." He found that all gases gave the same result.[6]

Even a cursory examination of Faraday's experimental researches in electricity, beginning in 1831, reveals that his rarefied gas studies came as the logical extension of systematic attempts to test the validity of his theory that all electrical phenomena depend on the action of contiguous parts. He says: "It would seem strange, if a theory which refers all the phenomena of insulation and conduction, i.e. all electrical phenomena, to the action of contiguous particles, were to omit to notice the assumed possible case of a *vacuum* ... I think I have observed the luminous discharge to be principally on the inner surface of the glass; and it does not appear at all unlikely, that, if the vacuum refused to conduct, still the surface of glass next it might carry on that action ... My theory, as far as I have ventured it, does not pretend to decide upon the consequence of a vacuum ... I have only as yet endeavoured to establish, what all the facts seem to prove, that when electrical phenomena, as those of induction, conduction, insulation, and discharge occur, they depend on, and are produced by the action of *contiguous* particles of matter, the next existing particle being considered as the contiguous one."[7] In other words Faraday felt that a vacuum would not conduct an electric current, but there was no way for him to experiment with a containerless vacuum.

Not much immediate attention was given to Faraday's discharge experiments and theoretical deliberations of 1838, except for an occasional reference to his discovery of the negative dark space – a phenomenon that continued for decades to evoke special puzzlement. A decade later, comparisons were drawn between laboratory experiments with attenuated gases in discharge tubes and other phenomena in the open air. William R. Grove (1811–1896), barrister by training, electrochemist by inclination, inventor and supplier of improved voltaic cells to Faraday, told his audience at one of the Friday Evening Lectures at the Royal Institution in January 1859: "Few subjects of physical investigation possess greater interest than the electric charge; its brilliant effects and mysterious characteristics offer powerful stimuli to curiosity and enquiry." Pursuit of these studies, he believed, would become important in reference to theories of electricity "and probably assist much towards the proper conception of other modes of force, or as they are termed, *imponderables*, heat, light, etc." Grove, like Faraday, believed that "ordinary matter is requisite for the transmission of electricity, and that if space could exist void of matter, there would be no electricity: thus supporting the views ... that electricity is an affection or mode of action of ordinary matter."[8]

There were stronger and more esoteric motivations for pursuing electrical discharge studies besides the old, somehow inaccessible problem of the nature of electricity. It was well known that the transmission of electricity through gases was impeded (required smaller electrode separations, and more intense electric currents) both at ordinary pressures and in highly evacuated discharge tubes. All the beautiful and puzzling phenomena showed up between the extremes. The objective was to establish the "easy path for the electrical force" in relation to good and poor conduction, density, and resistance of the gas to current flow. Such studies, Grove observed, "afford much assistance to the theory of the aurora borealis, a phenomenon, the appearance of which, the regions where it is seen, its effect on the magnet, and other considerations, have led to the universal belief that it is electrical ... [with] currents of electricity circulating to and from the polar regions of the earth." The height of these "beautiful phenomena," where the transit of electricity takes place, "would be just that at which the density of the air is such as to render it the best conductor." These matters all would then be approximated in the laboratory.[9]

Grove was voicing an opinion that had considerable support among investigators at mid-century when he wrote:

Thus by our cabinet experiments, light may be thrown on the grand phenomena of the universe, and the great questions of the divisibility of matter, whether there is a limit to its expansibility, whether there is a fourth state of attenuation beyond the recognised states of solid, liquid, and gaseous, as Newton seems to suspect, (30th query to the Optics,) and whether the imponderables are specific affections of matter in a peculiar state, or of highly attenuated gaseous matter, may be elucidated. The manageable character of the electrical discharge, and the various phenomena it exhibits when matter is subjected to its influence in all those varied states which we are enabled, by experiment, to reduce it, can hardly fail to afford new and valuable information on these abstruse and most interesting enquiries.[10]

Grove's preoccupation with voltaic cells and electrochemical investigations led him to conclude "that gases do not conduct in any similar
manner to metals or electrolytes." While pursuing the phenomenon of
polarization in gases, he remarked: "The dark spaces in the discharge
to which Faraday has called attention, may possibly be connected with
these [polarity] phenomena . . . I have observed, that in a well-exhausted
receiver . . . the discharge is throughout its course striated by transverse
non-luminous bands presenting a very beautiful effect."[11] The striae
mentioned by Grove were observed in experiments on gaseous mix-
tures and allotropic phosphorus, and were assumed to be associated not
with the electrical discharge as such, but with electrolysis accompanied
by electrolytic decomposition.

Grove's comment about striations and the reference to Faraday's
experiments caught the attention of the wealthy wine merchant and
munificent friend to science, John Peter Gassiot, F.R.S. (1797–1877).[12]
Gassiot promptly demonstrated that the dark bands, striae, and stratifications of an electric discharge were observable in a well-exhausted
receiver such as a Torricellian vacuum. Equipped, in his own home on
Clapham Common, with apparatus as good as any then to be found in
all of London, Gassiot was quick to cash in on experimental discoveries
that brought him the Bakerian lectureship for 1858.[13]

The striations, Gassiot was proud to report, were beautiful. They
were readily produced in a thoroughly cleaned receiver free from all
trace of moisture, in a vacuum "as perfect as can be attained with
the air pump." They are, he remarks: "figured as concave towards the
positive end, the concavity *decreasing* as the bands extend towards
the negative; at the center they become straight, and then gradually
concave towards the negative terminal *until* they arrive at the [Faraday]
dark space which separates the bands from the negative discharge."
Gassiot felt that "there must have been something wrong in the mode of
obtaining vacua [because they] exhibited such irregularities" from one
apparatus to another. "Nothing satisfactory has yet been ascertained as
to the cause of the stratification of light."[14]

Most striking was the action of a magnet on the stratifications: from
the positive end, the latter were "violently drawn down the tube as an
elongated spiral," but there were not "any signs of stratification in the
negative discharge." Gassiot was inclined to the opinion that these two
effects arose from different "distinct causes – the former from pulsations
or impulses of a force acting on [a] highly attenuated but a resisting medi-

um, the latter from interference." Only further experimentation would lead to "the elucidation of this novel and remarkable phenomenon."[15]

In an appended note and in his second communication, Gassiot noted that he had received "vacuum tubes of great delicacy," designed by Heinrich Geissler in Bonn, but that they were too complex for his own apparatus, or presumably "constructed for a different object from what I have been pursuing, and for which I purposely had mine made in the most simple form I could devise."[16] Grove had provided Gassiot with vital information; the Geissler and Plücker tubes from Germany were too delicate and had "reluctantly [been] laid . . . aside." Faraday was the prime mover and had provided the inspiration: "I cannot conclude this Note without expressing the deep sense of obligation which I owe to Faraday, who has, during the course of these investigations, not only afforded me the advantage of many important suggestions, but has also spared me much of his valuable time."[17]

Apart from Gassiot's Faraday-inspired experiments on the phenomenon of stratification and Grove's lofty words about luminary orbs and speculations concerning attenuated forms of matter, experimentation in England at mid-century, and beyond, provided essentially no new insights concerning the constitution of the gaseous discharge or the process by which an electrical discharge is sustained in gases at various degrees of rarefaction. To the extent that Faraday's electrical discharge experiments were pursued in England at all, some progress is evident in instrumentation: various intricately constructed glass discharge tubes, improved vacuum pumps, powerful sources of electricity, and more efficient induction coils. In due time such instrumental refinements would pay off generously – but elsewhere, namely on the Continent.

II. CONTROVERSY IN THE MAKING: AN OVERVIEW

By the 1860s the center of gravity for electrical discharge studies had been displaced to Germany. This transfer of disciplinary prowess was generated largely by two Faraday enthusiasts – Julius Plücker and his student, assistant, and collaborator Wilhelm Hittorf. For about three decades they had a corner on experimental studies connected with electrical discharge in attenuated gases. With the death of Plücker in 1868 in Bonn, one year after Faraday died, the center for these studies shifted to Münster, where Hittorf, in a series of classic papers (1869–1884), gave prominence and a new problem orientation to a subject that had become rather sterile and exhausted from a descriptive point of view. An examination of the contributions of Plücker and Hittorf provides the *terminus ad quem* for this paper. Nevertheless, a brief résumé of the post-Hittorf trajectory of the history of cathode ray investigations will serve, at this point in the narrative, to provide the requisite perspective on a controversy that only achieved resolution hesitantly and painfully by the end of the century.

Hittorf's focus on cathode rays, as the salient and theoretically most significant aspect of discharge phenomena, would set the stage for a new round of investigations that he no longer was privileged to participate in, but that did not escape the notice of Hermann von Helmholtz at the University of Berlin. The challenge was then passed on to two of Helmholtz's most promising doctoral candidates, Eugen Goldstein (1850–1930) and his close friend Heinrich Hertz (1857–1894).[18]

During a long career as experimentalist at the Potsdam observatory, most of Goldstein's research interests were given over to the study of electrical discharge in gases – a subject that was of keen interest to Helmholtz. In 1886, Goldstein discovered the so-called *Kanalstrahlen* that emerge from holes in the anode at low pressures in the discharge tube.[19] Goldstein's colleague, Heinrich Hertz, was drawn into the Helmholtz circle on arriving in Berlin in 1878. An assistant to Helmholtz at the Physical Institute of the University of Berlin from 1880 to 1883, Hertz – with the encouragement of Goldstein – became involved in cathode discharge experiments. Two years before his death, while at the University of Bonn in 1892, Hertz returned to the topic and published a seminal experimental paper on the passage of cathode rays through thin metal sheets.

After Hertz's premature death in 1894, his student Philipp Lenard (1862–1947) became director of Hertz's laboratory in Bonn. It was William Crookes's 1879 paper on radiant matter as a fourth physical state that originally had elicited Lenard's attention and criticism, and then led him to resume the cathode ray experiments that the ailing Hertz was unable to complete. By 1892, he had constructed a window in the cathode tube that enabled him to study the emergent rays outside of the tube, away from the discharge process.[20]

All of the above mentioned German scientists – Plücker's colleague Hittorf, Helmholtz's students Goldstein and Hertz, and Goldstein's student Hertz – championed the view that cathode rays correspond to a

wave-like phenomenon in the ether. By contrast Willy Wien, Jean Perrin, C.F. Varley, William Crookes, and J.J. Thomson took the view that
the cathode rays were particulate. The controversy drew out over a period of twenty-some years and came to a head and resolution with J.J.
Thomson's discovery, in 1897, that the electron, carrier of the electric
current in the discharge tubes, was a particle of discrete mass and charge.
In this paper, however, the subject is pursued only to the stage at which
Hittorf left it in 1884. As has already been indicated, the search for an
answer to the nature of cathode rays proper was carried on by others.[21]

III. PLÜCKER: MATHEMATICS ABANDONED, EXPERIMENTATION EMBRACED

Julius Plücker (1801–1868) was born in Elberfeld and educated in Bonn,
Heidelberg, Berlin, Paris, Düsseldorf, and Marburg. He was known as
a mathematician. Throughout his life he drew heavily on, was closely
attached to, and was generously supported by the French and English,
but summarily ignored by his German mathematician compatriots. He
was, by virtue of intellectual partisanship, an intransigent Francophile.
He spent most of his academic career in Bonn teaching and conducting
research in mathematics and physics. In his provisory position as director
of the physics *Kabinett* at the University, and with the responsibility for
delivering lectures on physics until the chair of physics would be filled,
Plücker decided abruptly in 1847 to quit mathematics and devote the
rest of his career to physics. A good number of his most important
contributions to mathematics and physics were published in French and
English.[22]

Plücker was one of the most innovative analytical mathematicians of
the 19th century. His contributions to fundamental questions in analytical and projective geometry, the theory of curves, conic sections, and
the arithmetization of analytical geometry found acceptance in France
and England in the school of Monge, Lazare Carnot, Bobillier, and
Poncelet. His excursions into a domain of mathematical analysis that
had been dominated by Descartes, Fermat, Monge, and Lagrange for
well over a century, brought laudation from France and contempt from
the syntheticists in Germany who were pursuing an algebra/analysis
programme. The aim of Jacob Steiner (1796–1863) at the University of
Berlin, who was honoured throughout Germany as a popular teacher and
brilliant geometer, was to discover the unity inherent in all branches of

mathematics. He harboured an intense aversion for analytical methods and was most critical of Plücker's *System der analytischen Geometrie*, published in Berlin in 1835.

In such a mathematics-hostile environment, it seems plausible to conjecture that intellectual embroilment and taciturn neglect in Berlin and Bonn were responsible, in part, for Plücker's self-imposed retreat from geometry to pursue experimental research in physics. It was only after his death that his mathematical prowess was recognized in Germany.[23] It is interesting to note that Felix Klein (1849–1925) at age 17 had become Plücker's assistent for two years and had completed his inaugural dissertation on a mathematical topic in 1868, the year Plücker died. Klein was determined to devote his life to physics and was appointed lecturer at Göttingen in 1871. He soon returned to mathematics, however, to develop a programme for the arithmetization of analysis.

Plücker idolized Faraday, studied all of his published works, communicated with him, visited him in London, and was determined to master the field of experimental physics by repeating, and hopefully extending, Faraday's experimental researches on the relation between electromagnetic and optical phenomena. His escape from mathematics into experimental physics was abrupt and unconditional – first into magnetic studies (1847–1857) and then into investigations on the discharge of electricity in rarefied gases (1857–1865).

It was only during the last three years of his life that Plücker returned briefly to mathematics by reworking a principle that he had begun twenty years earlier – an idea that space might be conceived, not necessarily as a totality of points, but as a composite of lines, a *Liniengeometrie*. His *New Geometry of Space* of 1868, revised and edited by Klein, "was known [in England] but not considered relevant to the crucial question about the reality of higher dimensional geometries" until so recognized in 1878 at the meeting of the British Association for the Advancement of Science by the mathematician William Spottiswoode (1825–1883).[24] It is noteworthy that Spottiswoode, who in 1870 was elected president of the London Mathematical Society, followed the pattern that Plücker had adopted in 1847. He dropped mathematics in 1871 to take up experimental research on the polarization of light and the electrical discharge in rarefied gases.[25] It seems to have been characteristic of the times in mid-century England that wealthy amateurs or persons who wanted to switch interests from one field to another, had a tendency to become engrossed in experimental investigations on the conduction of electricity in gases.

Around 1855, Plücker undertook to repeat Faraday's experiments with gases under conditions of increased rarefaction. His first two papers on the subject were published in 1857 and 1858.[26] His indispensable collaborator in these experiments was the eminent instrument maker and glass technologist Heinrich Geissler (1815–1879), referred to above in connection with Gassiot's work.[27] In 1855, in his workshop in Bonn, Geissler had constructed a manually-operated all-glass mercury air pump that was far superior to piston pumps of the type that Faraday and Gassiot had used. As a master glassblower, Geissler supplied small and thin, but sturdy and versatile "Geissler tubes" – as Plücker called them – "of the most different forms" that presented "an appearance of incomparable beauty."[28]

Both fused-in-glass and externally attached electrodes had been used for gaseous discharge experiments in the late 1830s, notably by John Gassiot. "Electric egg" electrodes also had been employed earlier by the Parisian instrument-maker H.D. Rühmkorff and his assistant Jean A. Quet.[29] Geissler's fused-in-glass platinum wire electrodes exhibited great advantage over other metal wires because a coefficient of expansion close to that of the glass permitted experimentation over an extended temperature range. Geissler's tubes, restricted to capillary size (radically tapered cross-section) in the middle of the discharge area, gave much brighter luminosity than heretofore and provided the close experimental access that was advantageous for carrying out spectral studies on the glow discharge. In order to get around the bothersome heat-induced scattering of metallic particles from platinum electrodes, Geissler discovered that aluminum, being less prone to vaporize, could be employed to cover the platinum electrode areas (except for a small tip) so as to prevent internal blackening of the tubes.[30]

These minutiae of electrode technology, and Geissler's ability to construct tubes that would allow Plücker and subsequent investigators to realize the most inordinately demanding thought experiments, helped to open up a "new branch of physics which led directly to the discovery of cathode rays."[31] The new findings reached far beyond what had been possible with the apparatus available to Gassiot. With higher vacua, a greater variety of specially constructed vacuum tubes, and a growing network of experimentalists who were knowledgeable about ways to secure access to these new instruments, the study of electrical discharge of rarefied gases mushroomed. Investigators perceived, then as now, that pushing the physical conditions of experimentation to the extreme – in this case the achievement of better vacua – favoured the discovery of new phenomena. Parallel with the expansion and growth of factual information came recognition of the immense and unexplained complexity of cathode ray phenomena. This pattern continued unabated into the 20th century and inevitably led to controversial interpretations of the nature of the rays and the mechanism of their production.

In his first paper, of 1857, on the action of the magnet on electrical discharge in rarefied gases, Plücker noted that Arago had predicted Davy's observation that the arch of light (*Lichtbogen*) produced by a powerful battery would be diverted by a magnet. Plücker discovered, as he had surmised, that the deflection also occurred in his Geissler tubes. Unexpected, however, was the observed division of the light-stream (*Lichtstrom*), namely "its decomposition at the negative electrode into an undulating flickering light, and the extension of the stream from the positive electrode [positive column] into a brilliantly illuminating fine point [the negative glow]." Different gases gave a "beautiful effect with the greatest certainty." Rarefaction manifested itself "suddenly by a remarkable alteration of colour." An analysis with the prism yielded "variously modified spectra" that were difficult to describe "inasmuch as the impression they produce on the eye depends upon the external illumination." Dark bands of the most varied shape formed at the "warmthpole" (cathode) with perfect regularity. The stratification (*Schichtung*) of the light was puzzling. The negative electrode and its immediate neighborhood were surrounded by an envelope of "variously coloured, finely stratified light."[32]

Conspicuous in the account is Plücker's poignant and running commentary about the salient, delicately variegated, and "beautiful" visual phenomena occurring along the full length of the tube: *schöner Effekt, besonders schönes Licht, schön geschichtetes Licht, schöne Ringe, schönes Spektrum, schöne Streifen*, etc. It evidently was a rewarding aesthetic experience for Plücker, and reminiscent of Faraday's accent on "beautiful" discharge phenomena two decades earlier. But one cannot resist puzzling about the motives for so precipitous a swerve from the life of pure mathematics to a pursuit, with passion, of empirical investigations nourished by beauty and curiosity.

There is no hint in Plücker's papers about possibly examining, analyzing, or even speculating about how mathematics or physical theory might come into the picture. Plücker had been snubbed by the Germans and was entranced by the non-German Faraday, who without even the most elementary competence in mathematics, and on the strength of imaginatively planned and executed experimental investigations, had turned the world of electricity and magnetism inside out. Plücker also had been inveigled by authorities in the administration at the University of Bonn to add physics to his teaching duties, until such time as a real physicist could be hired. He discovered that physics teaching was enjoyable and less tedious than mathematics, and so he began to fight tooth and nail for a laboratory that would be equipped with the special apparatus he would need for conducting experiments on the discharge of electricity in rarefied gases. We know too little about this stretch of his life to offer further speculations. After his death he was heralded as a great mathematician and put in place in the mathematical world from that perspective, without reference to the empirical side of his career.

The distinctive objective in Plücker's experiments was to study the effect of a magnet on discharge phenomena and to examine the spectral characteristics of the discharge. He discovered that "a great upright horseshoe magnet" drew the light in the vicinity of the negative electrode into "magnetic curves, or lines of magnetic force." These curves rendered "the distribution of the power of the magnet visible" in the way that iron filings strewn on a piece of paper arrange themselves in magnetic curves – "little magnetic elements placed with their attracting poles in contact," i.e., rays behaving "as a magnetic thread of perfect suppleness ... an electrical current twisted in an infinitely thin spiral." It was as if iron filings strewn into space and withdrawn from the action of gravity would arrange themselves around the light as they would around a magnet. Plücker was cautious on the interpretive side. "I have merely sought to make the nature of the phaenomenon intelligible, without in the least attempting to describe the nature of the magnetic light itself ... In consequence of our want of knowledge about magnetic light, and of the total want of analogous phaenomena, I performed many experiments in order to obtain magnetic light under other circumstances ... All such attempts, however, were fruitless."[33] Others before Plücker had noted the deflection of the electrical discharge by means of a magnet. Plücker first recognized the remarkable difference in the behaviours of the positive and negative discharges towards the magnet. He subsequently sought to explain the differences, and wrote papers on the subject, but was unable to come to a good understanding of the complexities involved.

Plücker's spectral examination of the discharge phenomena revealed, as he had "confidently expected," the presence of characteristic spectra for each gas. He assumed that the spectral patterns varied along the Geissler tube because "the ponderable matter which becomes luminous is differently distributed through the tube ... But in all cases, whatever may be the colour-impression produced on the eye, the distribution of the colours in the spectrum remains for the same gas *entirely of the same kind*; it is the intensity of the colour alone which changes in different degrees in different portions of the spectrum: so that when the eye (whose judgement is, moreover, considerably influenced by the external illumination) is at fault, still the nature of the gas ... is unfailingly determined by means of the spectrum."[34]

The electrical discharge investigations of Plücker in 1858 led to a solicitous and impressively detailed description of wonderful but puzzling configurations and hues. Not able to achieve the "requisite degree of accuracy" that he might have wished for in order to reach an understanding of how the current was being carried from one electrode to the other, he was prepared to strike out in another direction. Uppermost in his mind was what might be learned by spectral analysis concerning the chemistry of the electric glow: "The most difficult question which arises on the discharge of electricity through rarefied gases, is the chemical nature of the ponderable substance which gives rise to so infinitely varied phenomena of light. This question can only be safely discussed in connection with the prismatic analysis of the light which is produced – the more so as by this means every sudden or general chemical change in the substance is recognized ... The subject is one belonging, if I may use the expression, to Micro-chemistry. Conditions occur in it which differ from those under which chemical actions usually take place. It is only on the successful solution of these questions, that many not unimportant points for the molecular theory will be satisfactorily solved, such as – ... How are the spectra of a compound gas related to one another before and after its chemical decomposition by the current?"[35]

We see that at one stage in his rather straightforward, phenomenologically-focused investigations – continuing where others had left off – Plücker developed a yen to explore spectroscopically the chemical/physical nature of the glow discharge. The distinctions between physics and chemistry were not so sharp at that time. By moving in an alternative direction to the one he already had explored, he could anticipate uncovering a fresh set of hitherto unfamiliar phenomena that would bring new knowledge, questions, puzzles, and conceivably new insights into electrical discharge phenomena. As it turned out, his pursuits inescapably led again to problems that only could be illuminated experimentally. That entailed, as before, serious attention to experimental finesse. The result was more description and data, and more puzzles about what the data meant.

The records therefore show that the outcome was largely empirical; that the experiment had been motivated more by inquisitiveness than by theoretical guidelines. At least so it appears by hindsight. Nonetheless, it is evident that the knowledge that had been acquired approximated a certain level of comprehension as to what was going on in the Geissler tubes and had provided, as well, good insights on how to continue level-headedly and intuitively to acquire more knowledge by asking new questions. It is plausible to assume that exploratory theories; theory-pictures, or thought experiments, were lurking in the background, but concerning theory-talk there is no evidence.

As in other frontier areas of scientific investigation where new domains of knowledge are generated, cathode ray theory lagged far behind, while the facts of experience took precedence and set the stage, falteringly, for the promotion of theoretical comprehension.[36] In such cases the experiment-ladenness of theory is more in evidence than the theory-ladenness of experimentation. By contrast, the psychology of invention – scientific instruments are invented not discovered – is such as to keep pace with whatever stride is set by experimentation. In virtually all accounts of the electric discharge in gases, the concern with instruments – their invention, improvement, refinement – is central. Although it was said of Plücker by his contemporaries that he personally never attained great manual dexterity as an experimentalist, he was on the alert to recognize, encourage, and latch on to Geissler – as imaginative and ingenious an instrument maker as existed anywhere in the 1850s. In any case, Plücker engineered investigations that were more innovative, and in the long run more significant, than those of other experimentalists attacking the electrical discharge problem at the time.

In surveying the primary literature on cathode ray studies that were undertaken by several generations of investigators, it is credible to suppose that the drives, accomplishments, and thrills that carried the field along were linked primarily to the challenge of taking risks and exploring frontier domains of nature per se, without openly expressed commitments to potential theoretical outcome. The visual displays accompanying electrical discharge phenomena were, for Plücker, intrinsically "beautiful." Problems of interpretation, and ideas on potential ways to secure deeper (theoretical?) insights, had a way of becoming translated, not so much into theoretical solutions, as into variant puzzlements that pointed to alternative modes of experimental attack. Not every puzzle was attractive or manageable, but certain puzzles, perennially coming into focus, would exhibit a sturdiness that was embedded securely within the realm of nature. Such a puzzle had a life of its own, not to be snuffed out or sidestepped. It has been said that "problems worthy of attack show their worth by striking back."[37]

And so it was with Plücker. He pushed ahead into areas of cathode ray investigations that had not yet been explored. In seven classic papers (1858–1862), Plücker explored various facets of the discharge phenomena: size, shape, and composition of the discharge tube and the electrodes; effects of current intensity and degree of rarefaction of the various gases; characteristic spectra of gases and their current-induced decomposition products; the effect of the magnet on the discharge in various areas of the tube; fluorescence; thermal effects; and – notably – the nature of and laws governing the current carriers in relation to Faraday's electromagnetic investigations. At every stage of Plücker's probative search, Geissler's connoisseurship was enlisted in the creation of discharge tubes of ever-escalating complexity.[38] Virtuosity, experimentation, and innovation were handsomely synchronized; both Plücker and Geissler exploited the implications of the collaboration to full advantage.

On the basis of newly acquired experimental information, Plücker was prepared to correct previously recorded observations and suggest alternative interpretations. Many a time he was forced to admit that experiment had preceded theory. When the sheer complexity of the phenomena might have been disheartening, Plücker could reflect on the aesthetic features of the luminous discharge and their multifarious spectra. He was painfully aware of circumstances that conditioned the subjective judgment of the observer. Impurities in the gases were a nuisance, as were the chemical reactions of the gases with the electrodes, which produced "unpleasant blackening" of the glass and altered the concentration and therefore the degree of rarefaction in the Geissler tubes.

Plücker's analysis of the conduction of the current in the electrical discharge leaned heavily on Faraday's path-breaking discoveries. Plücker wrote: "The rotary motion of a magnetic pole around the [electrically] conducting wire, and of the conducting wire around a magnetic pole, are phaenomena which, on their discovery by Faraday, arrested the attention the more, because they were not connected by analogy with any previously observed phaenomena." Plücker believed that he could gain keen insight into his electric discharge experiments by adopting Faraday's model of interacting electric and magnetic forces: "We obtain a new point of view if we regard the conductor [of electricity] as *perfectly flexible*, and then inquire what would be the form of such a conductor as current-bearer under the influence of the magnet." In that case, "*equilibrium can only exist ... when the conductor assumes the form of a magnetic curve.*" For equilibrium to maintain over a given surface, "*the direction of the force acting at every point of the conductor must coincide with the normal to the surface at this point.*" Plücker called such curves "epipolar magnetic" (die epibolisch-magnetische Curven).[39] Plücker's account of the action of the magnet upon the electric current thus involved the adaptation of Faraday's views to a flexible, in place of a stationary, current.

During the course of his investigations, as Plücker became more addicted to, familiar with, and confident about his experimental investigations, he felt obliged to stake out some priorities: "I believe that I was the first to declare positively that the luminous appearance which accompanies electrical discharge through long tubes of rarefied gases is ... entirely and completely attributable to the traces of gas remaining in the tubes; further that the beauty and great diversity of such spectra for various gases offers a new characteristic for distinguishing them, and that any chemical alteration in the nature [decomposition] of the gas may be thereby at once recognized. This seemed to me to be the most important part of the subject, pointing, as it does, to a method of physico-chemical investigations of a new kind."[40]

The much used expression "absolute vacuum," to characterize the state of rarefaction that bars the passage of the electric current in Geissler tubes, annoyed Plücker: "An absolute vacuum, like a mathematical pendulum, is a fiction; and the practical question is only whether no electric discharge passes through the nearest possible approximation to an absolute vacuum that we may procure ... I agree with the opinion that ponderable matter is necessary for the formation of an electric current

... [and] that the light of the discharge-current, and the consequent corresponding spectrum of such gas-vacua, entirely depend upon the residual traces of the gas."[41]

In 1860 Plücker submitted an abstract of his experimental investigations and ideas to the Royal Society of London.[42] There are a number of reasons why he chose this route to publication. His British colleagues had supported his views with enthusiasm. For at least a decade Faraday had complained in letters to Plücker that he could not read German and therefore was cut off from learning about his work.[43] Meanwhile, the German scientific community was for the most part ignoring Plücker and his physical investigations.

As noted above, Plücker's experiments, which had been designed primarily to test the action of a magnet on the electrical discharge in rarefied gases, had led him straightaway, in 1857, to what he believed were more challenging and more significant issues, namely the study of characteristic spectra and the chemistry of the ponderable substances that constitute the glow in the Geissler tube. Now, in 1860, Plücker set himself the more ambitious objective of extending Faraday's laws of electromagnetism for fixed wires carrying a current to flexible electric currents in the discharge tube. The key idea was that the discharge, in general, could be accounted for by regarding "the discharge as a bundle of elementary currents, which under the influence of a magnet change their form as well as their position in the tube, *according to the well-known laws of electro-magnetic action.*" Two forms were discernable. The discharge could be concentrated into one form (path, arc, arch) "*if the arch be allowed to constitute a part of a line of magnetic force*" since by "theory there is no magnetic action at all exerted on any element of a linear electric current which proceeds along such a line [of magnetic force]." The other case of "electro-magnetic equilibrium ... takes place if the current proceeds along an '*epibolic curve*', i.e., along the curve ... whose elements, regarded as elements of an electric current, are perpendicular to the direction of the electro-magnetic force." This implied that "if neither of the two conditions ... be fulfilled, i.e., if the current cannot proceed either along a free magnetic or an epibolic curve, no *voltaic arch will be obtained*; the current will be disturbed and its light diffused."[44]

The other problem area in which Plücker made fundamental contributions to discharge phenomena relates to demonstrating important differences in the effect of the magnet on the luminous phenomena surrounding the anode and the cathode. In large tubes of cylindrical shape, the light around the negative wire was bent by the magnet into curves and surfaces. This was peculiar, "having no analogy with phenomena hitherto observed," but quite explicable with the help of the notions of "magnetic curves or lines of magnetic force ... shown by several philosophers, especially by Mr. Faraday." Plücker's arguments rests on an analogy to "a variable magnetic surface" that surrounds points on the negative wire, being produced by "a chain of infinitely small iron needles, absolutely flexible and not subject to gravity." In conclusion: "I think it most probable, that the luminous electric currents in question are double currents, – going from the wire to the glass and returning from the glass to the wire."[45]

As for the action of the magnet on the luminous area surrounding the positive wire, "where the origin of the current takes place ... striking phenomena were encountered when the two poles were fairly close (less than an inch apart) in a highly exhausted sphere. The whole area is "allmost [sic] uniformally illuminated by violet light, while the light of the positive electrode appears at one of its extremities ... [and] moves along an epibolic curve." According to Plücker, all phenomena in this class were "explained by the laws of electro-magnetic action" and Ampère's "molecular current" model of magnetism.[46] The electric discharge thus was seen to be a "double current" – two "electric currents returning on their own path" and separable by means of a magnet.[47]

In regard to the nature of the light itself, Plücker held that "*electric light does not exist; the light which we see belongs to the* [ponderable] *gas, rendered incandescent by the thermal action of the current*," and analyzable with the prism into characteristic spectra. Plücker was alert to the fact that his "primitive theoretical views [constantly] were [being] modified, reformed, or extended by subsequent experiments" and that his report referred "only to what I think *at present* to be the state of the question."[48] In the 1860 paper Plücker, more confidently than elsewhere, and more than anyone else at this time, was exploring electrical discharge phenomena along guidelines established by Faraday – and this in an area of physics that on the face of it was not a self-evident branch of mainstream electromagnetic investigation. Seven years after the initial excursion into electric discharge territory – where every innovative move was to hinge on experimental bravura – Plücker, with the assistance of Geissler, felt that he was in control of the best instruments available anywhere. But he decided that it was equally important to enlist the collaborative efforts of the chemist Hittorf from nearby Münster. After Plücker's death in 1868, Hittorf produced six classic studies on cathode rays." His work is examined below.

In 1864, Plücker and Hittorf submitted a joint paper to the Royal Society in English. It was a comprehensive report on the spectra of gases and vapours, designed to attack head on what had been denoted by Plücker in 1858 as "the most difficult [chemical] question which arises on the discharge of electricity through rarefied gases."[49] The objective of these investigations was to employ the electric current to acquire spectra of elementary bodies. It was anticipated that such spectra would furnish information not attainable at lower temperatures by flame analysis. With "the electric current, the heating power ... [could] be indefinitely increased by increasing its intensity." Two techniques were used: the application of a continuous electric current to a substance in the gaseous state at varying degrees of rarefaction, or, in the case of solids, the passage, through the two extremities of the conducting wire, of a strong spark from a large Leyden jar charged by a powerful Rühmkorff induction coil. In both cases, "the spectra are obtained the most beautifully and are the most suitable for examination in their minute details." The tube experiments "confirmed and supported in a striking way ... the theoretical conclusions of Dr. Faraday, that electricity being merely a peculiar condition of ponderable matter cannot exist without it, and cannot move without being carried by it."[50]

An important fact "as well with regard to theoretical conceptions as to practical applications" was that "certain elementary substances, which, when differently heated, furnish two kinds of spectra of quite a different character, not having any line or band in common." The remarkable feature was that "the passage from one kind of spectra to the other is by no means a continuous one, but occurs abruptly." In the case of nitrogen, for example, spark discharge gave "a beautiful richly coloured spectrum" not continuous but divided into bands that were resolved into dark lines (up to 34) on applying four prisms; Caveat: "but psychological effects of this description may be quite different: partly by our own will, partly by exterior circumstances."[51]

With the discontinuous discharge from an interposed Leyden jar, the spectrum of nitrogen (this was their test case) had no resemblance to the former "variously shaded bands, ... [being] replaced by brilliant lines on a more or less dark ground." What did these spectra reveal about nitrogen? "Certainly, in the present state of science, we have not the least indication of the connexion of the molecular constitution of the gas with the kind of light emitted by it, but we may assert with confidence that, if one spectrum of a given gas be replaced by quite a different one, there must be an analogous change of the constitution of the ether, indicating a new arrangement of the gaseous molecules ... [namely] a chemical decomposition ... We conclude that the change of the molecular condition of nitrogen which takes place if the gas be heated beyond a certain temperature by a stronger current, does not permanently alter its chemical and physical properties but that the gas, if cooled below the same limit of temperature, returns again to its former condition." The surmise was that the abrupt changes represented nitrogen in "the molecular and atomic states."[52]

In a similar way spectra were obtained for organic and inorganic gases and metals in the vapour state. Dependent on temperature, pressure, and electrical tension, the spectra nevertheless were seen to be distinctively characteristic for each substance in a given state. As in the case of spectra discovered by Fraunhofer for the arc and flame some thirty years earlier, this meant that electrical discharge spectra might be useful for substance identification.

The special investigations of Bunsen and Kirchhoff in the 1860s, which gave rise to methods of analysis of chemical elements by flame and spark, were given support when Plücker and Hittorf published their findings on the electricity-induced line and band spectra for gases under rarefied conditions. By varying the intensity of the electric current, and thus changing the temperature in the discharge tube, they observed and recorded changes in the line spectra. This discovery was later correlated with Norman Lockyer's observations of prominence spectra at total eclipse in 1868. Acclaimed in England, these advances were largely unacknowledged in Germany, being overshadowed by the more spectacular discoveries of Bunsen and Kirchhoff.[53] The 1870s, in general, saw an expansion in the study of the emission and absorption spectra that enveloped electric sparks, lightning, northern lights, solar protuberances, stellar light, and nebulae.[54]

Having focused primarily on Plücker's experimental investigations on the conductivity of electricity in gases, we now turn our attention to some of the contextual factors in Plücker's life; they help to illuminate his activities and methodology as a scientist. As already mentioned, Plücker owed a great deal by way of intellectual sponsorship to Faraday. A running correspondence with Faraday over a period of 15 years

(1847–1863) shows how much these two men esteemed each other's view as experimentalists.[55] Numerous references in the *Diary* through August 1855 show that Faraday was repeating Plücker's experiments and challenging some of them. Faraday recommended Plücker for election to membership in the Royal Institution in 1849 and the Royal Society in 1856. In 1868, as foreign member, Plücker was awarded the Society's coveted Copley Medal for his contributions to electrical discharge in gases.[56]

The Faraday–Plücker contacts had begun in November 1847, when Plücker wrote to Faraday and sent him a number of reprints on magnetism, diamagnetism, and magnecrystallic action – subjects in which Faraday had pioneered. There were some differences of opinion as to what the experiments showed and meant. Faraday had little to say in response to Plücker except that he could not read German and that the state of his health was deteriorating. As he wrote in June of 1848, "I am nearly 57 years of age – have worked long hours in my life and as to material strength am somewhat worn. In such cases a man may be patched but he cannot be remade."[57] By the late 1850s Faraday's mental acumen in fact had markedly deteriorated, and so it is to Plücker's letters, mainly, that one looks for clarification of the issues taken up in this paper.

Plücker's venture into electrical conductivity studies in 1867 was preceded by ten years of experimental investigations into the domain of magnetism in its various forms.[58] By 1857 Plücker's interests had turned to the conductivity of electricity in gases. In July he wrote to Faraday: "Lately ... I made a series of experiments in order to get an explanation of the stratification of light exhibited ... within certain rarified [sic] vapours ... The various experiments are beautiful ... but most beautiful is the effect, when the tubes are placed in the Electromagnet in different ways, as well axially as equatorially ... "[59] Faraday responded: "I am very glad that you are working on the stratified electric light – I hope that you will very shortly give us the fundamental explanation of the phenomenon. I cannot help thinking that it will aid us in developing some very important points about the nature of the electric discharge. We would rejoice to understand, truly, the first principles of that very striking electric action." Two weeks later Faraday added: "I have been obliged to give up thinking about the luminous current but whilst such as you and Gassiot and others work on the subject I know it is progressing." Faraday was especially interested in "the arch ... corresponding to the Magnetic

line of force and that other one of which you speak . . . that [seems] to be
directly at right angles to the course of the Electric current." Concerning
"the question of transmission of the discharge across a perfect vacuum
or whether a vacuum exists or not? is to me a continual thought and
seems to be connected with the hypothesis of the ether. What a pity that
one cannot get hold of these points by some *experiments* – more close
and searching than we have yet derived." It is clear that Plücker had
done a good job of conveying his ideas to Faraday in his papers (they
were now being published in English) and that Faraday's mind was by
no means yet in limbo.[60]

Plücker invited Faraday to Bonn. Faraday could not manage it: "Years
and their consequences limit our powers, and though I trust yours will
long run on successfully, mine are drawing nigh to their end."[61] The least
Plücker could do was to help Faraday become informed about the Bonn
experiments: "The discharge of Electricity through the tubes, exhibiting
the stratified light, cannot be the transport of light, or luminous matter
from one end of the tube to the other. There is, I think, within the tube a
distribution of ponderable matter produced by the discharge, that matter
becoming luminous by it, while the discharge is a dark one, as you call
it, from one luminous place to another . . . I showed the beautiful effect
they present, at the Meeting of Bonn . . . Since that time I observed a
quite new series of phenomena, which exhibit a very fine appearance. I
can, in a few words join no better account of them but by saying that I
am enabled by means of the electric light, to *render luminous your lines
of magnetic force.*"[62]

Six months later Plücker was producing "beautiful electric spectra by
conducting the discharge of Rühmkorff's Apparatus through a capillar
[sic] tube."[63] It took Faraday eight months to respond. Letter-writing
made his head ache; he was "fit for nothing now but small gentle acts
of thinking." He believed that "the luminous phenomena of the Electric
discharge . . . [were] so numerous so varied *so indicative* and yet . . . so
little understood in respect of their *law* or fundamental principle that I
cannot retain them in my mind . . . But though I cannot discuss these
beautiful . . . phenomena with you I can enjoy them and your success in
the development of them and I doubt not some day the whole beautiful
encircling cloud of luminous results will open up into perfect order and
intelligence and you will either produce that result or be a chief leader
in attaining it . . . In the meantime I commend myself to you as an old
worker in science that loves to look on the present bands of worker

[sic] & as far as he can keep up relations with them if it be only by reminiscences and the memory of the past times."[64] Plücker continued to give Faraday an account of his experiments. Faraday wrote: "Your results on the gas spectra are exceedingly interesting. What a wonderful branch of research that of luminous lines has become, and great honor belongs to Kirchoff [sic] and Bunsen."[65]

IV. HITTORF: DISCOVERY AND EXPLOITATION OF CATHODE RAYS

After Plücker died (1868), his erstwhile collaborator Hittorf confirmed the magnetic effects and the fluorescence; he also noticed that insulators or conductors placed in the path of the cathode beam stopped the cathode glow. This led to the observation that in an L-shaped tube, with electrodes at the extremities, the *Glimmlicht* ("rays of glow") was generated at the cathode and proceeded linearly from the cathode. These rays subsequently were referred to as *Kathodenstrahlen*, or "cathode rays." By varying the shapes of the vacuum tube, the location of the electrodes, and the degree of rarefaction of the gas, and by observing the effects of the rays on objects placed inside the tube, Hittorf was able to amass a great wealth of important information about cathode ray phenomena over a period of 20 years.

Wilhelm Hittorf (1824–1914) was born in Bonn, educated in Berlin and Bonn (with Plücker in 1846–1847) and spent his entire academic career at Münster.[66] His forte was electrochemistry and electroluminescence. In a series of classic experiments on the transport and mobility of ions in solution (1853–1859), he laid the foundations for the electrolytic solution theory of Arrhenius, van 't Hoff, and Ostwald. The work was not properly acknowledged until the end of the century.[67] As assistant to Plücker in the mid-1860s, Hittorf became involved in the study of electric discharges in gases and their spectra – work that after Plücker's death in 1868 led him to the discovery and characterization of cathode rays.[68]

Hittorf's analysis in the 1869 paper begins with a somewhat cynical evaluation of the state of knowledge concerning the mechanism of current conduction in gases: "The most obscure part of the present science of Electricity is undoubtedly the process by which the transmission of the current is effected in gaseous bodies. While for solid and liquid conductors, whether they are metallic or electrolytic, the relations between the facts have been collated, and have acquired a common bond in Ohm's law, our knowledge of the conductivity of gases, notwithstanding the endeavours of distinguished physicists, is of a decidedly fragmentary kind, and rests mostly on observations which are imperfect and isolated. The theory of the electrical spark, the longest known and the most striking of all electrical phenomena, can only be established when the condition of our knowledge is improved."

The problem, as Hittorf saw it, was that "gases at low temperature have an almost infinitely great resistance for electricity at such low tension as it is furnished by a voltaic element, and only begin to lose their insulation at red heat ... [since] at ordinary temperatures, the electricity must have a higher tension" if the most "characteristic" phenomena of the gaseous state are to be investigated under rarefied conditions. What was needed, in his opinion, were better air pumps to produce better vacua and barometers and manometers that would show when the gas (he suggested that air or nitrogen were optimum) had reached the requisite degree of rarefaction. Hittorf identified three parts to the problem: "The positive light, the dark space, and the negative glow-light."[69]

Hittorf's conception of a characteristic state of the gas and the identification of ways to achieve it set the stage for his production, recognition, and examination of cathode rays. His term for these rays was *Glimmlicht*. He recognized that the hurdles to be overcome were largely technical. He found it advantageous, for example, frequently "to change the shape and dimensions of the metallic ... electrodes," to make them of aluminum instead of platinum (as others mostly had done), and to fix them into the ends of the tube with sealing-wax instead of fusion so as to have them "easily done and undone." In a typical set-up the glow began at the anode and extended over the entire length of a tube 60 mm in width at a pressure of 0.33 mm (4×10^{-4} atm.). This glow-light took a rectilinear path such that each point on the cathode constituted the apex of an arc of rays.[70] Two observations followed: First, as was to be expected, "that all the rays of glow-light which proceed from the particles of the negative surface endeavour to join the positive light. If this is not possible, the glow ceases." What could not be foreseen was that "the positive light ... would find great difficulties for its formation in the vicinity of the kathode."[71]

Hittorf's summary conclusions were that: "Special conditions exist in the neighbourhood of the kathode which do not allow of the propagation of electricity in the same manner in which it takes place in the positive light, and prevent this current from entering the negative surface. It must therefore be supposed that the conduction there is due to the glow." The effect of the magnet was that the rays of negative light behaved like simple currents that flowed "from the neighbourhood into the kathode."

Hittorf's explanation was well in hand: "According to the law which Laplace deduced from the experiments of Biot and Savart, a pair of forces are at work between each particle of a linear current and the pole of an infinitely thin magnet, which are at right angles to the parts of the current and the pole, and whose intensity is inversely as the square of the distance between them, as well as directly proportional to the sine of the angle which the particles of the current form with its connecting line with the poles. Since, for a finite magnet, the tangent to magnetic curve which passes through the particle of the current represents the direction towards the pole, the force acting on the element of the current is at right angles to the plane which passes through it and the magnetic curve of its locus."[72] While working with Plücker in Bonn, Hittorf had been exposed to the conception of magnetically induced spirals moving in the axial direction, and negative rays that, in contradistinction to magnetic laws, coincide with the magnetic curves. Such views had enticed Hittorf to take up studies of the magnetic behaviour of the glow rays. He mentioned that: "The illness and unexpected death of my honoured teacher prevented me from bringing before him [my] experiments."[73]

On the pivotal question of the nature of "the rays of negative light," Hittorf reasoned that since "by their magnetic properties [they had] revealed themselves as simple currents which take the direction of the kathode, it can no longer be doubted that in gaseous media the propagation of electricity takes place in a two-fold manner." The positive light was seen to be analogous to conduction in metals and electrolytes, whereas the conduction in gases – that belongs to the glow-light "and deserves greater consideration than has hitherto been paid to it" – has its starting-point at the kathode and "propagates itself uniformly as rays in all directions in the gaseous medium, *and agrees therein with wave-motion*" (my emphasis).[74]

The task ahead was to "investigate the conditions for the glow-light." By formulating the problem in this way, that is, by deciding not to pay attention to all of the horrendously complex phenomena associated with electron discharge phenomena in order more exclusively to focus on the conditions for and nature of the *Glimmlicht*, Hittorf was establishing a new research programme in an old and somewhat weary enterprise that was muffled – choked up with too much empirical information. But Hittorf had done more than establish an agenda for future exper- imental investigations. He had coupled his discovery with the strong suggestion (growing out of that discovery) that the *Glimmlicht* was a wave-like phenomena in the ether, and not particulate. Thus began a thirty-year cathode ray controversy that wound up negating his wave theory interpretation in favour of the particulate view.

Hittorf assuredly knew that he had put electric discharge studies on a new course that was potentially of fundamental significance: "If I do not deceive myself, these comparisons are peculiarly favourable for form- ing conclusions as to the nature of the electrical current itself; it is not impossible that, as with the theory of heat, gases will enable us to rec- ognize most easily the essence of phenomena, and will liberate modern physics from the last of the imponderables, the electrical one."[75] Heat had been shown to be a wave-like phenomenon and not an imponder- able substance. Hittorf was confident that electricity, and therefore the *Glimmlicht*, would be shown to be a wave-like vibration in the ether.

Having said as much, Hittorf was disquietingly sensible of what it might take, technically, to resolve problems that still stood in the way. For example, since the conductivity of the gaseous particles was known to be "dependent in a far higher degree on the temperature than is that of metals and electrolytes," it would be necessary to determine resis- tances with "such feeble currents that the temperature of the medium is not appreciably changed, and must by other means reach that level by which the conduction can be recognized in our galvanometers." Hittorf continued: "It is easy to formulate this requirement, but difficult or even impossible to realize it completely. For gases only begin to lose their insulation for our more accurate measuring-instruments at red heat, a limit at which our means of maintaining a constant temperature are very restricted. The greatest drawback, however, lies in the fact that in those conditions of heat in which gases conduct, all other bodies that we know behave similarly, and no solid insulators exist. We cannot, therefore, restrict the current to a definite geometrically simple path in the gaseous medium, and in these measurements we shall remain far from that accuracy which can be obtained with metals. This view must not, however, lead us to neglect the subject. Even somewhat less delicate determinations may reveal facts of fundamental importance, and crude approximations, provided they are correct, may be of the greatest interest." But there were many more "drawbacks [that] ... decidedly preponderate."[76] There certainly was a new focus, but the experimental barriers, as before, were formidable. In fact, Hittorf never was able, for experimental reasons, to reach the goal that he had set for himself.

Although Hittorf proceeded in his investigations with considerable confidence, he wanted it to be known that his views were to be considered "as only preliminary studies and reconnaissances, which are necessary since the field is so unknown." Discouraged perhaps by technical difficulties, indecision about the best avenue of approach to the subject, and the need of equipment that he "could not procure," Hittorf waited for five years to return to his investigations on the conduction of electricity in gases.[77]

After a long interval Hittorf, as he said, "felt more strongly than before the necessity of being able to work with the current of the voltaic circuit, as well as with that of the induction-coil and of the electrical machine, for it has been most minutely investigated [since the late 1860s] in every direction." The first requirement, as Hittorf saw it, was to use a stronger galvanic current. He mentioned that Gassiot deserved the credit for having first demonstrated, in 1863, that "the enigmatic luminous phenomena [could] be produced by the galvanic circuit" but it had been accomplished with 3520 (later he mentions 3620) cells. With tubes of small dimensions and sufficiently rarefied gases, Hittorf discovered that 400 Grove's elements sufficed. Nevertheless, as suggested in 1869, Hittorf still felt that higher temperatures would solve some problems that otherwise were unmanageable: "The longer I busy myself with the discharge of electricity through gases, the more I am convinced that the enigmas which here present themselves will be solved when the electrical relations of matter in the third state of aggregation [presumably gases] at high temperature have been investigated. At a red heat all gases, as we shall see, lose the insulation which they possess at a low temperature, even for feeble tensions, and may be called conductors with just the same right as metals and electrolytes, even if the process itself is essentially different and its laws are completely unknown ... So long as these important facts are not more closely investigated, the basis for the subsequent construction of a theory will be wanting."[78] Having examined methods of elevating the temperature in the discharge tube by means of flame and heat-producing chemical reactions, in which "the formation of the glow-discharge is very beautifully seen," Hittorf comments: "all these means are available for the investigation of the discharge in rarefied gases. By using them we shall be able, with surprisingly low tensions of a few elements, to produce the glow discharge in rarefied gases."[79]

In the mid-1870s Hittorf was so preoccupied with teaching assignments and other academic responsibilities that it was not possible for him to devote his "scanty leisure to investigating the electrical conductivity of gases." With access to 1600 improved immersion batteries in 1879, he abandoned attempts to increase the temperature in the discharge area and took up voltaic methods: "The idea I had in view in fitting up this battery is manifest. With the help of this most efficient of all known sources of electricity, I hoped to produce the so-called glow-discharge in gases continuously, and thus obtain the great advantages for the investigation, which a continuous current offers, in contradistinction to a discontinuous one [with spark-discharge methods]."[80] In making this switch it was acknowledged that he had benefited from the work of many others and that he and Plücker had committed errors in their spectral studies by using very hot flames and chemical action rather than more intense voltaic currents.[81]

Anxious to clarify the nature of continuous discharge phenomena as in the *Spitzenlicht*, the "brush-light" that Faraday had described in his *Experimental Researches* (Nos. 1434 to 1447), Hittorf was keen to emphasize "that a transmission of particles of air accompanies the brush-light ... [and] is best of all demonstrated by the vane [Flugrädchen] which is caused to rotate rapidly by the reaction."[82] With the pointed end of a strip of metal as one electrode and the vane as the other, "it was possible to observe its behaviour at the ordinary pressure as well as at any required degree of rarefaction ... In order that the rotation may occur, glow-discharge must not take place in the intermediate space, but must remain confined to the immediate neighborhood of the point [as brush-light]: At the pressures under 70 mm [10^{-2} atm.] it was very easy, with the number of cells at my disposal, to cause and prevent the rotation at will, by suitably altering the resistance introduced into the circuit."[83]

What did these "radiometer" experiments tell about the conduction of electricity in gases? Hittorf believed that the glow-rays did not directly cause any rotation. In this he was in disagreement with William Crookes, who recently, said Hittorf, "without referring to my investigations," brought the ideas "before the Royal Society in London and the French Academy . . . [using] his radiometer as a fly-wheel, by fitting it up with aluminum plates, covered with mica on alternate sides."[84] We may note here that Crookes felt that he had demonstrated that the cathode rays were particulate and not wave-like in character. By contrast, Hittorf suggested that "it is the thermal radiation of the glass side . . . which first causes a rotation."

In his own experiments, Hittorf "used a light moving windmill . . . but it remained stationary when the whole bell-jar was filled with the glow-discharge." His conclusion was: "Reviewing all the facts in this paragraph, and having regard to the delicacy of fly-wheels and vanes, there is no doubt that in the glow-discharge there is no propagation of gas particles; and that, therefore, when it takes place, the transmission of the current everywhere, including the dark layers of the positive light and the dark space, is due to another cause." Faraday also held this view.[85]

In two final information-rich and congested papers, Hittorf presented the end-results of his investigations under the same heading he had used for 15 years: *Ueber die Electricitätsleitung der Gase*.[86] The caption, however, was not representative of the closely allied subject that had been explored. The investigations of electroluminescence and the *Glimmlicht* phenomena had spilled over into atomic and molecular spectra, the effects of the magnet, chemical reactions, flame and spark methods, the design of Geissler tubes, and the improvement of vacuum techniques, galvanometers, and condensors. The 1883–1884 papers, documented with thirty densely-packed tabular résumés, did not offer much more by way of interpretation than already had been presented in the earlier papers. Although Hittorf announced a final paper on the subject – perhaps to pull his ideas together – it apparently never saw publication.[87]

By the 1880s the study of gaseous discharge phenomena had become a hornet's nest of interlinked islands of experimental inquiry that enveloped branches of physics and chemistry such as crystal-, electro-, piezo-, radio-, chemi-, and thermo-luminescence and phosphorescence, as well as spectroscopy and astrophysics (notably auroral). More than any other investigator before 1890, Hittorf saw in his *Glimmlicht* studies the essential route to understanding the nature of the glow discharge phenomena produced by the electrical current in rarefied gases. Correlative to Hittorf's discovery of cathode rays came the inference of the wave-like/light-like disposition of those rays. Tentative and conjectural as that suggestion was, it nevertheless set into motion a legacy that for twenty years dominated the views of his German compatriots and initiated a controversy that ultimately led to rejection of the wave-like nature of cathode rays in favour of the particulate view. In 1897 it was shown – but not quite definitively and not accepted everywhere – that cathode rays were electrons.

Hittorf's work on electroluminescence was not widely acknowledged among his contemporaries. This, in part, was because his papers were so burdened with empirical details that important interpretations he had to offer were covered up by the data. Another factor in the neglect of his work was that so many papers were being published in the expanding domain of electrical gas studies – most of them being devoted to checking, duplicating, and demonstrating what already was known – that the effort to keep up in the field must have been rather formidable.

There is another explanation lurking in the background that is worthy of comment. Hittorf's public image had a low profile and revolves on his reticence at self-advertisement. Many persons who were close to him say as much and consider that others received the credit for much of what he had accomplished earlier and on the basis of less creditable evidence. It is noteworthy that his earlier work on ionic mobility (1853–1859) encountered the same neglect by his contemporaries, only coming into its own years later, when it was recognized that this work was crucial for the establishment of an electrolytic solution theory consistent with the new views on ionic dissociation.

V. CONCLUDING REMARKS

Essential and perhaps unconscious theoretical constructs and intuitively formulated hypothetical objectives are known to be at work, often unobtrusively, in the thought experiments and mental images of the scientific investigator. Intuition is used in this context to denote the direct inner perception that results from expert knowledge and superior experience in a scientific domain in which problematic and uncharted issues, phenomena, or conceptions are open for investigation. Experimentalists who bring knowledge and experience – that is, good intuition – to bear on a domain-related but unsolved problem, are at a decided advantage when it comes to insight and discovery. It is self-evident that intuition may not always provide the necessary or sufficient conditions for the

requisite insight. Nature commands its own complexities and regular-
ities. The history of investigations into electric discharge phenomena
in gases provides a seminal example of a frontier domain of science
in which theory-neutral expertise loomed paramount and on occasion
promoted insight and progress, in situations where theory manifestly
was absent or held in abeyance by the investigator.

Experiments, it rightfully is asserted, are theory-laden; yet that
emphasis often is uncritically invoked. Theories also are experiment-
laden – some more than others. In the case of 19th-century gaseous
discharge studies, history teaches that investigators, who were engaged
in creating knowledge in order to achieve understanding, were inundated
with experimental information for almost a century before satisfactory
explanations were reached in regard to the physical or chemical nature
of the constituents of the discharge, their mode of production, and the
mechanism of current conduction. The field was littered with apparent discoveries and embryonic attempts at explanation that invariably
evaporated with the acquisition of additional experimental evidence,
reanalysis of the data, or change of mental framework. At times the
would-be interpretations were too ambitious and too inclusive to be
put to decisive test. Then intuition, rather than "theory," comprised the
essential focus for subtle differentiation and interpretation of puzzling
experimental observations. Foregone and plausible but unproved explanations, on the other hand, functioned as barriers to advancement. They
either blocked the way to understanding or led to roundabout strate-
gies inferior to the path suggested by more straightforward, enlightened
empirical inquisitiveness, especially when sustained by superb instrumental craftsmanship.

It is understandable for an investigator to be so convinced of the outcome of an experiment or an unanticipated reality that discarded views
and conceptions block essential understanding long after favourite pre-
dispositions have been cleared away. Intrinsically complex phenomena
are then uniquely put in jeopardy by the heavy armory of preconceived
theoretical guidelines. In the case of electric discharge studies, it invariably proved to be the case that genuinely potent, unforeseen, and vexa-
tious puzzles provided, on their own, the essential signposts and instrumental means for deciphering an appropriate language of nature and
mapping the way to a more compelling puzzle – one, for example, that
would lead to a reshaping of the questions that had been asked, and then
to the design of experiments and instruments that would be essential

to pursue the newly formulated agenda. The Austrian dramatist Franz Grillparzer has captured the philosophy of the kind of puzzle-solving inherent in 19th century electric discharge studies: "Where riddles led me on to further riddles, to *them* the truth was quite precisely known."[88]

The emphasis given in this paper to the prominence and dominant role of experimentation in electrical discharge studies may be countered by saying that the inadequacy, irrelevance, or absence of theory does not imply that the experimental investigations were carried out in a theory-neutral mental setting. In fact, it is highly probable that a plurality of incipient theoretical hints and options on "what to do and how to proceed" was lurking "off stage." It also must be said that theories, such as these were, undoubtedly furnished points of departure, motivations, and transitory heuristics for the ongoing search for solutions to the problems encountered – even when they tended to vanish on subsequent investigation.

Two responses are offered. *First*: virtually all of our investigators were remarkably taciturn in print and correspondence about theoretical commitments and generalizations. Indeed, speculation was not much in evidence. *Second*: for better or for worse, there is a style of research on display here in which high premium is assigned to authentic and deep curiosity about nature and its inscrutable ways. This is coupled with traits of unsparing attention to fine-grained details, virtuosic tactile and visual sensitivity, craftsmanship, and above all a deferential and even celebratory acceptance of the endlessly exploration-worthy complexities of natural phenomena. Indeed, it could hardly have been otherwise in a field of research where, with every discovery, things became more complex. The boundary conditions for the investigations were tolerably well defined. There was a shared agenda in relation to the acquisition of knowledge, even in the midst of disagreement about the experimental results and interpretations. Astride such a mental disposition, the experiment-laden search for understanding became an end in itself, liberated from nervous concern about grand and final solutions. Conspicuous, above all, was the readiness to live with contradiction while playing an intellectually-directed experimental game.

The substance and focus of this paper has been to examine a style of scientific inquiry characterized by a narrowly identified domain of investigation – the conduction of electricity in gases – that attracted a genre of experimentalists who, within a specific intellectual disposition, were content to spend considerable time and effort exploring complex issues and puzzles where ultimate solutions were not readily forthcoming. Unlike the history of relativity, quantum physics, chemical thermodynamics, physical acoustics, or the conduction of electricity in solids, liquids, and solutions, the history of the conduction of electricity in matter in the gaseous state represents an area of scientific investigation in which advancement (the acquisition of knowledge) was uniquely tied to a brand of experimentation that was scarce on theory. Experimentation invariably was in advance of theory. Theoretical comprehension, generalization, and a measure of closure came only toward the end of the century.

Faraday's authority in the domain of electricity, coupled with his brief entrée into the field of electrical conductivity in gases in the 1830s, set the stage for the exhaustive gas discharge experiments of the mathematician Julius Plücker of Bonn and the physical chemist Wilhelm Hittorf of Münster. An historical examination of investigations along similar lines after 1880 – not included in this paper – would show that this branch of electrical studies was appropriated mainly by physicists from within the Helmholtz circle in Berlin: Eugen Goldstein, Heinrich Hertz, and Philipp Lenard. Following three decades of controversy concerning the nature of the "rays" produced in the electrical discharge – wave-like or particle-like? – it was demonstrated at the end of the century that the "cathode rays" are streams of electrons. This brought the century-old electric discharge phenomenon and its attendant puzzles to a point of closure and set the stage anew for exploration of the constituents, structure, and reactions of the atom and its nucleus.

We began with an account of Faraday's contribution to electrical discharge studies in gases. He will have the final word, because his attitude to the worth of experimental knowledge characterizes a point of view shared also by Plücker and Hittorf. Faraday many times referred to the impact that one particular book – the *Conversations on Chemistry* of Jane Marcet (1769–1858) – had played in his decision to take up chemistry. Unlike other contemporary textbooks, her book appealed to him because it had more details and was more technical. Paralleling Humphrey Davy's approach to the subject, Mrs. Marcet stood for chemistry as the key to unraveling the mysteries of nature. On the occasion of her death in 1858, Faraday wrote to August De la Rive (1801–1873), a Swiss physicist who had made a name for himself with a theory of electric discharge in rarefied gases as related to aurora borealis: "Do not suppose that I was a very deep thinker, or was marked as a precocious person. I was a very lively, imaginative person, and could believe in the 'Arabian Nights' as easily as in the 'Encyclopaedia'. But facts were important to me, and saved me. I could trust a fact, and always cross-examined an assertion. So when I questioned Mrs. Marcet's book by such little experiments as I could find means to perform, and found it true to the facts as I could understand them, I felt that I had got hold of an anchor in chemical knowledge, and clung fast to it."[89]

Department of the History of Science
Harvard University
U.S.A.

NOTES

[1] Jean Picard (1620–1682), instrument maker and one of the founders of the Académie Royale des Sciences, has been credited with having reported the phenomenon of "barometric glow" associated with electrical discharge in rarefied gases. See Juliette and René Taton, *Dictionary of Scientific Biography*, vol. 10, pp. 595–597. Joseph Priestley (1733–1804), in his search for new varieties of "airs," analyzed the products of electrical discharge in gases in the 1770s. Somewhat later Adriaan Paets van Troostwijk (1752–1837) employed the large electrical machine of Martinus van Marum (1750–1837) at the Teyler Museum in Haarlem for sparking and analyzing air. The German chemist Georg Friedrich Hildebrandt (1764–1816), early antiphlogistonist and reserved atomist, at the beginning of the 19th century reported on colours produced with electric discharge in rarefied gases. [Schweiggers] *Journal der Chemie* 1 (1811): 237.

[2] See, e.g., William Wheatstone, "On the Prismatic Decomposition of Electrical Light," *Communications to the British Association for the Advancement of Science* (at Dublin in 1835) (London, 1836), pp. 11–12.

[3] John Tyndall, "On Faraday as a Discoverer," Friday evening lecture, 17 January 1868, *The Royal Institution Library of Science*, Physical Sciences 2 (1970): 52 and 56.

[4] The definitive biography of Faraday is: L. Pearce Williams, *Michael Faraday. A Biography* (New York: Basic Books, 1965). See also the excellent short account in the *Dictionary of Scientific Biography*, vol. 4, pp. 527–540. Until about 1850 Faraday seems to have had limitless energy and enthusiasm for scientific investigations. Williams reports that "his friend and close associate at the Royal Institution, John Tyndall, a self-styled agnostic, wrote in ... his journal ... 'I think that a good deal of Faraday's week-day strength and persistency might be referred to his Sunday Exercises. He drinks from a fount on Sunday which refreshes his soul for a week.'" *Ibid.*, p. 527.

[5] Thomas Martin (ed.) with Foreword by William H. Bragg, *Faraday's Diary. Being the Various Philosophical Notes of Experimental Investigation Made by Michael Faraday*, 7 vols. and Index (London, 1932–1936). Quote: vol. 3, §3110–3140.

[6] Michael Faraday, *Experimental Researches in Electricity* (London, 1839), vol. 1, §1544–1548. The characteristic visible features that accompany electrical discharge in

a vacuum tube (from circa 10^{-1} to 10^{-6} atmospheres pressure) are complex, colourful, and tantalizing: spark discharge, negative glow, Faraday dark space, cathode glow, Crookes/Hittorf dark space, striated positive column, anode glow, Aston dark space, and fluorescence (begins at circa 10^{-5} atm.). In recent times the phenomenon of glow discharge has been recognized to be of theoretical importance for understanding the mechanism of atomic processes in current flow. See A. von Engel, *Ionized Gases* (London, 1965), pp. 217–221. The Faraday dark space phenomenon (that shows up at circa 6×10^{-3} atm.) is related to the depletion of charged particles by diffusion; at this rarefaction it becomes necessary to produce more particles in order to maintain constant current density.

[7] Faraday (1839), §1614–1615.

[8] W. R. Grove, "On the Electrical Discharge, and Its Stratified Appearance in Rarefied Media," (1859) *Royal Institution Library of Science*, Physical Sciences 1 (1970): 277–281. Quotes on pp. 277 and 280.

[9] Grove (1859) 280–281.

[10] Grove (1859) 281.

[11] W. R. Grove, "On the Electro-Chemical Polarity of Gases," *Philosophical Transactions of the Royal Society* 142 (1852): 87–101. See pp. 87 and 100.

[12] Edgar R. Morse, "Gassiot," *Dictionary of Scientific Biography*, vol. 5, pp. 292–293; William J. Harrington, "Gassiot," *Dictionary of National Biography* 7 (1885–1890): 935–936. As friend and experimental collaborator to Faraday, Gassiot is featured prominently in Faraday's diary, correspondence and publications, primarily in connection with investigations on the identity of static and voltaic electricity. Faraday, *Diary*, vol. 7 (1936), entries 23 Jan. 1838 to March 1858, sections 1–292; *Selected Correspondence of Michael Faraday*, L. Pearce Williams, ed., vol. 1 (1812–1848) and vol. 2 (1849–1866) (Cambridge, 1971). See General Index p. 1055.

[13] John P. Gassiot, "On the Stratifications and Dark Band in Electrical Discharges as Observed in Torricellian Vacua," *Philosophical Transactions of the Royal Society* 148 (1859): 1–16. A second communication on the subject is given in abstract form in *Proceedings of the Royal Society of London* 9 (1859): 601–605.

[14] Gassiot 148 (1859): 1–3.

[15] *Ibid.* 13–14.

[16] Gassiot 148 (1859): 14; 9 (1859): 601–602.

[17] Gassiot 148 (1859): 16.

[18] The belief that gas discharge studies would lead to fundamental insights concerning the connection between electricity and the structure of matter already was strong within the Helmholtz circle in Berlin in the late 1840s. The physical chemist Gustav Wiedemann (1826–1899) developed a lifelong friendship with Helmholtz at that time. At Karlsruhe in the 1870s, while Plücker and Hittorf in Bonn and Münster were engaged in their Helmholtz-motivated electric gas discharge studies, Wiedemann, in collaboration with Richard Rühlmann, made an attempt – that was only moderately successful – to disentangle the nature of the process involved in electric discharge phenomena in gases. Wiedemann and Rühlmann, "Ueber den Durchgang der Elektrizität durch Gase," *Annalen der Physik und Chemie* 145 (1872): 235–239, 364–399; 158 (1876): 71–87, 252–287.

[19] Goldstein's work was extended by Johannes Stark (1884–1957), who in 1900 demonstrated the Doppler effect for *Kanalstrahlen* and maintained that his discovery provided

proof for Einstein's special theory of relativity (1906) and the quantum hypothesis (1907). In 1919 Stark was awarded the Nobel Prize in physics "for his discovery of the Doppler effect in canal rays and the splitting of spectral lines in electric fields." Max Planck the same year received the Nobel Prize for his "discovery of energy quanta."

[20] Lenard was in charge of the publication of Hertz's three volume *Gesammelte Werke* (1894–1895). In 1905 he received the Nobel Prize in physics "for his work on cathode rays."

[21] An historical examination of the cathode ray controversy, as represented in the work of Goldstein, Hertz, Lenard, Crookes, Thomson, and their collaborators and contemporary critics and supporters, will be treated in subsequent papers.

[22] Plücker's *Gesammelte Mathematische Abhandlungen* (Leipzig, 1895) was edited by mathematician and crystallographer A. Schoenflies (1853–1928), whose chair in applied mathematics at Göttingen was created by Felix Klein. An evaluation of Plücker's mathematics is given by the noted Göttingen mathematician Alfred Clebsch (1833–1872) in "Zum Gedächtniss an Julius Plücker," pp. ix–xxxv. Plücker's *Gesammelte Physikalische Abhandlungen* was edited by F. Pockels (Leipzig, 1896). A résumé and appraisal of his physics is given by Göttingen physicist Eduard Riecke (1845–1915) in "Plücker's Physikalische Arbeiten," pp. xi–xviii. Other important secondary works include: Ad. Dronke, *Julius Plücker* (Bonn, 1871) – a short essay written with the help of documents supplied by the Plücker family. An incomplete list of Plücker's publications by Felix Klein is included. Dronke had been an assistant to Plücker in spectral studies: Plücker, *Philosophical Magazine* 18 (1859): 14. An inaugural dissertation by Wilhelm Ernst, *Julius Plücker* (Bonn, 1933), gives a more useful, because documented, portrayal and interpretation of Plücker's educational background, personality and contributions to mathematics and physics.

[23] Felix Klein, *Vorlesungen über die Entwicklung der Mathematik im 19. Jahrhundert* (Göttingen, 1926). Section on Plücker: pp. 119–126.

[24] Joan L. Richards, *Mathematical Vision. The Pursuit of Geometry in Victorian England* (Boston: Academic Press, 1988), p. 58. Evaluations of Plücker's views on space were not recognized in Germany. Minkowski wrote to Hilbert in 1893: "Schönflies will Plücker's Abhandlungen herausgeben, wozu, weiss ich nicht. In Plücker's wissenschaftlichen Nachlass ist, soweit ich hier [Bonn] für ihn auskunften konnte, seinerzeit Käse eingewickelt worden." L. Rüdenberg and H. Zassenhaus, eds., *Hermann Minkowski Briefe and David Hilbert* (Berlin: Springer, 1973), pp. 55–56. All of Plücker's mathematical papers, including *On a New Geometry of Space* (original long version 1865) and the French abstract (1867) are reproduced in the *Gesammelte Mathematische Abhandlungen*.

[25] *Encyclopaedia Britannica*, 11th ed., vol. 25, p. 736. See also Index in E.N. Harvey, and Obits in *Proceedings of the Royal Society of London* 38 (1883): 34 and *Nature* 27 (1883): 599.

[26] Plücker, "Ueber die Einwirkung des Magneten auf die elektrischen Entladungen in verdünnten Gasen," *Annalen der Physik* 103 (1857): 151–157 and 104 (1958): 113–128. English translation: "On the Action of the Magnet upon the Electrical Discharge in Rarefied Gases," *Philosophical Magazine* 16 (1858): 119–135; "Observations on the Electrical Discharge through Rarefied Gases," *Philosophical Magazine* 16 (1858): 408–418. German text: *Gesammelte Physikalische Abhandlungen*, nos. 30 and 31 (pp. 475–494); no. 32 (pp. 495–507). Where readily accessible, as in the above, the English translations

have been quoted. Otherwise the author has provided his own translations from the *Gesammelte Physikalische Abhandlungen*.

[27] Geissler, a descendant of master craftsmen in Thuringia, after some years of practice in Munich and Holland, established himself as a mechanic at the University of Bonn around the mid-1850s. Thence he became known far-and-wide in European scientific circles for his wizard-like inventive skills with glass and mechanical instruments. His close association with Plücker led to an honorary degree at Bonn in 1868. For more information on Geissler: Hans Kangro, *Dictionary of Scientific Biography*, vol. 5, pp. 340–341, and A.W. Hofman, obituary notice, *Berichte der Deutsche chemischen Gesellschaft* 12 (1879): 147–148. Geissler authored four papers, only one of which is of interest here: "Neue Erfahrungen im Gebiete der elektrischen Lichterscheinungen," *Annalen der Physik* 135 (1868): 332–335.

[28] Plücker, *Philosophical Magazine* (1858): 119.

[29] Rühmkorff (1803–1877), a German by birth, in 1855 established a workshop in Paris that supplied scientists with induction coils that he had invented. For many years these coils were used as the standard source of power in the operation of Geissler and Crookes discharge tubes. Bernard S. Finn, "Rühmkorff," *Dictionary of Scientific Biography*, vol. 11, pp. 603–604. Another noteworthy inventor and builder of instruments was the Paris-educated British chemist Warren De la Rue (1815–1889). His own fairly routine experiments on electric discharge in gases cover the period from 1868 to 1883.

[30] Heinrich Kayser, *Handbuch der Spektroskopie*, vol. 1 (Leipzig, 1906), p. 231.

[31] Hans Kangro, "Geissler," *Dictionary of Scientific Biography*, vol. 5, p. 341.

[32] Plücker, *Philosophical Magazine* (1858): 119–122.

[33] Plücker, *Philosophical Magazine* (1858): 408–411.

[34] Plücker, *Philosophical Magazine* (1858): 414–415.

[35] Plücker, *Philosophical Magazine* (1858): 417–418.

[36] See, e.g., Erwin Hiebert, "The Role of Experiment and Theory in the Development of Nuclear Physics in the Early 1930s," in D. Batens *et al.*, eds., *Theory and Experiment* (Dordrecht: Reidel, 1988), pp. 55–76.

[37] Piet Hein, *Grooks* (Copenhagen, 1966), p. 2.

[38] Plücker, *Gesammelte Physikalische Abhandlungen*, nos. 33–36, 38–40; pp. 508–598, 612–664. The first two (nos. 33 and 34) were translated into English: *Philosophical Magazine* 18 (1859): 1–7 and 7–20.

[39] Plücker, "Observations on the Electric Discharge," *Philosophical Magazine* 18 (1859): 1–7 and 7–20. Quote on pp. 1–2. The term "epipolic," Greek for the surface of a body, was introduced by John Herschel in *Philosophical Transactions* 136 (1845): 147.

[40] Plücker, *ibid.* 7.

[41] Plücker, *ibid.* 8–12.

[42] Plücker, "Abstract of a Series of Papers and Notes Concerning the Electric Discharge through Rarefied Gases and Vapour," *Proceedings of the Royal Society* 10 (1860): 256–269. *Gesammelte Physikalische Abhandlungen* (1890), no. 37, pp. 599–611. A French version of the influential 1860 paper was also published: "Analyse spectrale" in *Cosmos* in 1862; *Gesammelte Physikalische Abhandlungen* no. 40, pp. 657–664. During Plücker's mathematical career (1824–1847) almost half of his papers had been published in French simply because his mathematical interests and style of presentation were appreciated most by French mathematicians. On the other hand his physical investigations were taken up predominantly by the English; five of his 59 papers on physics

132 ERWIN N. HIEBERT

were published in English, two in Latin, the rest in German.

[43] See, e.g., letters from Faraday to Plücker nos. 358 (1847), 368 (184?), 383 (1848), 418 (1850), and 579 (1854) in the *Selected Correspondence of Michael Faraday*, L. Pearce Williams, ed. (Cambridge, 1971).

[44] Plücker, *ibid.* 599–600.

[45] Plücker, *ibid.* 601–602.

[46] Plücker, *ibid.* 602–604.

[47] Plücker, *ibid.* 605–606. Plücker, as he admits, had adopted the idea of "reciprocating currents" from Gassiot.

[48] Plücker, *ibid.* 611.

[49] J. Plücker of Bonn, For. Memb. R.S., and Dr. J.W. Hittorf of Münster, "On the Spectra of Ignited Gases and Vapours, with Especial Regard to the Different Spectra of the Same Elementary Substance," *Philosophical Magazine* 28 (1865): 1–29. Reproduced in English in Plücker's *Gesammelte Physikalische Abhandlungen*, pp. 665–700.

[50] Plücker/Hittorf (1865) 1–4.

[51] Plücker/Hittorf (1865) 6–7.

[52] Plücker/Hittorf (1865) 9–11.

[53] Actually Plücker's paper on electrical discharge in rarefied gases of 1859 predated the Bunsen/Kirchhoff announcement of 1860. Compare Plücker, *Annalen der Physik* 107 (1859) with Kirchhoff/Bunsen, *ibid.* 110 (1860).

[54] Clifford L. Maier, *The Role of Spectroscopy in the Acceptance of an Internally Structured Atom, 1860–1920*, Ph.D. Dissertation, University of Wisconsin, Madison, 1964. Published. An extensive survey of all aspects of 19th century spectroscopy is given in Heinrich Kayser, *Handbuch der Spektroskopie*, 8 vols. (Leipzig, 1900–1932).

[55] L. Pearce Williams, ed., *The Selected Correspondence of Michael Faraday*, vol. 1, 1812–1848; vol. 2, 1849–1866 (Cambridge, 1971).

[56] *Faraday Correspondence*, letter 404 in August 1849 and letter 627 in March 1856.

[57] *Faraday Correspondence*, letter 368.

[58] The two men first met in London in August 1848. Plücker spent three days in Faraday's laboratory and described some of his experiments "on the crystalline diamagnetic relations." See *Diary* entries for 16–25 August 1848. Tyndall reports that for "all September and October Faraday worked on the crystalline polarity of bismuth and on its relation to magnetic force" as reported in Faraday's 22nd series of the *Experimental Researches*. Bence Jones, *The Life and Letters of Faraday*, vol. 2 (London, 1870), pp. 238–241. In May 1849 Faraday wrote to congratulate Plücker (letter 391) "on the beautiful facts you describe. How wonderfully this branch of Science is progressing." In December 1849 (letter 411) Faraday mentioned the "growing of up facts." He hoped that when "properly understood" the various phenomena examined in terms of their magnecrystallic, crystallographic and optic axes would come under "one law [that] will include all these phenomena."

[59] *Faraday Correspondence*, no. 661.

[60] *Faraday Correspondence*, nos. 663 and 664.

[61] *Faraday Correspondence*, nos. 665 and 667.

[62] *Faraday Correspondence*, no. 675. The account here parallels Plücker's paper of 1858 on the action of light on the electrical discharge in rarefied gases.

[63] *Faraday Correspondence*, no. 698.

[64] *Faraday Correspondence*, no. 719.

[65] *Faraday Correspondence*, no. 781.

[66] For information on Hittorf: Gerhard Carl Schmidt, "Wilhelm Hittorf," pp. 1–18 in *Festschrift der 84. Versammlung Deutscher Naturforscher und Aertze* in Münster 1912; Alfred Coehn (Göttingen), "Wilhelm Hittorf," *Die Naturwissenschaften* 3 (1915): 41–43; G. Tamman, "Wilhelm Hittorf," *Nachrichten von der Königlichen Gesellschaft der Wissenschaften zu Göttingen*, Geschäftliche Mitteilungen aus den Jahr 1915 (1915): 74–78; H. Koner, *Deutsches Biographisches Jahrbuch* (Leipzig, 1925), pp. 41–44; Alfred Heydweiler, *Johann Wilhelm Hittorf* (Leipzig, 1915); and Gerhard C. Schmidt, *Wilhelm Hittorf* (Münster, 1924).

[67] W. Hittorf, *Ueber die Wanderungen der Ionen während der Elektrolyse* (1853–1859). The most important study on this subject is: Ollin J. Drennan, *Electrolytic Solution Theory: Foundations of Modern Thermodynamical Considerations*, Ph.D. Dissertation, University of Wisconsin, Madison, 1961.

[68] Wilhelm Hittorf, "Ueber die Elektricitätsleitung der Gase," *Annalen der Physik* 130 (1869): 1–31; 16 (1869): 197–234; Jubelband (1874): 430–445; 7 (1879): 553–631; 20 (1883): 705–755; and 21 (1884): 90–139. The first four papers except for pp. 617–632 (1879) were published in English translation in *Physical Memoirs* (London: Taylor and Francis, 1883), pp. 111–232. The quotes are taken mostly from the English. The illustrations that accompany Hittorf's published papers display a truly remarkable array of intricately constructed discharge tubes – 42 of them – used in the investigations, and frequently provide, as well, a vivid and helpful graphic image of the effects produced by magnets.

[69] Hittorf, *Physical Memoirs* (1869): 111–115. In *Annalen der Physik* 136 (1869): 6: "Die drei Teile, das positive Licht, der dunkle Raum, und das negative Glimmlicht treten bei jeder Dichtigkeit des Mediums auf, wenn die elektrische Ladung eine mehrfache Dauer annimmt."

[70] Hittorf (1869): 113–115. Extensive calculations, pp. 118–136, follow in an attempt to choose – in terms of these three elements and the various dimensional parameters of the tube – the best conditions for studying the stage at which good conductivity may be obtained at the smallest density for given capillary dimensions in the Geissler tube. Hittorf had been in contact with Geissler in 1865, and had given Plücker some of his tubes to be shown at the Paris exhibition in 1867 (p. 140).

[71] Hittorf (1869) 144.

[72] Hittorf (1869) 149.

[73] Hittorf (1869) 149–155.

[74] Hittorf (1869) 157.

[75] Hittorf (1869) 157.

[76] Hittorf (1869) 157–158.

[77] Hittorf (1869) 165–166.

[78] Hittorf (1874) 167–170.

[79] Hittorf (1874) 171–179.

[80] Hittorf (1879) 180–181.

[81] Hittorf (1879) 181–208. Mentioned are the works of Gassiot, Kirchhoff, E. Becquerel, Stokes, De La Rive and E. Wiedemann.

[82] Hittorf (1879) 221.

[83] Hittorf (1879) 221–222.

[84] Hittorf (1879) 224.

[85] Hittorf (1879) 224; Faraday, *Experimental Researches*, §1551.
[86] Hittorf, *Annalen der Physik* **20** (1883): 705–755 and **21** (1884): 90–139.
[87] Hittorf, *Annalen der Physik* **21** (1884): 139: "Schluss folgt."
[88] Erwin Schrödinger begins his *My View of the World* (Woodbridge, Conn., 1983) with this question: "Wo Rätseln mich zu neuen Rätseln führten, Da wussten *sie* die Wahrheit ganz genau."
[89] John Tyndall, "Faraday as a Discoverer," *The Royal Institution Library of Science*, Physical Sciences, vol. 2 (Amsterdam, 1970), pp. 52–53; L. Pearce Williams' article on Faraday, *Dictionary of Scientific Biography*, vol. 4, pp. 528–529; Evan Armstrong, "Jane Marcet and Her 'Conversations on Chemistry,'" *J. Chem. Educ.* **15** (1938): 53–57. Marcet's *Conversations on Chemistry; In Which the Elements of That Science Are Familiarly Explained by Experiments* was first published in two volumes in London and Philadelphia in 1806. In subsequent editions "late discoveries" by Sir Humphrey Davy and others were added. The last and 15th edition was printed in 1850.

ROBERT J. DELTETE*

GIBBS AND THE ENERGETICISTS

I. INTRODUCTION

The energeticists Georg Helm and Wilhelm Ostwald were enthusiastic in their praise of the thermodynamic writings of Josiah Willard Gibbs. Both admired the elegance and power of Gibbs's work and regarded it not only as energetic in character, but as exemplifying the course a rightly-conceived natural science should and would take. They disagreed, however, in their characterizations of Gibbs's general approach. Helm interpreted Gibbs as a phenomenalist, which was the approach to science he favored; Ostwald thought Gibbs a realist, which was the outlook he preferred. And since each regarded Gibbs's work in thermodynamics as reflecting the natural development of scientific inquiry, they also disagreed about the mandates of the *Weltgeist*: Helm saw science progressing in the direction of energetic phenomenalism, while Ostwald viewed it instead as progressing toward energetic realism.

The energeticists were wrong – in their understanding of Gibbs, at least – as I shall try to explain in this essay. Since Helm and Ostwald each regarded Gibbs as having furthered an approach to natural science initiated by Robert Mayer, I will begin, in Section II, with their very different understandings of his work. Section III, also preparatory, comments on their equally opposed views about energy. Section IV then describes Helm's reading of Gibbs and explains why it seems clearly mistaken. Section V, in somewhat more detail, does the same for Ostwald. Along the way, I try to imagine Gibbs's likely response to the energeticists. This is somewhat problematic, however, since Gibbs never commented on the project of the energeticists and commented only infrequently and indirectly on the goal of physical theory. Still, enough can be inferred from what he did say, and from the physics Gibbs produced, to conclude that he would have rejected energetic interpretations of his thermodynamics. Section VI, a brief epilogue, sketches the real motivation for his work.

135

A.J. Kox and D.M. Siegel (eds.), No Truth Except in the Details, 135–169.
© 1995 *Kluwer Academic Publishers.*

Since the subject of my essay is likely unfamiliar to many contemporary readers, a bit of context may prove useful. Let me begin with a broad brush, and then add a few relevant details.

The great unsettled question of late nineteenth-century physics was the status of the mechanical world view. For more than two hundred years – from Descartes, Huygens and Newton in the seventeenth century to Helmholtz, Hertz and Boltzmann at the end of the nineteenth – physicists had generally sought mechanical explanations for natural phenomena.[1] Indeed, as the last century drew to a close, Heinrich Hertz reaffirmed the classical goal of physical theory: "All physicists agree," he wrote, "that the problem of physics consists in tracing the phenomena of nature back to the simple laws of mechanics" (Hertz 1894, Vorwort). But when these words appeared in 1894, there was in fact no longer general agreement among physicists about the nature of their project. Many doubted, and some explicitly denied, that mechanics was the most basic science; other candidates for that honor – thermodynamics and electromagnetic theory, in particular – were seriously considered; and comprehensive alternatives to the mechanical world view were proposed and vigorously debated throughout the 1890s and early 1900s.[2]

Energetics was one of the alternatives. Tracing its origins to the founders of the law of energy conservation, especially Robert Mayer, and to the thermodynamic writings of Clausius, William Thomson, and Gibbs, energetics was an attempt to unify all of natural science by means of the concept of energy and of laws describing energy in its various forms. The energeticists believed that scientists should abandon their efforts to understand the world in mechanical terms, and that they should give up atomism as well, in favor of a new world view based entirely on the transfers and transformations of energy.

The emergence of energetics was largely, if not entirely, a German phenomenon. Its main proponents were Georg Helm, a Dresden mathematician and physicist, and Wilhelm Ostwald, the professor of physical chemistry at Leipzig. Both thought that the world view of modern science was moving toward a comprehensive theory of energy.[3] But, as previously noted, they disagreed about the general form that theory would take: Helm believed that it would be a phenomenalist theory, while Ostwald was convinced that it would be realist in character. Each thought, nevertheless, that the thermodynamic writings of Gibbs supported his position.

II. MAYER AND ENERGETICS

Helm and Ostwald included all of the pioneers of energy conservation among the founders of energetics, but they accorded a special place of honor to Robert Mayer. Sometimes, their reasons for doing so coincided. Each admired the boldness and independence of Mayer's thought, his skeptical attitude toward prevalent molecular and mechanical hypotheses, and the way he steadfastly opposed any attempt to reduce heat to a form of mechanical energy. Above all, each praised Mayer's insight that all natural phenomena are really energy transformations and his vision of a unifying science of energy.[4] But at the same time, they disagreed fundamentally about the content of Mayer's insight and the meaning of his vision.

To isolate the important issues, let me begin with Helm's evaluation of Mayer in his history of energetics. When Helm praised Mayer in 1898 for the clarity of his insight into fundamental principle, it was for conceiving the possibility of a science of energy that was a "pure system of relations," exemplifying a phenomenalism of the sort championed by Ernst Mach (Helm 1898, 20). Mayer had founded "a new world view," Helm claimed (Helm 1898, 214), that was both energetic and phenomenalist in orientation. Like Mach, Mayer was interested only in quantitatively describing and relating the data of experience, the phenomena. Eschewing metaphysical references to underlying substance or causes, he was satisfied to show that "a relationship exists in consequence of which one phenomenon decreases in favor of another, or increases at its expense" (Helm 1898, 26). But he went beyond Mach in suggesting that all our experience, and so all phenomena, are energy related. This was Mayer's "fundamental energetic idea" (Helm 1898, 29), the one Helm sought to promote in his own work.

This interpretation is fanciful and it conflicts with Helm's earlier reading of Mayer's intent.[5] But at present that is not my concern. (I shall comment later, when I come to Gibbs, on Helm's reliability as an historian.) Here I want to clarify the view Helm attributed to Mayer, because Helm *did* think that a "pure system of relations can be achieved by means of energetics"; this was the "fundamental energetic idea" that Helm sought to develop, defend, and promote in his history of the subject.[6] For reasons to be discussed shortly, I designate this Helm's "official position" on the goal of energetics, and I shall collect its main features under one heading, which I call the "Relations

Thesis." The Relations Thesis has epistemological, methodological and anti-metaphysical dimensions. First, it claims that we can only know phenomena and changes in phenomena, all of which – for Helm, as for Ostwald – are energetic in character. Second, it claims, in consequence, that the goal of natural science is to describe and relate energy phenomena in the simplest and most unified manner possible. Third, it rejects all inferences beyond the phenomena. Specifically, it rejects all efforts to substantialize energy or to reify energetic changes in terms of "migrations," "transitions," "transformations," "conversions," or what have you. When he wrote his history of energetics, Helm portrayed Mayer as the first significant advocate of the Relations Thesis.[7]

Ostwald disagreed. In his view, Mayer's most important contribution to energetics was to have ascribed reality and substantiality to energy as well as matter. *That* was the "essential insight" that Ostwald sought to promote and develop in his first writings on energetics,[8] but obstacles had made this difficult. Sometimes, Ostwald claimed that Mayer's insight had been obscured by subsequent developments of the energy concept, especially in thermodynamics, where energy tended to be regarded more as an interesting mathematical function, comparable to the potential function in mechanics, than as a physical reality. Usually, he put the blame elsewhere: "One may undoubtedly explain [general ignorance of Mayer's intent] as a consequence of the rapidly expanding mechanistic conception of nature," a way of thinking he found even harder to overcome.[9] Whatever the reason, Ostwald initially only wanted to recover and underline the importance of Mayer's basic idea, that energy is as real and fundamental as matter. Within a few years, though, Ostwald was converted to the way of "pure energetics" and began to defend in his writings the idea that *only* energy is substantial and real. "The more I reflected on the nature of energy," Ostwald wrote in 1891, "the clearer it became to me that matter is nothing but a complex of energy factors." Given that realization, he soon concluded that a genuine energetics had to do more than treat energy as "a real substance and not just as a mathematical abstraction"; it had to recognize energy a the ultimate substance and the only reality (Ostwald 1891, 566).[10] Opposing Helm's Relation Thesis was likely on Ostwald's agenda, therefore, when he later recounted the history of energetics. After insisting that his own development of the subject had not only opposed the "sterility of unbridled mechanism" but had sought to remove energy from "the realm of mathematical abstraction and to view it as the real substance of the world," Ostwald then proceeded to criticize Helm's initiatives as "a retreat to a position even less progressive than Mayer's" (Ostwald 1926, vol. 2, 157–158).

III. HELM AND OSTWALD ON ENERGY

Helm's position was less definite and consistent than Ostwald's criticism might suggest. In his writings on energetics, Helm vacillates between the ascetic phenomenalism of the Relations Thesis and some form of energetic realism, so that his intent in a given passage is not always clear. But two conclusions are reasonably secure. First, despite his later advocacy of the Relations Thesis, Helm always spoke of the internal energy of a system as if it were a substance. More precisely, he always attributed to a system, as a real possession, a definite internal (or intrinsic) energy, which was a function of its physical and chemical state.[11] Sometimes, he also seemed to be committed to a larger claim, which I shall call the thesis of "real presence."[12] This thesis claims that the internal energy of a system may be divided into components, each of which is really present in the system. Usually, however, he rejected as unfounded the idea of real presence, arguing that a system no more possesses a definite quantity of kinetic energy than it does of heat or volume energy.[13] So we may perhaps best summarize the *praxis* of Helm's history, in contrast to his official position, by saying that while he took for granted a substance view of internal energy, he opposed the idea of real presence. Hence, for example, his approval of Tait's criticism of Clausius: "We are quite ignorant of the condition of energy in bodies generally. We know how much goes in, and how much comes out, and we know whether at entrance or exit it is in the form of heat or work. But that is all" (Helm 1898, 121).[14] Helm did not object in principle to Helmholtz's distinction between "free" and "bound" energy as a conceptual or heuristic device, or to Rankine's between "actual" and "potential," or even to Clausius' between heat and internal work, but he generally rejected any realistic interpretation of energy components. The *appearance* of different forms of energy was a sign of intrinsic energy in transition, but these forms were not themselves really present in different amounts in the energy content of a body.[15]

Ostwald evidently disagreed, but his own considered position is also difficult to reconstruct. From the early 1890s, when he first began to write in earnest on energetic theory, he officially subscribed to a view of matter's relation to energy that might be called the Composition Thesis. On that view "material objects" (or "bodies" or "physical-chemical systems") are nothing more than energy complexes – spatially copresent and coupled clusters of energy. The Composition Thesis was doubtless central to Ostwald's conception of energetics; in fact, acceptance of it in some form or other constitutes much of what he meant when he spoke of his conversion to "pure energetics" (Ostwald 1926, vol. 2, 168–170).[16] In his more detailed discussions of energetic science, however, Ostwald usually employed a quite different view of matter's relation to energy. Then he frequently spoke of an object or system "containing" (or "possessing" or "having") energy of certain kinds in certain amounts, as if a system were not the same as, but something in addition to, its energy content. When he did this, moreover, he usually just assumed, without comment, that every system contains definite amounts of several distinct forms of energy (real presence) and that in each case the total energy content is given by the sum of the amounts of each form (really) present.[17] This view, into which Ostwald slipped whenever he attempted the mathematical development of energetic theory, might therefore be called the Containment Thesis.[18]

A study devoted to the energetic theories of Helm and Ostwald would require more attention to their various treatments of energy in practice, since that practice frequently appears to undermine or contradict the theory it is supposed to support. But as that is not my aim here, and to avoid complications in what follows, I shall take Helm and Ostwald (at their official words) as defending, respectively, the Relations and Composition Theses. In Helm's view, then, Gibbs followed Mayer in being a clear and consistent proponent of the Relations Thesis. In response, I shall argue that while Gibbs was evidently aware of the security as well as the power of phenomenological thermodynamics, he found the asceticism of phenomenalism physically unsatisfactory. On Ostwald's view, by contrast, Gibbs had gone beyond Mayer (like Ostwald himself) in conceiving of energy as the only reality and by embracing the Composition Thesis. In reply, I argue that Gibbs did not regard thermodynamics as essentially the study of energy and that he would have rejected the realism of the Composition Thesis as physically unfounded.

IV. HELM AND GIBBS

Like Ostwald, Helm thought the originality and importance of Mayer's insight had not been appreciated by his contemporaries, and to a large extent their reasons were the same. The main problem, in Helm's view, was that scientists were wedded to molecular and mechanical ways of thinking, which they were reluctant to give up. Nowhere was this more evident, perhaps, than in the subsequent development of thermodynamics and the effort expended to understand its laws. Here, Helm lamented, most physicists insisted that "the actual scientific foundation of thermodynamics had to be sought in the mechanics of atoms" (Helm 1898, 146). It was as if the laws of thermodynamics were taken to be only "rough estimates," useful for certain purposes, but ultimately unsatisfactory because they did not "open up a view into the mechanics of the interior of bodies." Helm agreed that "to someone for whom the highest goal of the theoretical knowledge of nature is the resolution of all change into the motion of atoms," thermodynamics probably appeared to be little more than a "bargain basement" theory, since its results were in fact the consequences of more basic causes (Helm 1898, 144). But he resisted that attitude as contrary to the spirit of energetics.

One can be more precise about the nature of that resistance, and in a way that sheds light on Helm's reading of Gibbs. In general, Helm applauded works that contributed to a phenomenological theory of energy, and criticized those that promoted molecular and mechanical theories of the same, conflated micro-mechanical theories with phenomenological ones, or valued the former sort of theory more than the latter. Of Clausius' 1850 memoir, for example, he wrote that it marked "a decisive turning point" in the history of energetics:

We have before us here for the first time the foundations of a system of theory that, without hypothetically going back to mechanics or even using mechanical analogies, can nonetheless make the same claim to unconditional and comprehensive validity as does mechanics itself: What Carnot and Mayer aspired to is here fulfilled. This energetic originality of Clausius' work emerges in a particularly striking way if one compares it to [J. M.] Rankine's memoir, published in the same year, which arrives at many of the same results, but which is based throughout on a mechanical hypothesis: Molecular vortices are conceived and, hypothetically, certain mechanical relations of these vortices are then interpreted as heat, others as temperature, in order to advance to the results (Helm 1898, 81).

Helm's evaluation of Clausius' memoir is selective and misleading; but the distinction Clausius began to formulate in that work between what we should now call the "general" and "special" theories of heat does help to locate what is essential to Helm's point of view. He enthusiastically praised the former theory, regarding it as fully energetic in spirit; but he regretted Clausius' later attempts to construct a special theory of heat – his excursions into the molecular realm and his efforts to provide his general theory with a mechanical basis.[19] To be sure, Clausius had been scrupulously careful to separate his work on thermodynamics from his ideas about molecular science, realizing that the latter were less likely to command general assent;[20] but others, such as Rankine, were perfectly willing to incorporate ideas about molecules and mechanisms into the very heart of thermodynamics. And Clausius, in any case, believed that an explanation of the laws of thermodynamics on the basis of molecular mechanics was possible, fundamental, and needed.

Helm did not share that point of view, and neither did Ostwald. But whereas Ostwald believed that the "subtleties of nature" sought by mechanical theorists were actually unveiled by a properly conceived energetics, Helm's considered view was that the search for such subtleties was not the proper task of science. If Ostwald was committed to energy as the only real substance of the world, and to the existence of distinct, irreducible forms of energy, Helm professed himself opposed to metaphysics of any kind. Of William Thomson and P. G. Tait's *Treatise on Natural Philosophy*, for example, he wrote:

This work has not escaped the ancient metaphysics of matter and motion; it has only put off discussing it until another time. But a rightly constituted theory of energy stands in need of no such discussion. Energetics, as a pure science of relations, does not require metaphysical speculations to support itself or to ground its applications (Helm 1898, 212).

Not surprisingly, therefore, Helm admired phenomenological thermodynamics and preferred a theory of energy modeled on that science. In the early parts of his history of energetics, in fact, he often simply identified the two (Helm 1898, Parts I–IV). While not wanting to deny that "the mechanical hypothesis" or the "molecular hypothesis" – two ideas he also frequently ran together – had sometimes yielded important results, he vigorously protested the tendency to interpret such hypotheses as more than conceptual or heuristic devices, lacking ontological import. Like Ostwald, moreover, Helm opposed attempts to defend molecular and mechanical hypotheses by means of "all sorts of artificialities," and sought to expose the confusion – all too prevalent, in his view – of atomism and mechanism with what was really essential to energetics. Having traced this confusion to the early writings of Helmholtz, especially to his famous memoir of 1847, Helm continued:

Robert Mayer completely avoids this confusion, and in England, too, under the steady influence of William Thomson, energetics developed more purely. In Germany the gradually increasing predominance of the mechanical hypothesis is very clearly revealed in the personal development of Clausius. His first work of 1850 sees in energetics [= thermodynamics] a new science joining mechanics on an equal basis, but the molecular hypothesis intrudes itself more and more into his later works. In the same way the entire course of development of the science in Germany from the mid 1850s to the mid 1880s appears as a falling away from the true clarity of Mayer's intuition (Helm 1898, 145).

Untouched by the German decline, however, Gibbs had held fast to Mayer's intuition. Helm had many reasons for praising Gibbs, but what most impressed him was the rigorously phenomenological character of Gibbs's work.[21] Nowhere, he thought, had the Relations Thesis found a more brilliant and productive development:

Completely free of any bias in favor of the mechanics of atoms, establishing with complete impartiality the strict consequences of the two laws [of thermodynamics], without any longing glances at and yearning for mechanics – thus the work of Gibbs suddenly stands before our gaze ... Here the great old idea of Mayer has come to life in mathematical formulae, free from all molecular-hypothetical adornment.

Helm can barely contain himself. Of Gibbs's three great papers on thermodynamics he exclaimed:

What a book, in which chemical processes are treated without the traditional chemical apparatus of atoms, in which the theories of elasticity, of capillarity and crystallization and of electromotive force, are set forth without all the usual devices of atomistic origin! Naked and pure, the true object of the theoretical knowledge of nature stands before us: the establishment of quantitative relationships between the parameters which determine the state of a material system during any changes subject to investigation. No wonder that people did not understand these works of Gibbs ... (Helm 1898, 146).

Again, however, Helm has selectively misread the history of science, attempting to bolster his own conception of energetics by claiming the authority of Gibbs for that point of view. He was certainly right in saying that Gibbs's writings were difficult to understand, but he was just as certainly wrong in claiming that Gibbs had no interest in a molecular and mechanical explanation of thermodynamics.

The clearest, most explicit statement of that interest is found in the remarkable preface Gibbs wrote in late 1901 for his treatise on statistical mechanics, a work that was "Developed With Especial Reference to the Rational Foundations of Thermodynamics" (Gibbs 1902, v–x). There we find him remarking that "the laws of thermodynamics, as empirically determined, express the approximate and probable behavior of systems of a great number of particles," and affirming the belief, in consequence, that the separate study of statistical mechanics "seems to afford the best foundation for the study of rational thermodynamics and molecular mechanics." In fact, he put forth the view that "the laws of thermodynamics may easily be obtained from the principles of statistical mechanics, of which they are the incomplete expression" (vi–vii).

This point of view was not new to Gibbs's thought: Gibbs spoke publicly on statistical mechanics as early as 1884;[22] and he taught courses on the subjected throughout the 1880s and 1890s, always with a view to laying a conceptual foundation for the laws of thermodynamics.[23] An obituary notice he wrote for Clausius in 1889 (Gibbs 1889) reveals the depth of his understanding of the subject and also his concern for the molecular approach to thermodynamics. There Gibbs compared the contributions of Clausius to the molecular interpretation of the second law with those of Maxwell and Boltzmann and made clear that he not only understood well their arguments but also approved of their orientation. Unlike Helm, he wrote approvingly of Clausius's interest in "the nature of the molecular phenomena of which the laws of thermodynamics are the sensible expression." He also praised Clausius for his "remarkable insight," for the "substantial correctness" of his ideas, and for his "very valuable contributions to the molecular science" (Gibbs 1889; [1906, vol. 2, 263–265]). The positive later evidence all suggests, therefore, that Gibbs thought thermodynamics to be reducible to mechanics, even if to some statistical version.[24]

Can we reasonably project this attitude backward into his earlier thermodynamics writings? This is more problematic, since Gibbs did not disclose his thoughts on the subject in these works. We may conjecture that he already thought a statistical approach to the second law of thermodynamics to be necessary, but that he did not approve of current approaches to the problem, such as Boltzmann's, and did not yet know how to extend the ensemble approach adumbrated in Maxwell's last work.[25] But the only direct evidence for this conjecture is a remark he made in the section of his memoir on heterogeneous equilibrium that developed that well-known "Gibbs paradox." There Gibbs noted that since it is possible in principle for the ordinary molecular motions in a gas to produce unmixing, with a concomitant decrease in entropy, one cannot absolutely rule out such a process, however unlikely it may be. "In other words," he wrote, "the impossibility of an uncompensated decrease of entropy seems to be reduced to an improbability" (Gibbs 1906, vol. 1, 167–168).[26]

This remark is not developed in Gibbs's thermodynamic writings; but given the objective of those works, it is not surprising. What we find instead are careful, if infrequent, references to the relations between classical, non-statistical mechanics and thermodynamics, for example, to the idea that the conditions for thermodynamic equilibrium may be regarded as a generalization of the condition for mechanical equilibrium.[27] A couple of passages may also suggest that Gibbs regarded classical mechanics as reducible to thermodynamics, since he explicitly states that real systems are thermodynamic in nature and seems to imply that the condition for purely mechanical equilibrium is merely a consequence of his own conditions for thermodynamic equilibrium.[28] But if Gibbs had thought that thermodynamics held the key to theoretical unity, he would have pursued the matter, and he did not. His next work on the fundamental equations of mechanics did not comment on the relation between mechanics and thermodynamics or even mention the equilibrium conditions of his famous memoir (Gibbs 1879). Indeed, he wrote no other major work on thermodynamics; with one exception, of no consequence here, his later publications in the field were elaborations of particular problems he had discussed in his memoir on heterogeneous equilibrium.[29] Gibbs apparently thought he had already said all that he needed to say about the fundamentals of the subject.[30] It was to statistical mechanics that he then turned to explain those fundamentals.

An indirect bit of evidence that he may have done so almost immediately is provided by E. B. Wilson, one of Gibbs's students. Commenting on the letter Gibbs wrote in January 1881 (Gibbs 1881), accepting the Rumford Medal of the American Academy of Arts and Sciences for his work in thermodynamics, Wilson wondered what Gibbs would have said about that work had he been present in person to receive it. He especially wondered what then occupied Gibbs's thoughts, since he was no longer at work on fundamental thermodynamic theory. "Was he concentrating his attention, as Clausius and Maxwell had done and as Boltzmann and Kelvin were doing, on the attempt to deduce thermodynamic behavior from dynamical properties of matter and possibly to find some equation expressing the thermodynamic functions of a body of variable composition other than perfect gases?" Wilson did not answer this question, but he evidently did not think it an unnatural one to ask.[31]

If we meet with only limited success in determining Gibbs's early position on the status of mechanics, his attitude toward molecular reasoning in thermodynamics is much more definite. He usually imagined a "mass of matter" under certain conditions and asked when it is in equilibrium. He referred to the components of the mass as different "substances," without regard for "any theory of their internal constitution." For purposes of general discussion, he wrote, "we may suppose all substances to be measured by weight or mass, [although] convenience may dictate the use of chemical equivalents" (Gibbs 1876–1878; in 1906 vol. 1, 62–64).[32] But Gibbs was not opposed in principle to the use of molecular arguments. In the interest of generality, he sought in his thermodynamic papers to do as much as possible without them. But when general molecular assumptions were needed to draw significant conclusions, he was not averse to making them. He even devoted a section of his large memoir to "certain points relating to the molecular constitution of bodies," wherein he explained phases of "dissipated energy," that is, states of stable equilibrium, and the action of catalytic agents in explicitly molecular terms (Gibbs 1876–1878; in 1906 vol. 1, 138–144).

Gibbs's attitude toward molecular hypotheses was therefore similar to that of Clausius.[33] Like Clausius, he tried to separate the general principles of his thermodynamics, and the consequences that could be drawn entirely from them, from special assumptions about the molecular constitution of bodies and their molecular motions. But also like Clausius, Gibbs seems to have had no doubt that matter really is molecular in nature and that a more adequate scientific theory would have to take account of this fact. In a lecture on thermodynamics given in 1899, he said: "We assume that heat has to do with motion of the particles of a body. We have little doubt that matter consists of very small discontinuous particles and there is no reason they should not move. In regard to molecular motion forces are conservative; there are no frictional losses."[34] There is also no reason to think that this statement did not express Gibbs's attitude a quarter of a century earlier. That is, in any case, the one he developed in his work on statistical mechanics, wherein each member of a given ensemble was regarded as a physical system that could be described by a set of generalized coordinates and momenta satisfying the laws of mechanics. For Gibbs, thermodynamics was a general theory of the equilibrium states of material systems and of the necessary and sufficient conditions of the stability of those states, but the needed mechanical and molecular explanation of stability was to be found in the distinctive properties of stationary ensembles.[35]

Helm was wrong, therefore, in his evaluation of Gibbs's thermodynamic work. Nowhere in his long commentary on that work in 1898 does Helm mention any of Gibbs's references to molecules, even in his discussion of the Gibbs's paradox (see Helm 1898, 162–163). The likely explanation for this incongruity is that while Helm approved of the greater part of Gibbs's memoir, which developed and applied a phenomenological approach consonant with the Relations Thesis, he opposed, and thus chose to ignore, any part attempting a molecular explanation. He evidently favored a phenomenological approach to energetics himself, and so emphasized the importance of that approach wherever he found it, even if it meant distorting the real intentions of his favorite authors, Mayer and Gibbs, in the process. Helm was not alone here: Ostwald did the same thing in defending his own conception of energetics.

V. OSTWALD AND GIBBS

Ostwald liked to recall the impact thermodynamics had had on him as a student. He had been initiated into the "Gedankenkreise" of the subject by Arthur von Oettingen, his physics professor at Dorpat, who also introduced him to Gibbs and encouraged him to apply thermodynamic reasoning to problems in chemistry.[36] But that was no simple task, as Oettingen himself had discovered; he found Gibbs "obviously significant, but difficult to approach," even for a physicist. The chemist Ostwald apparently fared no better. He later remembered the strenuous effort he had exerted to understand the principles of thermodynamics and to bring them to bear on his own work. "I soon realized," he wrote in 1913, "that my only recourse was to work my way through the dense thicket of the mathematical formulation of the second law, and to seek an understanding in this way."[37] But that effort apparently brought little reward, for there is little evidence in his early work that he had profited from it.[38]

Ostwald was motivated to try again in the late 1880s, when it became clear to him that thermodynamics and Gibbs's writings on the subject could help him answer basic questions that had long been of concern to him regarding problems of chemical affinity, rates of reaction, and the conditions of chemical equilibrium. The catalyst seems to have been his reading, in mid-1886, of Jacobus van 't Hoff's studies in chemical dynamics (Van 't Hoff 1884), which forcefully argued the applicability of thermodynamic reasoning to a wide range of chemical problems. Ostwald had been trying for more than a decade to focus the attention of chemists on the problems associated with chemical change, and in Van 't Hoff he immediately recognized a more successful ally. Starting from a work by the chemist Horstmann (Horstmann 1873), Van 't Hoff developed concise, quantitative expressions for rates of reaction, affinity, and the conditions for chemical equilibrium, and then applied them to a large number of different situations.

Ostwald was impressed.[39] Van 't Hoff's memoir amply demonstrated the power of thermodynamic reasoning in chemistry and apparently resolved him to restudy, carefully, the conceptual framework of thermodynamics. But the influence went deeper than that. First, Van 't Hoff's memoir revealed to Ostwald the importance of energy considerations in the study of chemical phenomena. Second, it reinforced his earlier suspicion that the key to the power of thermodynamics lay in the position it accorded to energy and its transformations. Finally, it confirmed his growing belief (curiously, given Van 't Hoff's avowedly molecular approach) that while theories based on detailed micro-mechanical hypotheses had made little progress with many problems in chemistry, energy-based approaches had been uniquely and dramatically successful. Armed and newly motivated, Ostwald was now ready to study energy for himself.[40]

This project brought him back to Gibbs. To understand that "most important of all aids to the development of the theory of chemical affinity," as he later referred to Gibbs's memoir on heterogeneous equilibrium (Ostwald 1926, vol. 2, 61), he resolved to study Gibbs's writings more carefully. But he then had difficulty in finding them and the same difficulty as before in comprehending them (Ostwald 1926, vol. 2, 63–64, 149). Ostwald later recalled that while he quickly recognized the "very great significance" of Gibbs's papers for the development of laws of chemical change, he discovered that the only way to study them was to translate them word by word, since the text was already so compact that no abbreviated summary of its content was possible. Having begun that time-consuming task, Ostwald was convinced that Gibbs's "long overlooked treasures" deserved a wider audience, and so wrote to their author suggesting that a German edition, which Ostwald would be pleased to prepare, would promote that end.[41]

Ostwald first made that proposal in April 1887, but it was not until more than three years later, after protracted negotiations, that he finally secured permission to translate Gibbs's thermodynamic writings.[42] Once begun, however, the task occupied him for the better part of a year, and its effect on him was deep and lasting. He wrote to Gibbs in August 1891 to send him first proofs and to comment: "The translation of your main work is nearly complete and I cannot resist repeating here my amazement. If you had published this work over a longer period of time in separate essays in an accessible journal, you would now be regarded as by far the greatest thermodynamicist since Clausius – not only in the small circle of those conversant with your work, but universally – and as one who frequently goes far beyond him in the certainty and scope of your physical judgment. Hopefully, the German translation will more quickly secure for it the general recognition it deserves."[43]

Ostwald was equally laudatory later. In his autobiography, he referred to Gibbs as "undoubtedly the greatest scientific genius produced by the United States,"[44] and described, in more detail, the impact of his close encounter with Gibbs's writings:

This work had the greatest influence on my own development; for while he does not particularly emphasize it, Gibbs works exclusively with quantities of energy and their factors, and shuns entirely all kinetic [molecular] hypotheses. He thereby obtained for his conclusions a certainty and permanence which place them at the uppermost limit of human achievement. Indeed, no mistake has been found in either his formulae or his conclusions – or, what is even more incredible, in his assumptions. There are many scientific works whose logic and mathematics are indisputable, but which are nevertheless worthless, because the postulates and assumptions used in them do not correspond to reality. In this respect, as well, Gibbs is perfect ...

The thorough immersion in these works in translating them was for me of considerable consequence. Although I could penetrate their mathematics only imperfectly, I nevertheless profited greatly from the clear objectivity with which [Gibbs] grasped individual problems and also from the circumspect manner in which he developed far-reaching consequences from established results. Moreover, I could not help but notice that the more than 200 equations stated and treated in his major work were almost without exception equations between *quantities of energy*. This observation, initially merely formal, became for me of the greatest importance, since it showed that this fundamental work could be characterized as a chemical *energetics* (Ostwald 1926, vol. 2, 61–62, 63–64; emphasis in original).

Later in his autobiography, in a section with the heading "Das Gesetz des Geschehens," Ostwald explained why. Referring to the intent of his own second "Studien zur Energetik" (Ostwald 1892a), which was published in June 1892, he wrote:

The central idea of a second paper was the extension of the second law [of thermody-namics], which until then had been stated only for processes in which heat was involved. This law says that heat never rises by itself from a lower temperature to a higher one, and the [energetic] extension says the following: For every kind of energy there is a quantity, comparable to temperature in the case of heat, which never rises by itself from lower values to higher ones, a quantity that may be called the 'intensity' of the energy in question ... Of particular interest to me was the question of the quantity of *chemical* intensity. It turned out that this highly important concept had been developed and regu-larly employed by W. Gibbs in his fundamental works under the name of the chemical *potential*. It can be seen from his choice of name, which previously had been used only for corresponding quantities of electrical and gravitational energy, that Gibbs realized these concepts to be of the same kind, although he does not seem to have emphasized it. But Gibbs presented the possibility of expanding the various aspects of the second law, which had been revealed by thermodynamic research where heat was concerned, to all of physics, that is to all happening (Geschehen), and of stating the general condition that must be satisfied *in order for something to happen at all*. For this there must be intensity differences in whatever energies are present (Ostwald 1926, vol. 2, 174–175; emphasis in original).

In short, Ostwald regarded Gibbs as a principled opponent of molec-ular reasoning, who had paved the way and charted the course to his own energetics. This conclusion, which Ostwald did not reach until late 1891 or early 1892, has little to recommend it. There is no doubt that he was profoundly influenced by Gibbs's thermodynamic writings in preparing their translation, but it is equally clear that he still large-ly misunderstood them. A brief description of the energetic theory to which the above remarks allude will to help to explain why.

The results of Ostwald's first energy studies were arguable, but limited in scope. An essay published in mid-1887 urged the use of Helmholtz's distinction between free and bound energy in the study of "energies of reaction," and suggested that the study of chemical affinity was essen-tially the study of "the transformations of chemical energy" (Ostwald 1887a; in Ostwald 1904, 9–11). Similarly, the inaugural lecture Ostwald gave at Leipzig later in the year concluded that while the establishment of laws governing mass relations in chemistry depended on the conser-vation of matter, the laws of chemical affinity had their basis in "the recognition that chemical processes are caused by transformations of persisting energy."[45] Soon, however, Ostwald cast a wider conceptual net. In the second edition of his outline of general chemistry, published in the fall of 1889, he developed further the two-part structure of his

Leipzig address. But he now claimed that transformations of energy were the causes, not only of chemical change, but of change in general. Every state of matter "is defined by the different amounts of distinct kinds of energy it contains," he wrote, a makeup that determines all the changes it can undergo. Moreover, it is the "intensity factors" of its energies that determine whether a body is in equilibrium with its surroundings, and if not how it will change. "In this property of energy," Ostwald asserted, "lies the cause of all happening, that is, of all change in the material world" (Ostwald 1889, 207–208).

Two years later, after his conversion to "pure energetics," Ostwald took the further step of embracing the Composition Thesis, claiming that "By matter we understand nothing but a spatially coincident occurrence of various forms of energy which are in a state of reciprocal dependence."[46] Recall that this position implied that "material objects" or "bodies" or physical-chemical "systems" are themselves just energy complexes – clusters of energy that are somehow coupled, either naturally or via extrinsic linkage. Conversely, these clusters are sufficiently distinct, that is, constitute sufficiently coherent collections, to be counted as separate individuals. It is to changes in such energy complexes that Ostwald now referred, at least officially, when he wrote that "everything that happens is in the last instance nothing but a change of energy."[47] Ostwald found the Composition Thesis implicit in Gibbs's writings. As he later wrote:

We . . . want to hazard the attempt to construct a world view exclusively from energetic material without using the concept of matter. The pronouncement that this task must be undertaken has often been made; and one finds, in one or another area, individual tendencies to representation in this sense. This postulate has even been developed practically with the widest compass for the new chemistry in the fundamental work of W. Gibbs, without, of course, having been explicitly formulated.[48]

The salient scientific particulars of the world view Ostwald sought to construct are as follows. Every object or system is constituted of energy of different, distinct forms.[49] Each form of energy, in turn, is resolvable into two factors: "Every quantity of energy may be represented as the product of two factors," Ostwald wrote, "of which the one will be called the *quantity of intensity* after the precedent of W. Gibbs and Helm, while I shall suggest the designation of *capacity factor* for the other one."[50] Since such factorization is a "universal characteristic of energy, which may be presupposed in the study of energy of every kind," he, like Helm, proposed a "Factorization Principle," calling it "the foundation of modern energetics."[51] In Ostwald's version, this principle took the form

(1) $E = ic,$

where E represents some form of energy, i is the magnitude of its intensity, and c is its capacity factor.[52]

Ostwald recognized a number of distinct forms of energy,[53] two of which deserve mention. First, he admitted – indeed emphasized the importance of – "volume energy" as one of the "spatial" forms of mechanical energy,[54] and factored it as

(2) $E_v = pV,$

where p is pressure and V is volume.[55] He also recognized "heat energy" as a distinct form of energy. Initially, he factored heat as $Q = Ts$, where T represents temperature and s specific heat.[56] Later, he factored it as

(3) $Q = Tc$

where c stands for either heat capacity or entropy.[57] (An important consequence of this factorization is that entropy has an official role in Ostwald's energetic theory only as the capacity factor of heat energy for isothermal changes.[58]) Ignoring the criticism that neither form of energy is a function of the state of a system,[59] Ostwald insisted that every system has a definite amount of volume energy as part of its makeup and that all but purely mechanical systems have definite amounts of heat as well.[60] The total energy composition of an object or system, in turn, is given by

(4) $E_T = \sum_n E = \sum_n ic,$

where the sum extends over every form of energy present and n is greater than one.[61]

To understand change, Ostwald directed his attention to the intensity factor of energy – "that property," he wrote, "upon which the motion or rest of energy, that is, everything which occurs, depends" (Ostwald 1892a, 371; also 366–367). Energetic intensities are the inherent tendencies of energy toward equilibration. Specifically, they are tendencies to pass over to regions of lower intensity, or to be transformed into other energies, unless prevented from doing so. Following Helm, Ostwald called this "essential and universal" fact about the behavior of energy the "intensity law."[62] This law says that intensity differences are a necessary condition of change (conversely, that equilibrium obtains when no differences are present) and that natural change never occurs from lower to higher intensity. To explain why the natural tendencies of energy are often prevented from being actualized, Ostwald introduced the notion of "compensated intensities." The basic idea here is that the intensity of one form of energy can balance the intensity of another, the resulting "quiet state" also being a state of equilibrium.[63] This must be going on all the time, Ostwald reasoned, since bodies composed of several forms of energy exist as stable individuals.[64] But the actual situation is more complex: the changes a body undergoes depend on the *relation* between the intensities of its energies and those of its surroundings. Ostwald concluded, therefore, that a system is in equilibrium with its surroundings if and only if the corresponding sets of energetic intensities are mutually compensatory. Conversely, the necessary and sufficient condition for change is the presence of non-compensated intensities, the fundamental idea expressed by his *Satz des Geschehens*: "In order for something to happen it is necessary and sufficient that non-compensated differences of energy are present."[65]

Ostwald expressed the condition of energetic equilibrium in several other ways, two of which need to be mentioned. In his first "Studien zur Energetik," he called this condition the "Principle of Virtual Energy Change" and stated it as follows:

In order for a system possessing any forms of energy to be in equilibrium, it is necessary and sufficient that, for any displacement of the system compatible with its state, the sum of the amounts of energy appearing and disappearing is equal to zero (Ostwald 1891, 567; italics omitted).

Ostwald attempted no analytical formulation of this principle, but said that as far as he could determine it "contains the entire theory of equilibrium, and permits equations of state of the briefest form to be found as soon as the nature of the system and the kinds of energies in it are given." He also remarked that it "plays an entirely essential role in Gibbs's fundamental paper" on heterogeneous equilibrium (Ostwald 1891, 567). In his second "Studien" a year later, Ostwald called one of his propositions expressing the condition for equilibrium the "Second Law of Energetics" and stated it for any possible change involving two forms of energy in this way:

If the conditions are such that in the event of a virtual change of the correlative factors just as much energy of type *A* vanishes on the one hand as must, on the other hand, come into being of type B by virtue of the assumed relationship, and vice versa, then the system is in a state of equilibrium (Ostwald 1892a, 377).

This time, however, he provided an analytical formulation:

(5) $\Delta A + \Delta B = 0$;

and extended it to changes involving any number of forms of energy:

(6) $\sum \Delta E = 0$ (1892a, 378).

The second law of energetics is allegedly a generalization of the second law of thermodynamics. The previous discussion, plus two additional items of background information, will allow us to see what Ostwald had in mind. The first item is Ostwald's view of thermodynamics. He regarded it as essentially the study of the reciprocal transformations of heat and mechanical energy, with the laws of thermodynamics governing such transformations.[66] The second item is his view of the second law. He regarded it as a law of energy, like the first, but a law confined to the behavior of heat. The proof of the second law depends on the characteristic behavior of heat energy, that it cannot by itself pass from a cold body to a warm one. But this assumption, which Ostwald sometimes called the "Clausius Postulate," is merely a perspicuous instance of the more general intensity law governing the behavior of all forms of energy.[67] The energetic essence and proof of the second law is, therefore, a simple matter. The Factorization Principle, Equation (1), yields Equation (3) for heat, which in turn yields the familiar $dQ = TdS$ upon differentiation at constant T.[68] This argument is intended to cut through the mathematical difficulties that had earlier perplexed Ostwald. It is also likely the basis of his remarks to Arrhenius and Boltzmann that finally, after eight years of reflection, he had grasped the content of the second law, and that it has nothing to do with the increase of entropy.[69]

To extend the second law of thermodynamics into the second law of energetics, however, Ostwald used the differentials of heat and volume energy at constant capacity. This gave him the equation:

(7) $dE = cdT - VdP,$

where c is now the heat capacity.[70] If we seek the equilibrium condition for such a change, the Principle of Virtual Energy Change tells us that $dE = 0$. Hence, the difference on the right-hand side of Equation (7) is zero, an example of the two-energy situation of Equation (5). Extension to any form of energetic change, involving any forms of energy, then yields the Second Law of Energetics, Equation (6).

Enough has been said, perhaps, to indicate the direction and character of Ostwald's thinking. How do they compare with Gibbs's ideas? Without recourse to very technical discussion, we may note immediately several important differences. To begin with, Gibbs is evidently not trying to construct an energetics, or "theory of energy as such," as Ostwald referred to his project.[71] For Gibbs, thermodynamics was not essentially a theory of energy, but rather the study of the equilibrium states of material systems and of the necessary and sufficient conditions of these states.[72] As noted earlier, Gibbs introduced assumptions about the nature of matter only when necessary, preferring instead an approach of the greatest possible generality. But there is no suggestion that he wished to reduce matter to a complex of energy factors in line with Ostwald's Composition Thesis.

This is not to say that energy did not play an important role in Gibbs's development of thermodynamics. It did; but it enters as an especially important property of the material systems that were the primary subject of his interest. "The comprehension of the laws which govern any material system is greatly facilitated by considering the energy and entropy of the system in the various states of which it is capable," Gibbs wrote at the beginning of his memoir on heterogeneous equilibrium (Gibbs 1906, vol. 1, 55). The reason, he explained, is that these properties of a material system allow one to understand the interactions of a system with its surroundings and its conditions of equilibrium. Gibbs said much the same thing in his letter accepting the Rumford Medal, which coincidentally also seems to have been his only public comment on the intent of his famous memoir.[73] "The leading idea which I followed in my paper on the Equilibrium of Heterogeneous Substances was to develop the roles of energy and entropy in the theory of thermo-dynamic equilibrium," he wrote, adding that his investigations had led him to "certain functions which play the principal part in determining the behavior of matter in respect to chemical equilibrium."[74]

Gibbs's remarks clearly indicate, moreover, that for him the concept of entropy was at least as important for comprehending the behavior of thermodynamic systems ("such as all material systems actually are") as the concept of energy. He put at the head of his memoir as a motto the dual statements of Clausius: "Die Energie der Welt ist constant. Die Entropie der Welt strebt einem Maximum zu";[75] and he began the

Abstract he prepared of the memoir soon after its completion with the following lines: "It is an inference naturally suggested by the general increase of entropy which accompanies the changes occurring in any isolated material system that when the entropy of the system has reached a maximum, the system will be in a state of equilibrium." This principle had been noted by physicists, Gibbs wrote; but its importance had not, he thought, been adequately appreciated: "Little has been done to develop the principle as a foundation for the general theory of thermodynamic equilibrium" (Gibbs 1878; in 1906, vol. 1, 354). Gibbs then stated the general conditions of equilibrium whose manifold consequences his memoir had developed.

These conditions make essential reference to entropy as well as energy. Beginning with the differential forms of the first and second laws of thermodynamics, which effectively define the state functions internal energy and entropy, Gibbs combined the two expressions to yield the general condition of equilibrium for any virtual change:

(8) $\delta U - T\delta S - \delta W \geq 0.$

If a system is isolated, so that there is no external work, this becomes

(9) $(\delta U)_s \geq 0$

or

(10) $(\delta S)_U \leq 0,$

for constant entropy and energy, respectively.[76] Ostwald, who had little regard for entropy, ignored relation (10); but he would have been attracted to relation (9). Indeed, we may conjecture that many of Ostwald's proposed conditions for equilibrium were attempts to interpret energetically, and thus to appropriate to energetics, this condition of Gibbs's. Ostwald's remark about the "essential place" of the Principle of Virtual Energy Change in Gibbs's memoir suggests as much, and so does the analytical form and indirect proof of the second law of energetics.[77] But, as Planck pointed out, Ostwald had misunderstood Gibbs's condition. Relation (9) does not say that the condition for equilibrium is $\delta U = 0$ *simpliciter*, as Ostwald seemed to think, but rather only for the reversible change of an isolated system at maximum entropy.[78]

Planck also pointed out other misunderstandings, two of which must be mentioned. In correspondence with Planck, Ostwald claimed the authority of Gibbs in defense of the Factorization Principle and of the idea that a system is composed of determinate amounts of energy of different forms.[79] To see why Planck thought that appeal misguided, we need to say a bit more about what Gibbs did.

In his first work on thermodynamics, published in 1873, Gibbs immediately combined the differential forms of the first and second laws for reversible processes to obtain a single "fundamental equation,"

(11) $dU = TdS - pdV,$

an expression containing only the state variables of the system, the path-dependent heat and work having been eliminated.[80] To treat the problem of chemical equilibrium, however, he had to modify this to include any change of internal energy due to a change in the mass of any of the chemical components. This he did for the simplest case of a homogeneous phase by writing Equation (11) in the form

(12) $dU = TdS - pdV + \sum_{i=1}^{n} \mu_i dm_i,$

where dm_i gives the change in mass of the ith independent chemical substance, $S_i, \ldots, S_n$, whose masses can be varied, and μ_i is what Gibbs called the "chemical potential" of the ith substance.[81] The chemical potential of any substance, in turn, is related to the energy, U, of the system by the equation

(13) $\mu_i = \left(\dfrac{\partial U}{\partial m_i} \right)_{S,V,m_k} \qquad (k \neq i),$

where the subscripts indicate that μ_i represents the rate of change of energy with respect to the mass of the ith component of the phase, the masses of all the other components being held constant along with the entropy and the volume.[82] Hence, the condition for equilibrium, under these circumstances, is that the chemical potential for each actually present component substance be constant throughout the whole of the system considered. In this requirement, Gibbs remarked, "we have the conditions characteristic of chemical equilibrium" (Gibbs 1876–1878; in 1906, vol. 1, 65). Finally, integration of Equation (12) gave him the energy of the system:

(14) $U = TS - pV + \sum_{i=1}^{n} \mu_i m_i.$[83]

Ostwald liked these conclusions, especially the one expressed in Equation (14), which he thought confirmed his version of the Factorization Principle and his view of the energy content of a system as comprised of distinct (really present) contributions – in this case, heat, volume energy and chemical energies. But there is no suggestion in Gibbs's thermodynamic writings that he thought of heat and work as distinct forms of energy, or that he thought a system to possess a determinate heat content or amount of volume energy. From the very beginning, in fact, he explicitly distinguished functions of the state of a system from path-dependent quantities (see Gibbs 1873; in 1906, vol. 1, 2–3). But he did not press the point, perhaps thinking it too obvious to belabor. Planck, who tried hard to understand Ostwald's version of energetics, recognized the importance of the distinction, but had repeatedly to insist on it in correspondence with Ostwald and elsewhere.[84] He also had to explain to Ostwald the conditions under which Gibbs could integrate Equation (12), none of which were mentioned in Ostwald's expressions [see Equation (4)] for the energy content of a system.[85] Gibbs began with a homogeneous mass having entropy S and volume V, and containing quantities m_i of the components S_i, and imagined adding quantities of a mass of the same composition and in the same state, that is, with T, p and μ_i unchanged, until the original mass had been doubled. The energy of the added mass is then equal to the energy of the original mass. There is no evidence, however, that Ostwald understood that this is what Gibbs had done.

VI. GIBBS'S THERMODYNAMICS

Gibbs never commented on energetics, so far as we know. There is nothing in his published writings, and he expressed no opinion on the subject in his correspondence with Ostwald. We cannot even be sure that he was aware of the interpretations the energeticists had given his thermodynamics. I have argued, nevertheless, that Gibbs would have rejected those interpretations – Helm's because it was too limiting and Ostwald's because it was physically unfounded. If we ask why Helm and Ostwald misunderstood Gibbs's work, part of the answer, surely, is that each read his writings with a certain view of the course and goal of science in mind, and so emphasized what seemed to conform to that view and ignored what did not. Another part, however, likely turns on the style Gibbs followed in his publications: he seldom told his readers about the specific problems that led to the work on which he was then reporting, much less inform them of any larger project he thought his results might further.[86] As a result, it is often not clear what motivated Gibbs or how he viewed the outcome.

It if was not, as I claim, an energetic project that motivated Gibbs's work in thermodynamics, one is naturally interested to know what did. The answer, apparently, was to develop his subject matter logically from a very simple and general point of view. This suggestion is confirmed by what seems to have been Gibbs's only public comment on his intent. "One of the principal objects of theoretical research in any department of knowledge," he wrote in his letter accepting the Rumford Medal, "is to find the point of view from which the subject appears in its greatest simplicity."[87] Other aspects of Gibbs's style indicate what this entailed for him. The author of a commemorative essay on Gibbs's father wrote the following of the elder Gibbs: "Mr. Gibbs loved system, and was never satisfied until he had cast his material into the proper form. His essays on special topics are marked by the nicest logical arrangement" (see Wheeler 1952, 9). The same could equally have been said of Gibbs himself. He always sought to take a very general approach to the subjects that engaged him, to develop his ideas rigorously and systematically from first principles, and to avoid whenever possible any special assumptions. In short, Paul Tannery's evaluation of one of Pierre Duhem's works could easily have been made of any of Gibbs's: "To draw all logical consequences from a very general principle, to show clearly what it contains and what it does not, and to specify the points where experiment must intervene to bring in something really new – such is the aim [Duhem] pursues, and undoubtedly he will thus contribute in large measure to the organization of current science."[88]

This is a good summary of what Gibbs tried to do in his main study on thermodynamics. His memoir on heterogeneous equilibrium begins by stating very simple and general conditions for the equilibrium and stability of a material system, and then proceeds to work out, very carefully, the consequences of those conditions for diverse situations (see Wheeler 1952, 75). He is aware that for practical purposes the consequences are often more useful than the general statements from which they are derived; but he preferred to begin with the latter rather than the former, "believing that it would be useful to exhibit the conditions of equilibrium of thermodynamic systems in connection with those quantities which are most simple and most general in their definitions, and which appear most important in the general theory of such systems" (Gibbs 1878; in 1906, vol. 1, 355–356). It therefore seems fair to say that Gibbs was motivated in this work, as in others, by the search for a simple, very general standpoint that would allow him to retrieve accepted results in a rigorous manner and to develop new ones.[89] The reason for logical rigor,[90] we may conjecture, is that he valued it as the best means of "showing clearly what a principle contains and what it does not" and in order "to specify the points where experiment must intervene to bring in something really new."

That theory – even simple and general theory – is constrained by experiment is something of which Gibbs was evidently aware.[91] A prescient example is found in the Preface he wrote to his work on statistical mechanics. Gibbs proposed his statistical theory as a "branch of rational mechanics" that did not "attempt to frame hypotheses concerning the constitution of material bodies." In Section IV, I argued that Gibbs was no defender of phenomenalism, that he was persuaded of the molecular structure of matter; but in his statistical mechanics, where one might otherwise expect him to do so, he declines to offer any theory of that structure. The reason is that Gibbs does not see how to formulate a general theory that is compatible with experimental findings. In fact, he says, "In the present state of science, it seems hardly possible" to construct such a theory and that, in consequence, one would be "building on an insecure foundation." He therefore concludes: "Difficulties of this kind have deterred the author from attempting to explain the mysteries of nature" and have forced him to be contented with a more modest aim (Gibbs 1902, vii–viii). Ostwald, in particular, might have learned something from this approach.

Department of Philosophy
Seattle University
U.S.A.

NOTES

* I would like to thank Russ McCormmach, Pat Hillegonds, and Reed Guy for helpful comments on earlier drafts of this essay
1 See Klein (1970) 53–54; (1972a); (1974).
2 See, for example, McCormmach (1970) and Jungnickel and McCormmach (1986),

vol. 2, ch. 24.

[3] For example, Ostwald (1893a), 40–43; (1896b), Vorwort; and Helm (1898) 144–146.

[4] See Helm (1887) 15, 23–26; (1898) 16–28; Ostwald (1892a) 363; (1893a) 40–41; (1895a) 162–164 [(1904) 231–233]; (1910) 79, 84, 91.

[5] Compare Helm (1887) 14–15, with (1898) 16–19.

[6] See Helm (1898) 206, 216, 361–366.

[7] Helm (1898) 20, 22, 225, 293–294, 322.

[8] See Ostwald (1887b) 13–14, 20; [(1904) 192–193, 200]; (1891) 566; (1893a) 41–43; (1895a) 162–164, 164–165; [(1904) 231–233, 234–235]; (1926) vol. 2, 154.

[9] Ostwald (1887b) 13–14; [(1904) 192–193]; (1891) 566; (1893a 10, 40–44; (1895a) 161–162; [(1904) 229–231]; (1926) vol. 2, 156.

[10] Also (1892a) 375–376, 385; (1893a) 4–6; (1895a) 155, 158–159, 161, 164–165; [(1904) 220–221, 226–227, 229, 233–235]; (1896b) 18–20; (1902) 163–167; (1908) 102; (1924) 144; (1926) vol. 2, 153–162.

[11] See Helm (1887) 14, 15–16, 34, 42–43, 66, 71; (1898) 226, 296–299.

[12] See Helm (1894) 16, 24–28, 42–43, 58, 60, 70–73.

[13] For example, Helm (1895a) XII; (1898) 224, 296–298.

[14] Other criticisms of Clausius are more consistent with the Relations Thesis. Of the first of Clausius' most famous statements, Helm remarked: "With the pronouncement 'The energy of the world is constant,' the firm footing of the energy law has been abandoned, which is in fact nothing but an empirical relationship between measurable quantities we find present in any natural process; and for this sacrifice absolutely nothing is gained except an empty saying" (Helm 1898, 114–115). Helm extended this evaluation, in another place, to Clausius' second famous pronouncement, commenting that while both statements had no doubt encouraged the study of energy and entropy more than "sober claims which try to give expression to the true importance of these concepts," they were, in fact, nothing but "metaphysical abberations" (125).

[15] See Helm (1898) 111–112, 121–122, 187–188, 298.

[16] This section of Ostwald's *Lebenslinien* is entitled "Wesen der Energetik." See also Ostwald (1896a).

[17] For example, Ostwald (1892a) 367–368, 371–378, 380; (1893a) 16, 30–35, 41–42, 47, 485–490.

[18] The tension in Ostwald's thought is especially noticeable and acute when he discusses the "substances" involved in chemical reactions. On the composition view, these should just be clusters of energy, but Ostwald treats them as *Stoffmengen* which possess energy – chemical energy included. See Ostwald (1893a) 500–517; (1893b); (1893c).

[19] See Helm (1898) 120–126, 342–343. Also (1887) 53–56.

[20] Clausius (1850) 371–372, 374, 378; (1854) 481–482; (1862) 74; [reprinted in (1867) 16, 19, 21, 108, 206]. Also Klein (1969).

[21] See Helm (1898) 146–147, 151–152, 164, 172. Also (1894) III–IV, 124–125; (1895a) XII–XIII; (1895b) 29; (1896) 648.

[22] See Gibbs (1884). Also Wheeler (1952) 154; Klein (1987) 282 and (1989) pp. 12–13.

[23] See Klein (1987) 283 and (1989) p. 13.

[24] See, also, Wheeler (1952) 38–39.

[25] See Maxwell (1879); in (1952) vol. 2, 713–741; esp. 720–726. We may conjecture that Gibbs did not follow Boltzmann's approach because it relied on too many special molecular hypotheses. See Wheeler (1952) 120, 124, 155–157, 170.

[26] Boltzmann quoted this remark in his dispute with Zermelo on the proper interpretation of the second law of thermodynamics as evidence that Gibbs regarded the law as statistical in character. See Boltzmann (1896) 779. Also Wheeler (1952) 81–82.

[27] Gibbs (1876–1878), in (1906) vol. 1, 89–92; (1878) in (1906) vol. 1, 354–356, 358note.

[28] See Gibbs (1876–1878), in (1906) vol. 1, 55–56; (1878) in (1906) vol. 1, 354–355.

[29] See Gibbs (1906) vol. 1, 372–417. The anomaly discussed "An Alleged Exception to the Second Law of Thermodynamics" (404–405); but Gibbs's rejections of the exception appeals to his earlier thermodynamic theory and does not advance it.

[30] Shortly before his death, Gibbs began to prepare several supplementary chapters for a republication of his thermodynamic papers, but none of his notes suggest a novel development of theory. See (1906) vol. 1, 418–434. A decade earlier, he politely declined Ostwald's insistent request that he add some form of commentary to the German translation Ostwald was then preparing of them. See Ostwald (1961) 96, 102. Also Gibbs's letter to the publisher Veit & Comp., Ostwald (1961) 110.

[31] Wilson (1936); in Donnan and Haas (1936) vol. 1, 57. See 54–55 for Gibbs's letter, 55–57 for Wilson's comments.

[32] Significantly, "Stoffemenge" is Ostwald's later translation of Gibbs's "mass of matter" (see, for example, Ostwald (1892b) 75); but it is clear from Gibbs's use of the term that "masses of matter" are neither energies nor factors of energy.

[33] See Gibbs (1902) xii, 165–167; also Wheeler (1952) 121, 157.

[34] As recalled by Wilson (1936); in Donnan and Haas (1936) vol. 1, 21.

[35] See Gibbs (1902) ch. XIV. Also Klein (1970) 129–137. Ever aware of the limitations of his work, Gibbs did not claim for his own statistical theory the status of a complete explanation. See (1902) 166. Also Bumstead (1903); in Gibbs (1906) vol. 1, xxii–xxiii; and Klein (1972b) 392.

[36] Ostwald (1926) vol. 1, 177–180, 196–197; vol. 2, 61–64, 148–149.

[37] Ostwald (1926) vol. 1, 179–180. Ostwald is commenting on his years as a professor in Riga, so 1881–1886.

[38] Ostwald admitted that his "strenuous work of conceptual analysis" had yielded no "filterable precipitates in the form of published scientific papers." Still, he thought that early reflections on thermodynamics had "prepared the ground for the later development of *energetics*" (Ostwald 1926, vol. 1, 179–180).

[39] See Ostwald (1926) vol. 1, 177, 235–236; vol. 2 20–25. Also (1924) 138–139, 145.

[40] From his own recollections. See Ostwald (1926) vol. 1, 178–179; also (1924) 145.

[41] Ostwald (1926) vol. 2, 61–62. Ostwald to Gibbs, 26 April 1887; in Ostwald (1961) 89.

[42] The story of Ostwald's efforts may be reconstructed from Ostwald (1961) 89–110. Gibbs was not helpful. Ostwald very much wanted him to comment on his writings or at least to write a Preface to them. Gibbs declined to do either. He did not even send the photograph Ostwald had requested until after the work had appeared.

[43] Ostwald to Gibbs, 9 August 1891; in Ostwald (1961) 99.

[44] Ostwald (1926) vol. 2, 62–63. Also (1910) 379–380.

[45] Ostwald (1887b); in (1904) 200 (without italics). See, also, (1888) 277. Ostwald later wrote that the struggle to grasp the conceptual structure of thermodynamics, and its bearing on the problems of chemistry, had so occupied his thoughts that he had decided to address his Leipzig audience on the question of "energy and its transformations"

(1926 vol. 2, 149–150).

46 Ostwald (1892a) 385. Also (1891) 566; (1892a) 375–376; (1893a) 4–6; (1895a) 158, 146–165 [(1904) 224, 233–235].

47 Ostwald (1892a) 363; (1893a) 46, 48–49; (1893b) 234–235.

48 Ostwald (1902) 165–166. Gibbs's memoir on heterogeneous equilibrium is the only work cited explicitly in the lecture, titled "Das Energetische Weltbild," from which the quotation is drawn.

49 Ostwald (1889) 208, 214, 254, 265, 311; (1891) 567; (1892a) 376; (1893a) 7–8, 11–12, 17–18, 471–472; (1893b) 232–233; (1893c) 50; (1895a) 166 [(1904) 236–237].

50 Ostwald (1891) 571 (emphasis in original). Also (1893a) 45, for a similar reference to Gibbs, and more generally (1892a) 367–368; (1893a) 16, 485–488.

51 Ostwald (1892a) 368; (1893a) 44, 46–47.

52 Ostwald (1891) 571–572; (1892a) 368; (1893a) 43–44. The capacity factor of a given energy is said to be "the amount of energy which, with a given intensity i, is present in a system" (1892a, 367; also 1893a, 16, 485). Helm's principle factors the differentials of energies, thus $dE = i dc$. See (1887) 61–62.

53 See Ostwald (1892a) 369 for a preliminary table.

54 "For our purposes, the most important form of spatial energy is that which depends on volume, since this comes into consideration almost exclusively for questions of thermodynamics and chemical equilibrium" (Ostwald 1893a, 30). Other forms of "spatial" energy are "distance" energy and "surface" energy. Later, Ostwald also included "form" energy, which enables a system to maintain its shape or form. See (1902) 167–168.

55 Ostwald (1891) 566; (1892a) 369; (1893a) 12–14, 17–18, 24–27, 30–35, 37–38.

56 Ostwald (1889) 208, 244. Ostwald did not say which specific heat, whether that at constant pressure or constant volume, and apparently did not realize the need for a distinction.

57 Ostwald (1892a) 370, 380–382; (1893a) 484–485. (1891) 577 recognized only heat capacity.

58 Ostwald (1892a) 370, 382; (1893a) 49–50, 485, 490, 494–496. Also Ostwald to Planck, 12 October 1891 and 29 February 1892; in Ostwald (1961) 348–349. Ostwald frequently confused specific heats (and heat capacities) with entropy, whose presence in thermodynamics he neither welcomed nor understood.

59 Planck to Ostwald, 20 March 1892, 22 April 1892, and 25 June 1893; in Ostwald (1961) 41–45; Helm to Ostwald, 4 June 1893; in (1961) 76–77. Also Planck (1896) 73–75; (1958) vol. 3, 384–385.

60 Ostwald (1892a) 367–368, 371–376, 380; (1893a) 485–490. Also the exchange of letters between Ostwald and Planck in June and July 1893 (Ostwald 1961, 46–59). Ostwald conceded that, in general, the absolute amount of energy of a given form in a system cannot be determined experimentally, but he spent a good deal of effort trying to solve the problem for particular cases.

61 There are no single-form energy systems, according to Ostwald, since such "systems" (given the Composition Thesis) would have no boundaries. (1892a) 377; (1893a) 24–25, 28, 32–33, 46–48.

62 Ostwald (1891) 571; (1892a) 367–368; (1893a) 46. Compare Helm (1887) 53, 55, 58–59, 61–62.

63 Ostwald (1893a) 25–27, 34–36, 47–48, 512–514. Ostwald has changed his mind, likely due to recent correspondence with Planck, and will later change it again. I cannot

trace his vacillation on this point here, although I think it worth doing.
[64] Ostwald (1892a) 377–378; (1893a) 24–25, 28.
[65] Ostwald (1893a) 31–36, 46–49. The Satz des Geschehens is said to be "the most general natural law known to science" (1893a, 48; also 46).
[66] See Ostwald (1889) 244–252; (1893a) 12, 484–486; (1893b) 232–233.
[67] Ostwald (1889) 245; (1893a) 45–46, 477–478, 492–493.
[68] Ostwald (1893a) 484–494, esp. 490–494; also (1892a) 380–382; (1893a) 562–563, 1023–1024. Ostwald recognized that this equation holds only for reversible processes (1893a, 482, 484), but he seldom mentioned irreversibility – even in discussions of entropy, where it would have been appropriate. See (1893a) 482–484, 494–497.
[69] Ostwald to Arrhenius, 11 June 1892; in Ostwald (1969) 110; Ostwald to Boltzmann, 2 June 1892; in Ostwald (1961) 9–10. Ostwald was evidently pleased with himself. In an address in 1894, he explained that while the first law seemed obvious, a "mysterious darkness" had always hung over the second one. He continued: "I strove for seven years to penetrate [the second law] without coming any closer to understanding it. Of course, the calculations that are connected with it may be understood, but the deeper intuitive meaning that lay hidden in these calculations ... this meaning was not revealed to me ... I also believe that for approximately 15 or 20 years, many suffered with me the same complaint. Today, I do not believe I suffer from it any longer. I believe I have grasped the essential content of the second law and have acquired the dearly won freedom in its use that my modest powers permit" (Ostwald 1894; [(1904) 360–361].

In 1892 and 1893, Ostwald explained irreversibility and dissipation as entirely the fault of radiant energy. Because of its peculiar nature, radiant energy cannot be coupled to energy of other forms, and so cannot take part in reciprocal transformations. Its formation is therefore undirectional and always results in a loss of "moveable energy." See (1892) 370–371, 384–386; (1893a) 1006–1022.
[70] Ostwald (1892a) 382; (1893a) 498–499. Or, at least, c should have been heat capacity. All of Ostwald's theoretical remarks on the factorization of heat energy explicitly state that heat capacity is its capacity factor for variable temperature. In practice, however, he used entropy, as in Equation (7), presumably so that he could obtain relations, such as the Clausius–Clapeyron equation, that he already knew to be correct.
[71] Ostwald (1891) 565; (1893a) 10.
[72] Gibbs (1876–1878); in (1906) vol. 1, 55–56; (1878); in (1906) vol. 1, 354; (1881); in Donnan and Haas (1936) vol. 1, 55 and Wheeler (1952) 89. Also Klein (1977) 334–335; (1989) p. 5; (1990a) 42, 44; (1990b) 61.
[73] See Wheeler (1952) 88, 101.
[74] Gibbs (1881); in Donnan and Haas (1936) vol. 1, 55 and Wheeler (1952) 89.
[75] Gibbs (1906) vol. 1, 55. The original statements are found in Clausius (1865) 400.
[76] Gibbs (1876–1878); in (1906) vol. 1, 56; (1878); in (1906) vol. 1, 354. Gibbs showed that (9) and (10) are equivalent and that each is a necessary and sufficient condition for equilibrium. See (1906) vol. 1, 55–62, 354–356.
[77] Ostwald's proof of the necessity and sufficiency of his second law of energetics as a criterion of equilibrium seems evidently to be modeled on Gibbs's indirect proof of the necessity and sufficiency of relation (9) as a condition of equilibrium. See Ostwald (1892a) 377–379.
[78] Planck to Ostwald, 27 April 1892; in Ostwald (1961) 43. Planck tried on several occasions to understand Ostwald's various statements of the conditions of equilibrium.

See the exchange of letters in July 1891 (in Ostwald 1961, 34–37), where he questioned the Principle of Virtual Energy Change and a variant from the Introduction to Ostwald's (1893a): "For a system which contains several forms of energy to be in equilibrium or stationary, the sum of lost and gained energy must be equal to zero for any change compatible with the conditions of the system" (1893a, 25; without italics; also 34–35). Also see an exchange on the Second Law of Energetics in June and July 1893 (in Ostwald 1961, 46–53).

[79] Ostwald to Planck, 21 March 1892; in Ostwald (1969) 349–350. Ostwald had earlier arranged to have a copy of his translation of Gibbs sent to Planck. See Planck to Ostwald, 20 March 1892; in Ostwald (1961) 41–42.

[80] Gibbs (1873); in (1906) vol. 1, 2–3. I have altered Gibbs notation, here and in what follows, to conform with more familiar conventions. For a discussion of Gibbs's early papers, see Klein (1977).

[81] See Gibbs (1876–1878); in (1906) vol. 1, 62, 65, 86, 92–96.

[82] Gibbs (1876–1878); in (1906) vol. 1, 89, 93. See Haas 1936); in Donnan and Haas (1936) vol. 2, 21–22.

[83] Gibbs (1876–1878); in (1906) vol. 1, 87. See Butler (1936); in Donnan and Haas (1936) vol. 1, 87–91, for an explanation.

[84] Planck to Ostwald, 20 March 1892, 22 April 1892, and 25 June 1893; in Ostwald (1961) 41–45. Also Planck (1896) 73–75 and (1958) vol. 3, 384–385. Planck cogently argued that any quantity of energy, which in Ostwald's energetics "represents substance *par excellence*," must be a function of the physical and chemical state of a system, else the Principle of the Conservation of Energy would be undermined.

[85] Planck to Ostwald, 27 April 1892; in Ostwald (1961) 43.

[86] For example, Pierre Duhem wrote: "We must resign ourselves to ignorance about the philosophical ideas that no doubt presided over the birth of physical theories in Gibbs's mind" (quoted in Klein 1990b, 63). See also Klein (1980) 156–158; (1989) p. 4. An exception is the Preface to Gibbs (1902), which is uncommonly revealing.

[87] Gibbs (1881); in Donnan and Haas (1936) vol. 1, 54 and Wheeler (1952) 88. Gibbs made a similar remark about the goal of his work on statistical mechanics: "I do not know that I shall have anything particularly new in substance, but shall be contented if I can so choose my standpoint (as seems to me possible) as to get a simpler view of the subject." Gibbs to Rayleigh, 27 June 1892; quoted in Klein (1972b) 391–392; (1987) 284; and (1989) p. 14.

[88] Of a work by Duhem written in 1893; quoted in Jaki (1984) 279–280. We know that Duhem, too, was a large admirer of Gibbs. See Klein (1990b).

[89] See Gibbs (1881); in Donnan and Haas (1936) vol. 1, 55 and Wheeler (1952) 89.

[90] Bumstead (1903) rightly speaks of the "logical austerity" of Gibbs's work, remarking that "his logical processes were really of the most severe type" (in Gibbs 1906, xvi, xxiv). Also, Wheeler (1952) 170.

[91] See, for example, Gibbs (1881); in Donnan and Haas (1936) vol. 1, 55 and Wheeler (1952) 89–90.

REFERENCES

Boltzmann, Ludwig (1896). "Entgegnung auf die wärmetheoretischen Betrachtungen des Hrn. E. Zermelo," *Annalen der Physik* **57**: 773–784.

Bumstead, Henry A. (1903). "Josiah Willard Gibbs," in Gibbs (1906), vol. 1, pp. xi–xxvi.

Butler, J. A. V. (1936). "The General Thermodynamic System of Gibbs," in Donnan and Haas (1936), vol. 1, pp. 61–179.

Clausius, Rudolph (1850). "Über die bewegende Kraft der Wärme welche sich daraus für die Wärmelehre selbst ableiten lassen," *Annalen der Physik* **79**: 368–397, 500–524.

Clausius, Rudolph (1854). "Über eine veränderte Form des zweiten Hauptsatzes der mechanischen Wärmetheorie," *Annalen der Physik* **93**: 481–506.

Clausius, Rudolph (1862). "Über die Anwendung des Satzes von der Aequivalenz der Verwandlungen auf die innere Arbeit," *Annalen der Physik* **116**: 73–112.

Clausius, Rudolph (1865). "Über die verschiedenen für die Anwendung bequeme Formen der Hauptgleichungen der mechanischen Wärmetheorie," *Annalen der Physik* **125**: 353–400.

Clausius, Rudolph (1867). *Die mechanische Wärmetheorie* (Braunschweig: Vieweg).

Donnan, F. G. and A. Haas, eds. (1936). *A Commentary on the Scientific Writings of J. Willard Gibbs*, 2 vols. (New Haven: Yale University Press).

Gibbs, Josiah Willard (1873). "Graphical Methods in the Thermodynamics of Fluids," *Transactions of the Connecticut Academy of Arts and Sciences* **2**: 309–342. Reprinted in (1906), vol. 1, pp. 1–32.

Gibbs, Josiah Willard (1876–1878). "On the Equilibrium of Heterogeneous Substances," *Transactions of the Connecticut Academy of Arts and Sciences* **3**: 108–248 (1876); 343–524 (1878). Reprinted in (1906), vol. 1, pp. 55–353.

Gibbs, Josiah Willard (1878). Abstract of "On the Equilibrium of Heterogeneous Substances," *American Journal of Science* **16**: 441–458. Reprinted in (1906), vol. 1, pp. 354–371.

Gibbs, Josiah Willard (1879). "On the Fundamental Formulae of Dynamics," *American Journal of Mathematics* **2**: 49–64.

Gibbs, Josiah Willard (1881). "To the American Academy of Arts and Sciences," January 10, 1881. Reprinted in Donnan and Haas (1936), vol. 1, pp. 54–55 and Wheeler (1952), pp. 88–89.

Gibbs, Josiah Willard (1884). "On the Fundamental Formula of Statistical Mechanics with Applications to Astronomy and Thermodynamics" (Abstract), *Proceedings of the American Association for the Advancement of Science* **33**: 57, 58. Reprinted in (1906), vol. 2, p. 16.

Gibbs, Josiah Willard (1889). "Rudolf Julius Emanual Clausius," *Proceedings of the American Academy of Arts and Sciences* **16**: 458–465. Reprinted in (1906), vol. 2, pp. 261–267.

Gibbs, Josiah Willard (1902). *Elementary Principles of Statistical Mechanics. Developed with Especial Reference to the Rational Foundations of Thermodynamics* (New Haven: Yale University Press).

Gibbs, Josiah Willard (1906). *The Scientific Papers of J. Willard Gibbs*, 2 vols., H. A Bumstead and R. G. Van Name, eds. (New York: Longmans, Green, and Co.).

Haas, Arthur (1936). "The Thermodynamic Principles as Extended and Perfected by Gibbs," in Donnan and Haas (1936), vol. 2, pp. 1–57.

Helm, Georg (1887). *Die Lehre von der Energie, historisch-kritisch entwickelt. Nebst Beiträgen zu einer allgemeinen Energetik* (Leipzig: A. Felix).

Helm, Georg (1894). *Grundzüge der mathematischen Chemie. Energetik der chemischen Erscheinungen* (Leipzig: W. Engelmann).

Helm, Georg (1895a). "Überblick über der derzeitigen Zustand der Energetik," Beilage zu den *Annalen der Physik* **55**: III–XVIII.

Helm, Georg (1895b). "Über den derzeitigen Zustand der Energetik," *Verhandlungen der Gesellschaft deutscher Naturforscher und Ärtze*, II, 1: 28–33.

Helm, Georg (1896). "Zur Energetik," *Annalen der Physik* **57**: 646–659.

Helm, Georg (1898). *Die Energetik nach ihrer geschichtlichen Entwicklung* (Leipzig: Veit & Comp.).

Hertz, Heinrich (1894). *Die Principien der Mechanik in neuem Zusammenhang dargestellt*, P. Lenard, ed. (Leipzig: J. A. Barth).

Horstmann, August (1873). "Theorien der Dissociation," reprinted in *Abhandlungen zur Thermodynamik chemischer Vorgänge* (Leipzig: J. Barth, 1903), pp. 26–41.

Jaki, Stanley, L. (1984). *Uneasy Genius: The Life and Work of Pierre Duhem* (Dordrecht: Martinus Nijhoff).

Jungnickel, Christa and R. McCormmach (1986). *Intellectual Mastery of Nature: Theoretical Physics from Ohm to Einstein*, 2 vols. (Chicago: University of Chicago Press).

Klein, Martin J. (1969). "Gibbs on Clausius," *Historical Studies in the Physical Sciences* **1**: 127–149.

Klein, Martin J. (1970). *Paul Ehrenfest: The Making of a Theoretical Physicist* (Amsterdam: North-Holland).

Klein, Martin J. (1972a). "Mechanical Explanation at the End of the Nineteenth Century," *Centaurus* **17**: 58–82.

Klein, Martin J. (1972b). "Gibbs, Josiah Willard," in *Dictionary of Scientific Biography*, C. G. Gillispie, ed. (New York: Scribner's), vol. 5, pp. 386–393.

Klein, Martin J. (1974). "Boltzmann, Monocycles and Mechanical Explanation," in *Philosophical Foundations of Science*, R. J. Seeger and R. S. Cohen, eds. (Dordrecht and Boston: D. Reidel), pp. 155–175.

Klein, Martin J. (1977). "The Early Papers of J. Willard Gibbs: A Transformation in Thermodynamics," in *Human Implications of Scientific Advance. Proceedings of the XVth International Congress of the History of Science. Edinburgh, 10–15 August 1977* (Edinburgh, 1978), pp. 330–341.

Klein, Martin J. (1980). "The Scientific Style of Josiah Willard Gibbs," in *Springs of Scientific Creativity: Essays on the Founders of Modern Science*, R. Aris, H. T. Davis, and R. H. Stuewer, eds. (Minneapolis: University of Minnesota Press, 1983), pp. 142–162.

Klein, Martin J. (1987). "Some Historical Remarks on the Statistical Mechanics of Josiah Willard Gibbs," in *From Ancient Omens to Statistical Mechanics: Essays on the Exact Sciences Presented to Asger Aaboe*, J. L. Berggren and B. R. Goldstein, eds., Acta Historica Scientiarum Naturalium et Medicinalium, vol. 39 (Copenhagen: University Library), pp. 281–289.

Klein, Martin J. (1989). "The Physics of J. Willard Gibbs in His Time," in *Proceedings of the Gibbs Symposium. Yale University, May 15–17, 1989*, D. G. Caldi and G. D. Mostow, eds. (New York: American Mathematical Society), pp. 1–21.

Klein, Martin J. (1990a). "The Physics of J. Willard Gibbs in His Time," *Physics Today* (September, 1990): 40–48.

Klein, Martin J. (1990b). "Duhem on Gibbs," in *Beyond History of Science: Essays in Honor of Robert E. Schofield* (London and Toronto: Associated University Presses).

McCormmach, Russell (1970). "H. A. Lorentz and the Electromagnetic View of Nature," *Isis* **61**: 459–497.

Maxwell, James Clerk (1897). "On Boltzmann's Theorem on the Average Distribution of Energy in a System of Material Points," *Transactions of the Cambridge Philosophical Society* **12**: 547–550. Reprinted in (1952), vol. 2, pp. 713–741.

Maxwell, James Clerk (1952). *Scientific Papers of James Clerk Maxwell*, 2 vols., W. D. Niven, ed. (Cambridge: Cambridge University Press, 1890). Reprinted New York: Dover, 1952.

Ostwald, Friedrich Wilhelm (1887a). "Die Aufgaben der physikalischen Chemie," *Humboldt* **6**: 249–252. Reprinted in (1904), pp. 3–12.

Ostwald, Friedrich Wilhelm (1887b). *Die Energie und ihre Wandlungen. Antrittsvorlesung, gehalten am 23. November in der Aula der Universität Leipzig* (Leipzig: W. Engelmann). Reprinted in (1904), pp. 185–206.

Ostwald, Friedrich Wilhelm (1888). "Über die Dissoziationstheorie der Elektrolyte," *Zeitschrift für physikalische Chemie* **2**: 270–283.

Ostwald, Friedrich Wilhelm (1891). "Studien zur Energetik," *Berichte über die Verhandlungen der Sächsischen Akademie der Wissenschaften zu Leipzig* **43**: 271–288. Reprinted in *Zeitschrift für physikalische Chemie* **9** (1982): 563–578.

Ostwald, Friedrich Wilhelm (1892a). "Studien zur Energetik II. Grundlinien in der allgemeinen Energetik," *Berichte über die Verhandlungen der Sächsischen Akademie der Wissenschaften zu Leipzig* **44**: 211–237. Reprinted in *Zeitschrift für physikalische Chemie* **10** (1892): 363–386.

Ostwald, Friedrich Wilhelm (1892b). *Thermodynamische Studien von J. Willard Gibbs*, trans. W. Ostwald (Leipzig: W. Engelmann).

Ostwald, Friedrich Wilhelm (1892c). "On the General Laws of Energetics," *Report of the British Association*: 661–662.

Ostwald, Friedrich Wilhelm (1893a). *Lehrbuch der allgemeinen Chemie. Zweite ungearbeite Auflage. II. Band, I. Teil: Chemische Energie* (Leipzig: W. Engelmann).

Ostwald, Friedrich Wilhelm (1893b). "On Chemical Energy," *Journal of the American Chemical Society* **15**: 231–238.

Ostwald, Friedrich Wilhelm (1893c). "Über chemische Energie," *Verhandlungen der Gesellschaft deutscher Naturforscher und Ärtze*, II, 1: 49–55. Reprinted in (1904), pp. 207–219.

Ostwald, Friedrich Wilhelm (1894). "Johann Wilhelm Ritter," Rede, gehalten auf der ersten Jahresversammlung der Deutschen Elektrochemischen Gesellschaft am 5. Oktober 1894 in Berlin. Reprinted in (1904), pp. 359–383.

Ostwald, Friedrich Wilhelm (1895). "Die Überwindung des wissenschaftlichen Materialismus," *Verhandlungen der Gesellschaft deutscher Naturforscher und Ärtze* I, 1: 155–168. Reprinted in *Zeitschrift für physikalische Chemie* **18** (1895): 305–320.

Ostwald, Friedrich Wilhelm (1896a). "Zur Energetik," *Annalen der Physik* **58**: 154–167.

Ostwald, Friedrich Wilhelm (1896b). *Elektrochemie. Ihre Geschichte und Lehre* (Leipzig: Veit & Comp.).

Ostwald, Friedrich Wilhelm (1902). *Vorlesungen über Naturphilosophie, gehalten im Sommer 1901 an der Universität Leipzig* (Leipzig: Veit & Comp.).

Ostwald, Friedrich Wilhelm (1904). *Abhandlungen und Vorträge allgemeinen Inhaltes (1887–1903)* (Leipzig: Veit & Comp.).

Ostwald, Friedrich Wilhelm (1908). *Grundriss der Naturphilosophie* (Leipzig: P. Reclam).

Ostwald, Friedrich Wilhelm (1910). *Grosse Männer* (Leipzig: Akademische Verlagsgesellschaft).

Ostwald, Friedrich Wilhelm (1924). "Wilhelm Ostwald," in *Philosophie der Gegenwart in Selbstdarstellungen* (Leipzig: B. G. Teubner), vol. 4, pp. 127–161.

Ostwald, Friedrich Wilhelm (1926). *Lebenslinien: Eine Selbstbiografie*, 3 vols. (Berlin: Klasing & Co., 1926–1927).

Ostwald, Friedrich Wilhelm (1961). *Aus dem wissenschaftlichen Briefwechsel Wilhelm Ostwalds, I. Teil: Briefwechsel mit Ludwig Boltzmann, Max Planck, Georg Helm und Josiah Willard Gibbs*, H.-G. Körber, ed. (Berlin: Akademie-Verlag).

Ostwald, Friedrich Wilhelm (1969). *Aus dem wissenschaftlichen Briefwechsel Wilhelm Ostwalds, II. Teil: Briefwechsel mit Svante Arrhenius und Jacobus Hendricus van 't Hoff*, H.-G. Körber, ed. (Berlin: Akademie-Verlag).

Planck, Max (1896). "Gegen die neuere Energetik," *Annalen der Physik* **57**: 72–78.

Planck, Max (1958). *Physikalische Abhandlungen und Vorträge*, 3 vols. (Braunschweig: F. Vieweg und Sohn).

Van 't Hoff, J. H. (1884). *Études de dynamique chimique* (Amsterdam: F. Muller & Co.).

Wheeler, Lynn P. (1952). *Josiah Willard Gibbs. The History of a Great Mind*, 2nd ed. (New Haven: Yale University Press).

Wilson, Edwin B. (1936). "Papers I and II as Illustrated by Gibbs' Lectures on Thermodynamics," in Donnan and Haas (1936), pp. 19–59.

JED Z. BUCHWALD

HEINRICH HERTZ'S ATTEMPT TO GENERATE A NOVEL ACCOUNT OF EVAPORATION

Although Heinrich Hertz is known primarily for his discovery in 1888 of electric waves and for his canonical formulation of field theory over the next several years, he had spent much of the 1880s searching for other ways to produce novel effects in the laboratory.[1] These investigations, which did not have much impact on his contemporaries, nevertheless provide uniquely revealing insight into Hertz's characteristic methods of thinking and of working in the laboratory. Among the earliest of the areas that he tried to evolve into something novel was evaporation, which became interesting to him as a byproduct of a brief investigation he did early in 1882 on hygrometry (Hertz 1882a). He had developed a new method for measuring absolute or relative humidity by weighing a hygroscopic substance. Such a substance absorbs water from the air until its vapour pressure becomes equal to the partial pressure of the unsaturated vapour "actually present in the air." Hertz built a simple device to measure relative humidity that employed a torsion-balanced glass rod carrying on one of its ends tissue-paper saturated with calcium chloride. Shortly afterwards Hertz became interested in the process of evaporation itself, as a vehicle for discovery.

What happens, Hertz wondered, when a liquid evaporates "in a space which contains nothing but the liquid and its vapour" (Hertz 1882b), when, that is, no other pressure acts on the liquid's surface beyond that of its own vapour?[2] (The kind of evaporation, or condensation, that occurs in Hertz's hygrometer takes place under conditions in which the total atmospheric pressure vastly exceeds the vapour pressure.) In a space filled solely with vapour far from equilibrium with the liquid, Hertz argued, two things could happen. Either (1) the liquid will continue to evaporate at a rate limited solely by the speed with which heat can be conveyed to its surface, or else (2) every liquid has an intrinsic *maximum* to its rate of evaporation, no matter how rapidly heat can flow through it. If, as Hertz believed, the second possibility holds true, then "the rate of evaporation will depend upon a number of circumstances, but chiefly upon the nature of the liquid, *so that there will be for every liquid a*

171

A.J. Kox and D.M. Siegel (eds.), No Truth Except in the Details, 171–189.
© 1995 *Kluwer Academic Publishers.*

specific evaporative power" (Hertz 1882b, p. 187; emphasis added). Every liquid will, that is, have a property or *quality* that determines its evaporative character. Liquids might then interact with one another in ways that depend upon their respective "evaporative powers," upon qualities of the substances, *per se.*

Suppose with Hertz that several liquid surfaces evaporate into a given enclosed space. If, Hertz claimed, the evaporation rate were unlimited, then, supposing that heat can be supplied rapidly enough, "all liquid surfaces in the same space must assume the same temperature; and this temperature as well as the amounts of liquid which evaporate are determined by the relation between the possible supply of heat and the different areas." Or if, as Hertz believed, the rates are intrinsically limited, then

there may be surfaces at different temperatures in the same space, and the pressure and density of the vapour arising must differ by a finite amount from the pressure and density of the saturated vapour of at least one of these surfaces: the rate of evaporation will depend upon a number of circumstances, but chiefly upon the nature of the liquid; so that there will be for every liquid a specific evaporative power.

Hertz's cryptic remarks require considerable explanation in order to grasp why an unlimited rate requires the vapour to be saturated and all evaporating surfaces to have the same temperature, whereas limited rates do not. These consequences hardly leap unaided to mind.

Hertz's conclusions depend upon the properties of a saturated vapour, and upon what might happen in the absence of equilibrium between a vapour and its liquid. When, he clearly knew, a liquid and its vapour coexist in equilibrium in some region, then the state of the vapour must lie somewhere along the vapour–pressure curve (Figure 1) that divides the pressure–temperature space into liquid and gaseous regions. Raising the temperature causes evaporation and increased pressure; lowering the temperature causes condensation and decreased pressure. But in quasi-static conditions the system sits always on the vapour–pressure curve. Under these circumstances the vapour is said to be *saturated.* If the pressure at a given temperature were, say, suddenly increased – swiftly enough to preclude a rapid return to equilibrium – then the vapour would become *supersaturated* and much of it would condense. A rapid decrease in pressure produces *superheated* vapour; evaporation then occurs until the system returns to the vapour–pressure curve. At equilibrium, then, the system sits on the vapour–pressure curve, the liquid and vapour have

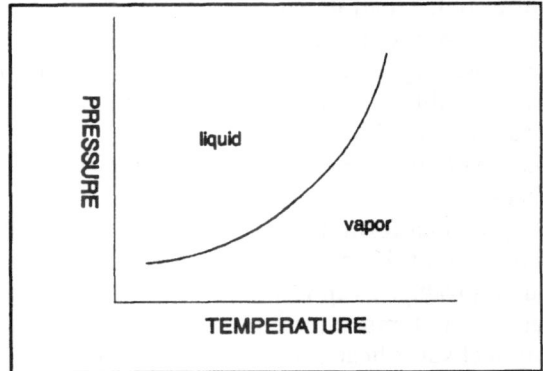

Fig. 1. Vapour–pressure curve.

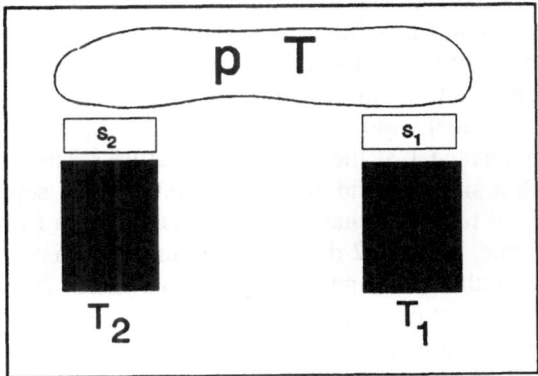

Fig. 2. Hertz's system.

the same temperature, and the pressure on the liquid surface and due to the vapour equals the vapour pressure.

To uncover what Hertz had in mind, consider Figure 2. Here we have two heat sources, one at the high temperature T_2, the other at the lower temperature T_1. During evaporation heat flows through the sources to their surfaces, which are separately represented in the figure as s_1 and s_2, and which may have their own temperatures (T_s^1 and T_s^2). The vapour in the surrounding region has some temperature T and pressure p. Suppose that the high temperature source is hotter than its surface, which is hotter than the surface of the low temperature source, which is in turn cooler than its surface ($T_2 > T_s^2 > T_s^1 > T_1$). Heat accordingly flows *from* source 2 to its surface, where vapour forms. At the low temperature surface heat flows *to* the source from condensing vapour and from conduction. For the system to reach a steady state, the rate of evaporation from 2 must equal the rate of condensation on 1, which requires that there be no net heat flow into or out of the vapour between surfaces.

Saturated vapour can form in such a system provided that the evaporation rate has no limit. If so, then *all* of the heat that flows to surface 2 from its source produces vapour; none remains to accumulate at the surface, increasing its temperature. However, the actual rate at which vapour formation absorbs heat must not be greater than the rate at which the heat formed by condensation at the low temperature surface can be removed by conduction to its reservoir. If both surfaces have the same temperature – whatever that might be – then a steady state may occur and saturated vapour form. Suppose for example that both surfaces drop to the low temperature T_1. Then, in effect, the high-temperature source simply feeds heat into a single-temperature system: surface 2 uses the heat to form saturated vapour at its (now low) temperature, which then condenses on surface 1 at the same temperature.[3] This system differs from one with a single liquid surface in that it has a separate, higher-temperature heat reservoir that can be drawn upon to form vapour. As the vapour forms, surface 2 drops while surface 1 rises. (In a single-source system nothing changes, though both systems contain saturated vapour.) This is a sort of evaporation engine.

If, *per contra*, the evaporation rate were limited, then the common surface temperature necessary to form saturated vapour might never be reached. To see this, suppose that the surfaces initially share a temperature that is intermediate between the source temperatures. If T_2 is large enough, then the rate at which heat flows to surface 2 might be greater than the finite evaporation rate can absorb. And if T_1 is low enough, then it may suck heat from its surface faster than can be provided to the surface by the condensing vapour generated at surface 2. As a result the temperature of surface 2 will increase towards T_2, and that of surface 1 will decrease towards T_1, until the heat flows balance (by decreasing the fluxes between the surfaces and their respective reservoirs): the system might accordingly be unstable at a common surface temperature. To observe such an effect clearly requires sufficiently high and low temperature sources, but how high and how low cannot be calculated beforehand.

Although Hertz did not do so, we can elucidate his reasoning with a bit of simple algebra. Consider the heat fluxes through the two surfaces on the supposition that the surface temperatures are intermediate between T_1 and T_2:

flux across S_2: $+F_2^F$ due to heat flow from T_2
$\qquad\qquad$ $-F_2^e$ due to evaporation
$\qquad\qquad$ $+F_2^c$ due to condensation

flux across S_1: $-F_1^F$ due to heat flow from T_1
$\qquad\qquad$ $-F_1^e$ due to evaporation
$\qquad\qquad$ $+F_1^c$ due to condensation

We can use this to establish equilibrium conditions, i.e., conditions under which the net flux through each surface separately vanishes, and, therefore, such that the surfaces have the same temperature. The surfaces would, in essence, form two parts of the same surface:
conditions for equilibrium

1. $F_2^F - F_2^e + F_2^c = 0$ and so $F_2^e = F_2^F + F_2^c$
2. $-F_1^F - F_1^e + F_1^c = 0$ and so $F_1^e = -F_1^F + F_1^c$

We see that for equilibrium to subsist, the rates of heat flux due to evaporation for each surface (which measures the rate of evaporation) must equal the sum of the corresponding rates due to flow and condensation. This at once implies that the system cannot reach equilibrium unless the evaporative rates are essentially unlimited, since the reservoir temperatures (and so the heat fluxes from them) can be made arbitrarily large or small. But if the rates are intrinsically unlimited, then the surfaces must have the same temperature, and the vapour must therefore be saturated. If the pace of evaporation cannot exceed a certain maximum, then for some reservoir temperature the heat flux to or from the surface will exceed the rate at which the heat can be turned into, or condensed from, vapour.

Hertz accordingly had to find some way to determine, for a given liquid, the temperature t at its *surface*, the pressure P upon it, and the height h of the liquid layer that evaporates in unit time. He evidently planned originally to measure at the high-temperature surface only, seeking to

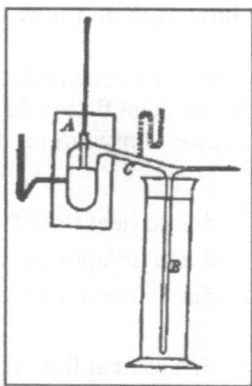

Fig. 3.   The first evaporation device.

find whether its temperature corresponds to a saturation pressure that differs from the pressure of the surrounding vapour (thereby implying a limit to the rate of evaporation). He thought, we shall presently see, that he could obtain the vapour pressure without directly measuring it. The temperature raised problems, because the very large temperature gradient from the surface towards its reservoir meant that "if we dip a thermometer the least bit into the liquid it does not show the true surface temperature," but Hertz was convinced that he could overcome these problems.

He built the apparatus, using mercury as the liquid, which he drew in Figure 3. In his words:

Into the retort A, placed inside a heating vessel, was fused a glass tube open above and closed below; inside this and just inside the surface of the mercury was the thermometer which indicated the temperature. To the neck of the retort was attached the vertical tube B, which was immersed in a fairly large cooling vessel, and could be maintained at $0°$ or any other temperature. By brisk boiling and simultaneous use of a mercury pump all perceptible traces of air were removed from the apparatus. The rate of evaporation was now measured by the rate at which the mercury rose in the tube B.

Uninterested in an absolute measure for evaporation, Hertz intended to operate entirely with relative values, measuring solely the liquid's height in his apparatus under different conditions. Similarly, he did not care to know the pressure P in absolute measure; he needed only its relative value. Indeed, he at first thought that he could avoid measuring the pressure altogether, on the following grounds: whether the mercury surfaces in A and B differ in temperature or not, the pressure on either surface "could not exceed the pressure of the saturated vapour at the lower temperature," namely the saturation pressure corresponding to $0°$.[4] This meant that he would not have to measure P because he could control it directly, by heating the low-temperature source.

Hertz rapidly uncovered an oddity:

... when the temperature [of the high source] began to exceed $100°$, and the evaporation became fairly rapid, the vapour did not condense in the cold tube B, but in the neck or connecting tube at C. This became so hot that one could not touch it; its temperature was at least $60°$ to $80°$. This cannot be explained on the assumption that the vapour inside has the exceedingly low pressure corresponding to $0°$; for in that case it could only be superheated by a contact with a surface at $60°$, and could not possibly suffer condensation.

If the vapour pressure was not governed by the low temperature source, B, then perhaps it was governed by the higher temperature of A. He attached a manometer at the neck, C, and measured the pressure at different rates of evaporation – that is, at different source temperatures of A. "But," he reported, "this did not show any change from its initial position when the rate of evaporation was increased." This truly startled him, because it seemed to mean that vapour pressure somehow did not depend solely on the states of the active surfaces.

"I began to doubt," he wrote, "not whether these magnitudes [surface temperatures and pressure] were necessary conditions, but whether they were sufficient conditions for determining the amount of liquid which evaporates." But as he thought the problem through, it occurred to him that the "vapour pressure" should be distinguished from the actual pressure on the evaporating surface: since the vapour is ejected with a certain speed it must exert a reaction force on the surface, and this must be added to the intrinsic vapour pressure, whatever the latter might be. Nor would such a reaction be at all negligible: the vapour moved so fast at higher source temperatures that "when the drops of mercury on the glass attained a certain size they did not fall downwards from their own weight, but were carried along [by the vapour] nearly parallel to the direction of the tube." The vapour kinetic energy might itself produce a great deal of heat, and this could account for the otherwise-surprising high temperature of the neck, with the temperature of the vapour still being determined by the cooler reservoir. Far from being super*heated*, the vapour is at it were super*saturated* by the pressure kick it gets from the ejecting surface. Contact with the neck then raises the latter's temperature towards the point on the vapour–pressure curve that corresponds to saturation at this higher pressure, but droplets form as the vapour resaturates at its own temperature. The neck produces condensation by *slowing* the supersaturated vapour, not by cooling it through conduction.

But was this reaction pressure sufficient to produce the necessary kinetic energy? To find out, Hertz attached a manometer directly to the reservoir A, and he measured the pressure at several reservoir temperatures (and so rates of evaporation). "It turned out that there was a very perceptible pressure," amounting to "2 to 3 m when the thermometer stood at 160° to 170°." Or, it would be more complete to say, the pressure that Hertz had measured would be substantial *if* the vapour were so cold that its intrinsic (vapour) pressure could be ignored, leaving only the reaction force to press the surface. He did not at this time wonder whether the *vapour pressure* itself at these temperatures might have produced the reading, because the invariance with evaporation rate of the pressure measured in the neck had already convinced him that the vapour state was not governed by the high-temperature source.

These several developments hardly jump to the eye in Hertz's account, because he wrote about them shortly afterwards and not during the process of discovery. Nevertheless, Hertz's early articles, in print and in manuscript, bear unmistakeable marks of the complicated paths he followed to discovery. To recapitulate, when he found to his surprise that the vapour condensed at a high temperature, Hertz first thought that the vapour might take its state from the high source. This did not prove out, because, when he measured the pressure, he found that it did not vary with the high temperature (rate of evaporation). Deeply puzzled, and thinking that perhaps something altogether odd was involved, Hertz turned back to his original view that the vapour is governed by the low-temperature source and no doubt began to look for something else that could account for the anomalous condensation in the neck. He found a possible reason in the rapidity of the streaming vapour: its temperature might be low, as he had initially assumed, but its speed might be so great as by itself to contain energy that could be retrieved as heat (presumably by friction between vapour and neck). To test the possibility, he measured the pressure on the high-temperature surface itself, instead of the vapour pressure in the neck, and found it to be "very perceptible," perceptible enough to represent a powerful recoil kick from the speeding vapour. That result convinced him that a careful experimental distinction had to be made between the vapour pressure proper and the net

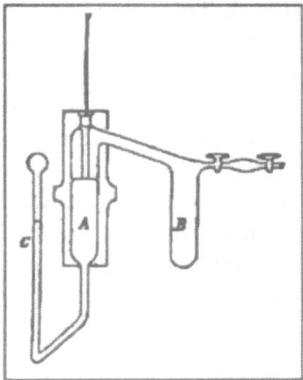

Fig. 4.   First redesign.

pressure on the evaporating surface. Significantly, Hertz did not report measuring this surface pressure at various temperatures to see whether it did indeed vary with the rate of evaporation – as it should have, but as the pressure in the neck (whose magnitude he did not describe) had *not*.[5] He interpreted the existence of the pressure as warranting his conclusion that the rapidity of the vapour's egress precluded using the apparatus as he had hoped, not as evidence that the state of the vapour differed from his initial supposition.

The original apparatus had failed to work properly, Hertz now felt, because the rapidly streaming vapour made it impossible to control the native vapour pressure, though Hertz believed it to be that of saturated vapour at the temperature of the lower source. He decided to alter the apparatus in order to slow the stream, which required keeping it under pressure. That way the measured pressure on the surface would also be the native pressure of the gas. The goal of the experiment would then be to see whether that pressure corresponds to saturated vapour at the temperature of the surface. If not, then the evaporation rate had to be limited.

His redesigned apparatus appears in Figure 4. The two most striking differences from the first experiment are the connection of an extension tube, C, to reservoir A, and the removal of the cooling bath from the condensing tube B. The vapour still condenses in B, but since Hertz would no longer directly control the low temperature source, it made no difference what the temperature there might be. His attention concentrated entirely on reservoir A. The extension tube permitted measuring the pressure on A by reading the difference in height of the liquid in A and C, while the absolute height of the liquid in A over time measured the rate of evaporation. Because mercury's density varies markedly with temperature, and tube C was not itself heated to the temperature of A, Hertz had to correct his pressure measures by amounts some of which "were much larger than the quantity whose value was sought." Nevertheless, Hertz asserted, the pressure measure could be relied on to about 0.1 mm, and the evaporation measure to about 0.02 mm.

The surface temperature was "the most uncertain element." Of it he remarked:

I thought it was safe to assume that the true mean temperature of the surface could not differ by more than a few degrees from the temperature indicated by the thermometer when the upper end of its bulb (about 18 mm long) was just level with the surface; and it seemed probable that of the two the true temperature would be the higher. For the bulk of the heat was conveyed by the rapid convection currents; these seemed first to rise upwards from the heated walls of the vessel, then to pass along the surface, and finally, after cooling, down along the thermometer tube. If this correctly describes the process, the bulb of the thermometer was at the coolest place in the liquid.

The bulb of the thermometer was, as it were, immersed in the same stream of liquid that, beginning deep within the fluid mass, creeps up the tube's walls and bathes the surface itself. Confident that his device correctly measured the surface temperature, Hertz "carried out a large number of experiments at temperatures between 100° and 200°, and at nine different pressures'; he altered the pressure by admitting air into the device through valves located near the condensing tube B. The results were precisely what he had hoped for: "The observed pressure P was always smaller than the pressure P_t of the saturated vapour corresponding to the temperature t," and by a considerable amount indeed – at *circa* 180° the difference amounted to over 7 mm of mercury, which was vastly larger than any possible error in measuring the pressure. But what about the temperature? To lower the pressure of the corresponding saturated vapour by 7 mm requires a temperature 30° lower than Hertz had measured. No measuring error in pressure could have occurred; "nor do I believe," he wrote, "that the second [temperature error] could."

When Hertz first wrote these last words, he was apparently convinced that he had been able reliably to elicit a limit to the evaporative rate of mercury. The words remain in both the manuscript and in the printed article. Indeed, it seems quite likely that Hertz originally intended to close the experimental part of his article at this point, concluding with a 'theoretical' discussion of how to calculate the limits given certain data. These 'theoretical' remarks do indicate that a new effect should exist, but between the two he now inserted a discussion of yet another (third) series of experiments, ones that he had certainly intended solely to provide further evidence for his claim to have fabricated something novel. (He would presumably not have described these experiments in the detail that he did if they had merely corroborated his previous results.) These experiments manifestly contradicted Hertz's claim to discovery, and they forced him thoroughly to redraft what he had planned (and probably begun) to write when the second set of experiments had worked to give him something new.

Utterly convinced that his temperature measurements simply could not be in error by nearly 20%, he could have gone into print with these results. Why didn't he? One thing pressed him on. Not doubt that he had produced novelty – of that he was certain. Rather, he felt that he needed to provide more secure quantitative measures in order to be able accurately to specify the limiting conditions on the evaporative rate. Contemporary demands, particularly in the German physics community, for exact measurement would have made that seem to be essential. His second set of experiments could be improved upon to that end in two respects. First of all, they required very large corrections to be made for mercury's expansion in determining the pressure. These corrections had required "a careful application of theory and . . . special experiments," which, Hertz was himself convinced, made the result reliable to 0.1 mm. Hertz accordingly sought to manipulate the device into a form that did not require elaborate corrections to get the correct pressure. Second, and much more important, even though Hertz was certain that his temperature measures could not have been off by 20% (i.e. by 30°), nevertheless, they might admittedly have been inaccurate by "a few degrees."

In the second mutation of his original device, Hertz produced the apparatus of Figure 5. The first obvious change is the absence of tube C, which had required measuring corrections because it lay outside the heated vessel. Instead, the two arms now both lie within the vessel, and are, moreover, symmetric. Pressure measurements could be read off directly from the height difference between the mercury in the arms. But the temperature required an entirely separate device based on equating the heat that flows to the surface in unit time to the heat absorbed during that time in evaporation. In Hertz's words:

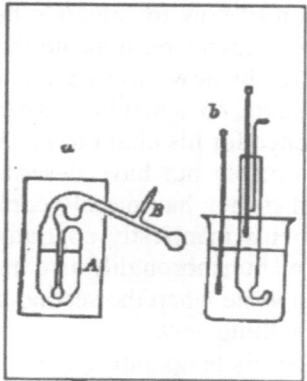

Fig. 5.   The second mutation.

This [surface] temperature is equal to that of the bath [reservoir], less a correction which for a given apparatus is a function of the convection current only which supplies heat to the surface. The known rate of evaporation gives us the required supply of heat; from this again we can deduce the difference of temperature when the above-mentioned function has been determined.

Hertz had observed large-scale convection currents in his earlier apparatus and had concluded that most of the heat conveyed to the surface came to it in this way. He needed a device untroubled by evaporation to pin down the heat flow. The right-hand diagram in Figure 5 illustrates it. In Hertz's words:

A piece of the same tube from which the manometer was made, was bent at its lower end into the shape of the manometer limb. This was filled with mercury to the same depth as the manometer tube; above the mercury was a layer of water about 10 cm deep, and in this a thermometer and stirrer were placed. This tube was immersed up to the level of the mercury in a warm linseed-oil bath, the temperature of which was indicated by a second thermometer. A steady flow of heat soon set in from the bath through the mercury to the water. The difference between the two thermometers gave the difference between the temperatures of the bath and of the mercury surface; the increase in the temperature gave the corresponding flow of heat.

Hertz's clever device measures the mercury's surface temperature by putting it in contact with another liquid, which, when stirred, is in thermal equilibrium with the surface. Further, instead of immersing a second thermometer in the mercury proper, Hertz placed it in the surrounding heat reservoir (here of linseed-oil), thereby obtaining the temperature at the source of the convection current that feeds the surface.[6] In this way, he found that a layer of water 117 mm high would be heated through 0.48° per minute by a 10° difference between bath and surface temperature.

The apparatus yielded an (evaporation) rate of 0.057 mm of mercury per minute and a pressure difference between the two arms of the manometer of 0.26 mm for a temperature of 118° in the paraffin bath. Given from elsewhere how much heat (in unit weights of water) is necessary to vaporize (unit weight of) mercury at 180° (and given as well the ratio of the specific gravity of mercury to that of water), Hertz concluded that the heat involved here would raise 117 mm of water (on the same base) through 0.48° – which is why he had sought for this specific difference in his temperature-measuring apparatus. Consequently, the evaporating surface (in the leftmost tube in Figure 5) must have been 10° colder than the paraffin bath in which the manometer sat, making its temperature 108°.

The mercury surface in the rightmost limb of the manometer, however, bounds an enclosed region, and consequently produces saturated vapour at the temperature of the encompassing bath, that is at 118°. Now if the evaporative rate were unlimited, then each surface would be in contact with saturated vapour at its respective temperature of 118° on the right and 108° on the left. If this were the case, then (using values for the vapour pressure of mercury that he himself had generated in a separate series of experiments), Hertz noted, the pressure difference across them would be 0.27 mm. This was in fact only 0.01 mm more than he had actually measured – leaving only an experimentally-insignificant amount to represent the effect of a limited evaporation rate. Hertz felt that only one conclusion was possible given the great care that he had taken in temperature and pressure measurement: "the positive results obtained by the earlier method had their origin partly, if not entirely, in the errors made in measuring the temperature."

Hertz's confession that he had *failed* to produce something new contrasts remarkably with his assertion, following his description of the second series of experiments, that the *positive* results he had obtained relied on pressure and temperature measurements that were not subject to substantial doubt. In his words, again, "The first-mentioned error [pressure] could not have occurred; *nor do I believe that the second [temperature] could*" [emphasis added]. Both statements – that the temperature measures in the second set of experiments are reliable, and that they must not have been – appear in the same published article, as well as in the manuscript (which scarcely differs from the printed version). Both cannot be correct, nor would Hertz likely have held both consciously and simultaneously. We have evidence here for one aspect of the neophyte Hertz's laboratory work-habits.

The article as printed divides into (but was not so divided by Hertz) five distinct parts: first, an introduction that raises the issue of a new effect and elliptically discusses the results of the experiments; second, the initial experiments that immediately revealed flaws in his assumptions; third, the experiments that seemed to reveal the new effect; fourth, the more elaborate experiments that contradicted the results of the second set; finally, 'theoretical' considerations which have the effect of justifying the ultimately negative results, without giving up belief in the effect's existence. The manuscript for the printed article, now held at the Science Museum in London, has no physical marks of discontinuity between these several parts. And yet we have just seen that statements in parts three and four concerning temperature measurement manifestly contradict one another. It seems reasonable to conclude that the article as it was finally written contains in its second and third parts substantial vestiges of a *first* article, now lost, in which Hertz had announced success. After, or perhaps even while, he was writing this lost account, he decided that the force of the paper would be considerably strengthened if he could provide something more than evidence that the new effect simply exists – if he could, that is, pin it down quantitatively to something better than "a few degrees": "But I could not conceal from myself," he wrote in recollection of what had prompted him to undertake the new experiments, "that the results, from the quantitative point of view, were very uncertain." He accordingly built the apparatus of part four to do so, and he then discovered, undoubtedly to his great surprise and consternation, that the two sets of experiments did not tally at all with one another. We can be nearly certain that he must have checked his results very carefully, but could see no way out: his experiments had *failed* to bring the effect to laboratory life.

Despite Hertz's intense, indeed overpowering desire to produce novelty, and despite his initial belief that he had done so, his own experimental results forced him to back down. He pursued his initial success in order to be certain that he could convincingly counter objections that might be raised to his claim. Hertz was, in the end, utterly convinced that, despite his strong hopes, his devices simply could not be made to produce what he wanted from them. They were not infinitely malleable; they could not be forced into an appropriate mold. Here, then, we have a striking instance of a scientist who deeply wanted to produce a certain result – and who thought he had initially succeeded in doing so – deciding that he had failed. Hertz's devices, one might say, spoke unequivocally to him, and their words could not be denied. He knew just how reliable his apparatus was; hours and days of experience, and weeks of skilled manipulation, had taught him its capabilities and its limits. He could not argue away the new, depressing results, without making his entire workshop experience with apparatus he had himself built a mirage. Perhaps the abstract, noumenal *Nature* of scientific realism did not foil Hertz's designs. Equipment worked with knowing skill certainly did.

APPENDIX: HERTZ'S RETROSPECTIVE ACCOUNT OF EVAPORATION'S THEORETICAL LIMITS

Hertz felt that he had been unable to achieve a "lucid theory" for evaporation, even though his published article contained two distinct excursions in that direction, one of which moved almost entirely outside the laboratory. The first, and much the longer of the two, was probably written (or at least thoroughly rewritten) after Hertz concluded that he had been mistaken in thinking that his experiments had ever revealed an evaporative limit. These remarks were designed to show why Hertz's failure had been inevitable; they were intended to indicate that none of the devices that he used could possibly have revealed the limit, even if it did in fact exist. Obviously, Hertz could not have had anything quite like these considerations in mind when he began experimenting, for if he had, then there would have been no point in going further. These arguments were constructed after his failure to transform the experiment into something else – if not into success, at least into something not entirely negative.

Hertz envisioned "two infinite, plane, parallel liquid surfaces kept at constant, but different temperatures." Vapour between will move "from the one surface to the other," conserving its heat and establishing an interaction between the two surfaces. During this passage, Hertz argued, "it follows from the hydrodynamic equations of motion" that the state of the vapour, including its velocity, remains constant. The vapour, that is, has some particular native pressure and temperature p, T. These values may be different from the temperatures T_1, T_2 of the surfaces and from the pressure P that the vapour exerts on each surface. The latter, Hertz asserts, must be the same for both surfaces, presumably because the force exerted on the vapour at one surface must be the same as the force that the vapour exerts on the other surface.

Hertz's theory – or, better, the reason for the failure of his experiments – hinges on the magnitude of the difference between p, the native vapour pressure, and P, the pressure between vapour and surface. The difference is due entirely to the fact that the vapour streams away from the surface with a certain speed; the difference $P - p$ must therefore be the force that accelerates the vapour to that speed. Suppose the vapour (of density d) has, after acceleration, speed u, and that in unit time a weight m passes from one (unit) surface to the next. Then, clearly:

$$m = ud.$$

We can use this to establish a link to the pressure P, which is what Hertz measured. In his words:

For let us suppose the quantity m spread over unit surface, the pressure upon one side of it being P and on the other side p, and its temperature T maintained constant. It will evaporate just as before; after unit time it will be completely converted into vapour, which will occupy the space u and have the velocity u. Hence its kinetic energy is $(1/2)mu^2/g$; this is attained by the force $P - p$ acting upon its centre of mass through the distance $u/2$,[7] so that an amount of work $(P - p)u/2$ is done by the external forces. From this follows the equation $P - p = mu/g$; or, since $m = ud$, $m^2 = gd(P - p)$.

Hertz wanted to use this result to assign limits to the relations between the six quantities that he had at his disposal, namely T, T_1, T_2, p, P, d, u, and m. That way he could calculate from observational data what the corresponding evaporative limit ought to be. But first he had somehow to establish that there *ought to be* such a limit. This was not easy, but Hertz succeeded in formulating an argument based on two "assertions which, according to general experience, are at any rate exceedingly likely to be correct" – they do not derive in any way at all from theory. They are:

1. "If we lower the temperature of one of several liquid surfaces in the same space while the others remain at the original temperature, the mean pressure upon these surfaces can only be diminished, not increased."

2. "The vapor arising from an evaporating surface is either saturated or unsaturated, never supersaturated. For it appears perfectly transparent, which could not be the case if it carried with it substance in a liquid state."

These two rules from "general experience" suffice to establish that the rate of evaporation must have a limit for a given temperature. From the first we know that the surface pressure P must be less than the pressure p_M of saturated vapour at the temperature of the high source, T_1; from the second we know that the actual density d of the vapour is less than the density d_p of *saturated* vapour at the same pressure p. Hertz could therefore write the following inequality:

$$m < \sqrt{g d_p (p_M - p)}.$$

Now when the pressure p vanishes then so do the other pressures, so that the quantity under the root goes to zero with p. It also goes to zero when p goes to p_M. In between these two points, Hertz noted, it must reach some maximum, which is then the theoretical maximum for the evaporation rate at a given high temperature T_1. No matter how low the temperature of the other reservoir may be, the evaporation rate cannot exceed this value.

Of course Hertz had not observed any limit at all, and the point of going through this bit of 'theory' – based on a pair of decidedly untheoretical rules – was to show why he had not succeeded. To do that he needed to compute numbers, which meant he had to be able to calculate volumes and temperatures from pressures. To do so, he assumed, without elaboration, that the Boyle law holds for the vapour; the pressure–temperature relation could be derived from thermodynamics and known experimental values. The calculation gave values for the limit that were vastly greater than the evaporation rates that Hertz had achieved (e.g., at 180° the calculated limit was 20.42 mm/min, whereas Hertz had achieved only 1.67 mm/min), which gave him the confidence to remark of his experiments that "they do not show definitely the existence of the deviation from this rule which probably occurs, and which is of interest from the theoretical point of view."[8]

Hertz's second excursion into theory provided a reason for *why* a limit might exist at all. He based it on the kinetic theory of gases, arguing in effect that the evaporation rate cannot exceed a quantity determined by the "mean molecular velocity of the saturated vapour corresponding to the temperature of the surface." His argument was, however, extremely undeveloped, amounting to little more than a sketch.

Dibner Institute for the History of Science and Technology

Cambridge, MA 02139
U.S.A.

NOTES

[1] Buchwald (1990) and (1992) provide background to Hertz's work and career.

[2] If measurable evaporation occurs then the system has not yet reached equilibrium since, when it does, the rate of condensation equals the rate of absorption.

[3] Alternatively, though Hertz did not initially consider the possibility, the surface temperatures might lie halfway between T_1 and T_2 (all else being equal), since then surface 1 would transfer the heat produced by condensation to its reservoir at exactly the rate that surface 2 sucks it in from its source to produce vapour.

[4] Because, he initially believed, the common surface temperature that would be achieved *if* he evaporation rate were unlimited would be close to that of the low-temperature reservoir. If the rate had a limit then the surface-temperatures could differ, but, Hertz evidently thought, the vapour would stay close in temperature to that of the low (condensing) source.

[5] The surface pressure must vary with the evaporation rate because it is due predominantly to the 'kick' as streaming vapour leaves the surface. The pressure measured in the neck should not vary because it is determined by the state of the vapour, which in turn is governed by the low-temperature surface.

[6] In the evaporation apparatus Hertz correspondingly measured the temperature of the reservoir that surrounds the manometer, and not the temperature within the mercury itself. Because, he believed, the heat was supplied by convection along the walls of the containing vessel – i.e. at the boundary of the liquid – measurements taken in the mercury could not be reliably used.

[7] Hertz does not justify the factor 1/2 here, nor is it a simple matter to see what the factor should be. The force P presumably acts only very near the surface since it is by assumption the force exerted by the surface on the vapour. The force p comes into action just as soon as the vapour leaves the surface. We must somehow imagine that the vapour is shot from the surface by P but is slowed down by p. After the action ceases – when the ejected vapour experiences p on either side – it has speed u. Hertz convinces himself that the *distance* over which a net force acts on the vapour "centre of mass" is actually $u/2$, at the end of which the "centre of mass" has speed u. Hertz may have reasoned that, at the surface proper, the speed is zero, whereas a distance u away from the surface the speed is u, so that the mean speed of the vapour mass that fills the space u is $u/2$, and consequently that after unit time the "centre of mass" of the vapour will be $u/2$ away from the surface.

[8] Note the peculiar phrasing here: "they do not show *definitely*" – in the original, "sie zeigen aber nicht mit Bestimmtheit." The experiments in fact show *nothing at all* with regard to such a limit.

REFERENCES

Buchwald, Jed Z. (1990). "The Background to Hertz's Experiments in Electrodynamics," in *Nature, Experiment, and the Sciences*, T. H. Levere and W. R. Shea, eds. (Dordrecht, Kluwer), pp. 275–306.

Buchwald, Jed Z. (1992). "The Training of German Research Physicist Heinrich Hertz," in *The Invention of Physical Science: Essays in Honor of Erwin N. Hiebert*, M. J. Nye, J. L. Richards, and R. H. Stuewer, eds. (Dordrecht, Kluwer), pp. 119–145.

Hertz, Heinrich (1882a). "On a New Hygrometer," in Hertz (1896), pp. 184–185.

Hertz, Heinrich (1882b). "On the Evaporation of Liquids, and Especially of Mercury, in Vacuo," in Hertz (1896), pp. 186–199.

Hertz, Heinrich (1882b). "On the Pressure of Saturated Mercury-Vapour," in Hertz (1896), pp. 186–199.

Hertz, Heinrich (1882b). *Miscellaneous Papers*, transl. D. E. Jones and G. A. Scott (Leipzig: J. A. Barth).

REFERENCES

FREDERIC L. HOLMES

CRYSTALS AND CARRIERS: THE CHEMICAL AND PHYSIOLOGICAL IDENTIFICATION OF HEMOGLOBIN

I

I am very glad to be included in this volume of articles celebrating a landmark in the professional life of Martin Klein. As a historian of the life sciences, with a few footholds in history of chemistry, I have been uncertain that I could make a contribution fitting the scholarly interests of a preeminent historian of physics. Those who know him, however, know that Martin's professional interests extend far beyond the field in which his own work has centered. For several years he and I taught together a seminar in the history of science designed for a small group of Freshmen in Yale College who intended to concentrate in science. That experience intensified my appreciation for the breadth, depth, and precision of his thought that I had already admired in his published writings and his lectures. I also saw at close range his enthusiasm for critical, imaginative intellectual activity in whatever domain it occurs. I hope that what I have to offer here in tribute to his achievements, to his integrity, and to his long friendship, will meet at least his basic standards for scholarly endeavor.

The topic for my paper originated in another occasion, an invitation to participate in a conference on hemoglobin held during the summer of 1993 at the Wellcome Institute for the History of Medicine. The complexity of the subject, and my interest in probing beneath the surface results of individual investigations critical to its historical development, led me to write about it at greater length than could be fitted into the confines of such a conference. I am, therefore, grateful for the special opportunity to present my story in full here. Although centered in the history of physiological chemistry, the story is not bounded by any of the sciences that contributed to the resolution of the problems that arose during its course. Two of the prominent participants, and at least one of the critical methods applied, are best known within the history

191

A.J. Kox and D.M. Siegel (eds.), No Truth Except in the Details, 191–243.

of physics. The story provides a strong example of the permeability of disciplinary boundaries within the sciences, and for the unities that should connect our subspecialties within the history of science. For me personally, it is an opportunity to display a few of the many points of contact between the fields within which I have worked and the large field to which Martin Klein has devoted his distinguished scholarly life.

Historians of science have written extensively on the history of hemoglobin during the nineteenth century.[1] I shall not survey that history comprehensively here, but will concentrate on a more particular historical question: how did hemoglobin come to be understood as a combination of a protein and an iron-containing pigment, whose physiological function was to transport oxygen from the lungs to the tissues of vertebrate animals? The story has, as we shall see, many layers. The dénouement was dominated by a well-delineated series of investigations conducted over the course of the decade 1857–1867 by Felix Hoppe-Seyler, whose outcome established the capacity of hemoglobin to "bind oxygen loosely," and thereby to make arterial blood an "oxygen carrier," able to pick that substance up in the lungs, and to give it off at the capillary walls to the organs of the body.[2]

The ability of Hoppe-Seyler to take up such an investigation rested, however, on several long lines of previous investigation, including especially (1) experiments on the respiratory gaseous exchanges extending from the late eighteenth century to his own time, (2) the gradual refinement of chemical methods for extracting and characterizing organic substances, and their application to proteins, (3) the development of methods capable of revealing the presence of gases in the blood.

During the course of Hoppe-Seyler's prolonged investigation, several concurrent advances enabled him to bring his study to a successful conclusion in ways that would not have been feasible at the time he started. These included (1) the introduction of more accurate quantitative methods of gas analysis applicable also to the study of the blood gases, and (2) the emergence of spectroscopic chemical analysis, which made it feasible just at this time to identify the absorption spectra of pigmented substances in solution. These methods, devised in part for problems in physics and chemistry, were applied almost immediately to the study of physiological questions. These rapid transfers demonstrate that hemoglobin was already in the mid-nineteenth century a broadly multidisciplinary research object.

If the determination of the basic functional properties of hemoglobin was made possible, at close range, by the availability of increasingly precise analytical methods, their significance depended, on the other hand, on a much broader shift in biological viewpoints during this era. Functions traditionally associated with the organ systems and fluids of the body were being relocated, at least in principle, within the elementary units of the body tissues. This gradual reorientation reflected in particular the steadily expanding impact of the cell theory in the decades following the epoch-making work of Theodor Schwann.

The role of hemoglobin in the transportation of oxygen could hardly have been established prior to the 1850s, because methods adequate to do so were not yet in place. By the 1880s so much converging evidence existed that it is hard to imagine its role *not* being understood by then, even if Hoppe-Seyler had not undertaken his systematic study of the subject. The history of the specification of that role raises, therefore, fundamental questions concerning the extent to which the individual scientist can exert a creative force in the construction of scientific knowledge, and the extent to which she or he is only the agent for an inexorable line of scientific advance.

The dynamic of scientific investigation during the mid-nineteenth century was much like that of our own day. Although the overall scale of research was much smaller, individual investigators worked in specially equipped laboratories in well-organized institutes, often as heads of research groups. They were highly professional specialists in specific fields and subfields. They contributed scientific papers regularly to specialized journals that published rapidly, and they competed for priority. When a particular set of problems became visible as a promising area for investigation, multiple investigators in laboratories spread across Europe tended to move into it. In our historical accounts we must often limit ourselves, for simplicity, to the most prominent discoveries and the most celebrated scientists. That should not obscure the fact that these peak events and personalities were, then as now, surrounded by less outstanding investigations and investigators, who nevertheless contributed collectively to sustain many robust investigative streams.

II

The problem area out of which the functional definition of hemoglobin eventually emerged was rooted in the theory of respiration of Antoine

Lavoisier. By 1790, when he presented his most comprehensive memoir on the subject, Lavoisier had come to view respiration as a slow combustion of carbon and hydrogen that produced carbonic acid and water. This process was the source for both the heat released in the body and the work produced by an animal. By this time Lavoisier also realized that there was no decisive experimental evidence to support his simple initial assumption that the combustion takes place in the lungs, where he had supposed that the inspired oxygen comes into immediate contact with the combustible material. It would be possible, he wrote, "that the carbonic acid gas that is disengaged during expiration is formed ... during the course of the circulation by the combination of the oxygen of the air with the carbon of the blood."[3]

Lavoisier's uncertainty was prolonged for nearly 50 years, as various investigators sought inconclusively to decide between these alternative locations for respiration. Various attempts to extract oxygen gas from arterial blood and to associate the color change from venous to arterial blood with the removal of carbon, or other chemical changes, remained unconvincing until well into the 1830s.[4]

At the time Lavoisier formulated his theory of respiration, chemists recognized several basic constituents of blood. With the microscope it could be seen to contain "a great number of globules," which appeared red when intact, but broke down into colorless fragments. When removed from the body, blood separated spontaneously into "two parts, the clot and the serum." The serum contained, in addition to water and neutral salts, "a peculiar matter of great importance." This substance, which contained nitrogen and "coagulates and hardens long before it boils," was called albumin, or albuminous matter, because of its resemblance to egg white. The clot could be further "separated into two very distinct substances." One of them was a "white fibrous matter" which was, accordingly, named fibrin. The other matter, when dissolved, gave the fluid a red color and contained a great "quantity of iron."[5] Lavoisier never attempted to identify the matter burned in respiration with any of these tangible components of the blood. Usually he referred simply to "a portion of carbon and hydrogen" furnished by the blood, occasionally to a hypothetical "carbonated hydrogen" separable from the blood as a "viscous humor."[6]

In the first decades of the nineteenth century, some of the leading chemists of that era of rapid change in general chemistry also studied extensively the composition of animal matter. The most important of these studies were conducted by Jöns Berzelius, who published in 1813 a summary of the principal results of the analyses he had performed over the preceding years on the composition of animal fluids. Largely qualitative, but as rigorous as contemporary methods allowed, these analyses applied a selective repertoire of solvents and reagents to separate and to ascertain the chemical properties of each of the constituents that could be differentiated. In his study of the blood Berzelius concentrated on three substances – fibrin, albumin, and the "coloring matter [that is] suspended, not dissolved," in the fluid. From their respective reactions to heating and with alcohol as well as dilute and concentrated acids and alkalis, from the characteristic yellow compound they formed with nitric acid, he concluded that all three "resemble one another so closely that they can be considered modifications of one single substance." He named them, therefore, the "albuminous parts of the blood."[7]

The "coloring matter" was, however, readily distinguishable from fibrin and albumin, not only because of its red color, but because it contained a "certain quantity of iron oxide." Berzelius devoted special attention to the question of what form this iron oxide took, partly in response to a claim by French chemists that the coloring matter consisted of a solution of sub-phosphate of iron in albumin. After trying all of the reagents known to precipitate or otherwise react with salts of iron oxide, however, and finding that none of these characteristic reactions occurred, he concluded that "these experiments prove that iron in coloring matter is not contained in such a manner that it can be detected by our best reagents," unless the coloring matter itself is first completely destroyed. "The mode of combination of the iron with the coloring matter," he added, "is and will probably remain for a long time unknown."[8]

As this passage suggests, Berzelius was acutely aware of how limited the reach of his methods was, judged against the complexity of the problems with which blood confronted him. "After all these experiments," he wrote "one could ask, what is the difference between venous blood and arterial blood? To that question I have no reply."[9] Only in a footnote did he allow himself to ponder the question. The most striking difference between the coloring matter and the other albuminous constituents of the blood, he wrote,

is found in the property of the coloring matter to absorb oxygen and thereby to undergo a color change. Serum absorbs very little oxygen, and only in proportion as it decomposes. Does the iron contained in the coloring matter confer that property upon it? That is probable; but we can never arrive at an exact knowledge of these phenomena without first analyzing the elements of the animal kingdom with the most scrupulous exactitude. It will be then, and only then, that we can form conjectures; for now they are useless.[10]

Despite these scruples, Berzelius's query offered the prospect that the unanswered questions about the relation between the blood and respiration could be brought to a sharper focus by directing them not just at the overall differences between arterial and venous blood, but at the distinctive properties of the iron-containing coloring matter. For more than two decades chemists and physiologists could do little to exploit this insight, however, because they could not settle the most basic question: whether the carbonic acid that is exhaled is "only produced in the lungs, as the inspired oxygen oxidizes there a part of the carbon of the blood, or whether when the venous blood reaches the lungs it already contains preformed carbonic acid that is merely discharged" there.[11] They recognized that the difficulty was technical rather than conceptual. Multiple efforts to detect carbonic acid in arterial and venous blood had so far produced only "striking contradictions."[12]

Representative of multiple efforts to solve this problem, viewed by the 1830s as critical to further progress, were experiments carried out between 1831 and 1834 in Heidelberg by two outstanding scientists of their time, the chemist Leopold Gmelin and the physiologist Friedrich Tiedemann. A decade earlier Gmelin and Tiedemann had collaborated to produce a landmark investigation of the chemical changes in digestion.[13]

Gmelin and Tiedemann drew samples of blood from the femoral artery and vein of a dog, transferred them quickly (protected from contact with the air) to inverted tubes filled with mercury, placed the tubes inside the chamber of an air pump, and evacuated the chamber. Neither arterial nor venous blood discharged any permanent gases. This result favored the theory, long associated with Lavoisier, but actually the brainchild of his younger associate Armand Seguin, that a hydrocarbonaceous substance oozes from the blood into the lungs, to be directly oxidized there. Gmelin and Tiedemann regarded this theory as implausible, however, and searched for an alternative. Perhaps, although the blood contained no free carbonic acid, that substance might be present but bound to an alkali. To test their idea, they added some concentrated acetic acid to fresh samples of arterial and venous blood, and were then able to extract carbonic acid gas in the air pump. Venous blood yielded more than arterial blood. On the basis of this result they proposed "fragments of a theory," whereby the various secretory and nutritive processes of the body discharge acetic or lactic acid into the blood. In the lungs these acids displace the bound carbonic acid, which is then exhaled. Acknowledging that many difficulties stood in the way of the further development of their theory, they reserved the solution of these problems for future research.[14]

At about the same time that Gmelin and Tiedemann were concluding that carbonic acid was not present in free form in the blood, Gustav Magnus was exploiting a different method to find it there. Trained in chemistry at the University of Berlin under the direction of Eilhard Mitscherlich, Magnus had then studied in Stockholm in 1827 and 1828 with Berzelius, whom he afterward regarded as his mentor. Back in Berlin, Magnus gave lectures in chemistry and in technology, but was appointed in 1833 extraordinary professor of technology and physics. He maintained a private laboratory, where he encouraged students to conduct their own investigations. Early in 1834 one of them, a medical student named Bertuch, was able to confirm claims made by two English physicians, named Stevens and Hoffmann, that they could drive carbonic acid gas out of blood by means of hydrogen gas. Shortly afterward Bertuch died of smallpox. Magnus thought the question so important that he immediately continued the experiments.[15] By late May he reported to Berzelius that

carbonic acid gas can, indeed, be driven out of human venous blood by hydrogen gas, or nitrogen, even when one uses blood that has been received directly from a vein. It now seems to me an interesting question, whether one can obtain by means of these gases just as much carbonic acid from the blood as by means of oxygen; because until this is shown, one can draw no conclusions about respiration. Then it would still be left to show, why in the experiments conducted up until now, one had not been in a position to discover in the blood the pre-formed carbonic acid.[16]

The reason Magnus thought it crucial to show that the two "indifferent" gases displaced the same quantity of carbonic acid that oxygen did was that only in that way could he prove that atmospheric air in the lungs does not form carbonic acid by oxidizing the carbon of the blood. In the following months he succeeded in showing, by parallel experiments on "the same blood," that the quantities of carbonic acid received with hydrogen and with atmospheric air were "as close to the same as one can expect in experiments of this type."[17] By September he wrote Berzelius, "I can now say with certainty that carbonic acid is preformed in the blood," and he believed that he was also in a position to show "why this gas had eluded previous observers."[18]

There were two reasons, Magnus thought, why "most investigations had revealed no carbonic acid in the blood." First, the standard way to drive absorbed gases out of liquids was to boil the liquids. This method could not be carried out with blood, however, because the prolonged heating coagulated it. Second, the experiments with the air pump had probably failed because "the rarefaction of the air had not been driven far enough," and because the space in which the gas was to be collected was too small to admit much of it.[19] Although convinced that his experiments with hydrogen gas had already revealed that Gmelin and Tiedemann's experiments with the air pump had led to "false conclusions," Magnus did not yet "dare" to publish his results, because they had so far yielded "no new facts," only refuted other claims. If carbonic acid existed in the venous blood and was not formed directly in the lungs, he realized, and consequently

the oxygen could not be consumed immediately in the lungs itself for the oxidation, then the removal of the carbonic acid there can only result from the fact that oxygen is absorbed and displaces the absorbed carbonic acid. The most direct proof of the correctness of this assertion seemed to me to be to show that arterial blood contains oxygen, and this seemed to be only possible if one collects the air contained in the blood and can then analyze it.[20]

The best way to do this, he saw, was to devise a method that would enable him to reduce the pressure of the air in contact with the blood much further than Gmelin and Tiedemann's air pump had done, and to provide a larger total air space into which the air could escape. At first he attempted to reach these goals by allowing the air to ascend into the empty space of a barometer tube closed at the top with a stopcock, above which another tube filled with mercury was attached. After setting up many arrangements that he later termed "useless," and trying many "useless experiments," he finally turned to a different form of apparatus that worked very well. A pear-shaped vessel closed with a stopcock at the top and open at the bottom into a dish containing mercury was inserted into the evacuation chamber of an air pump, with its upper end projecting through an air-tight seal (see Figure 1). He placed the sample of blood in the top of the vessel when it was filled completely with mercury (supported by the pressure of the atmosphere). When he exhausted the air pump until the pressure was only 1 inch of mercury, the pear-shaped vessel being otherwise entirely empty, the gases of the blood expanded into this relatively large space. Then he screwed a tube filled with mercury to the top of the vessel, opened the stopcocks so that this mercury descended to the bottom of the vessel, and let air back into the receiver of the air pump. Mercury rising into the pear-shaped vessel

Fig. 1.   Gustav Magnus's vacuum apparatus for extracting blood gases. *Annalen der Physik und Chemie* **40** (1837): 594.

pressed the blood gases into the now-empty tube attached at the top. Magnus then closed the stopcocks and removed the tube containing the gases for analysis. He could repeat the operation, attaching new tubes at the top, until he had collected all of the gas that he could extract from the blood by this method.[21]

It took Magnus until the spring of 1837 to overcome all his experimental difficulties and "deliver" finally a "complete proof for the correctness of his assertions."[22] Table I summarizes the outcome. From both arterial and venous blood taken from a horse, he was able to obtain carbonic acid, oxygen, and nitrogen. In general the quantities of carbonic acid were higher in the venous blood than in the comparable sample of arterial blood, the oxygen higher in the arterial blood. As inspection of his table easily shows, however, this was not true in every case.[23]

On April 12, 1837, Magnus returned from the veterinary school with a sample of arterial horse blood "for the last decisive experiment," and found on his table a *Habilitationschrift* written by Theodor Bischoff for a professorship at Heidelberg. Bischoff, who had previously been unknown to Magnus, had also found that carbon dioxide can be displaced from venous blood with hydrogen or with the air pump: Gmelin

FREDERIC L. HOLMES

TABLE I

	Cubikcentimeter	
Blut von ein. Pferde	125 gaben 9,8 Luft	5,4 Kohlensäur. 1,9 Sauerstoff 2,5 Stickstoff
Venöses Blut von demselben Pferde, am 4. Tage, nach der Entziehung des arteriellen aufgefangen	205 gaben 12,2 Luft	8,8 Kohlensäur. 2,3 Sauerstoff 1,1 Stickstoff
Dasselbe Blut	195 gaben 14,2 Luft	10,0 Kohlensäur. 2,5 Sauerstoff 1,7 Stickstoff
Artielles Blut von einem sehr alten, aber gesund. Pferde	130 gaben 16,3 Luft	10,7 Kohlensäur. 4,1 Sauerstoff 1,5 Stickstoff
Dasselbe Blut	122 gaben 10,2 Luft	7 Kohlensäur. 2,2 Sauerstoff 1 Stickstoff
Venöses Blut von demselben alten Pferde nach 3 Tagen aufgefangen	170 gaben 18.9 Luft	12,4 Kohlensäur. 2,5 Sauerstoff 4,0 Stickstoff
Arterielles Blut v. einem Kalbe	123 gaben 14,5 Luft	9,4 Kohlensäur. 3,5 Sauerstoff 1,6 Stickstoff
Dasselbe Blut	108 gaben 12,6 Luft	7,0 Kohlensäur. 3,0 Sauerstoff 2,6 Stickstoff
Vernöses Blut von dem- selben Kalbe nach 4 Tagen aufgefangen	153 gaben 13,3 Luft	10,3 Kohlensäur. 1,8 Sauerstoff 1,3 Stickstoff
Dasselbe Blut	140 gaben 7,7 Luft	6,1 Kohlensäur. 1,0 Sauerstoff · 0,6 Stickstoff

had written a preface declaring that he had witnessed Bischoff's experiments and was now convinced that the earlier conclusions he and Tiedemann had reached were "false." Although Bischoff had concluded that there was no carbonic acid in arterial blood, the inferences he reached concerning respiration were similar to the views at which Magnus had arrived. "As unpleasant as this incident is," Magnus wrote Berzelius 8 days later, "I know nothing better to do than to make my experiments known, and to describe the situation as it really stands." He had no intention to make a priority claim. "What is new in my work everyone will easily see, and I hope it will soon be printed."[24]

In June Berzelius replied that

This property of the blood to absorb oxygen and give off carbonic acid is an extremely interesting matter. Your dissatisfaction with the fact that someone else has found that blood exhales carbonic acid is groundless. This is a matter that has already been maintained and disputed, which is in itself of little weight, but becomes important in connection with the observation of the absorption of oxygen as a gas, without the latter forming a chemical combination with the blood. This fact, which is your own, forms the crown of these beautiful observations.[25]

I have gone into these details concerning the events leading up to the publication of Magnus's landmark paper for two reasons. First, it illustrates that already in the 1830s areas of central concern in physiology and chemistry were competitive enough so that an investigator who encountered delays in his work, or postponed publishing partial results until he could complete a longer project, risked being overtaken. Second, Berzelius's opinion of the novelty and significance of Magnus's "observations" is revealing of their impending influence in the wider scientific community. Why did both Magnus and Berzelius believe that the oxygen Magnus extracted from the blood was present in it as a gas, not in chemical combination? This account of the local context within which Magnus pursued this investigation makes clear that that view was not an inference drawn from the experimental data, but an expectation built into the structure of the problem Magnus set out to solve. He required a mechanism to explain the displacement of carbonic acid from the blood in the lungs, to replace the prevailing view that the inspired oxygen *produces* the carbonic acid there. The fact that oxygen, hydrogen, and nitrogen were all capable of driving carbonic acid out of samples of blood persuaded him that the best available explanation for the physiological discharge of carbonic acid was that oxygen entering the blood in the lungs acted as one dissolved gas displacing another from solution.

In his published discussion of the experiments Magnus pointed out that the quantities of gas extracted were "obviously only a small portion of the total gas contained in the blood." If it had been possible to use a larger pear-shaped vessel (to increase the space into which the gases were evacuated), one could "obviously have received more gas in the same time." Moreover, there were irregularities due to the fact that he had been unable to repeat the same number of times in each experiment the displacement of the gases into the tube attached to the top of the vessel. It was striking that, despite variations in the results, they "nevertheless were in agreement in the respect that the quantity of oxygen taken up was nearly the same as that of the exhaled carbonic acid." Had it been possible to extract all of the gases, he was confident that one would find in the arterial blood "exactly as much more oxygen as there was [more] carbonic acid in the venous blood."[26] Until then one could not provide certain proof that the exhaled carbonic acid is replaced by a corresponding amount of oxygen; but the results already attained "proved to satisfaction" that "the formation of carbonic acid does not first occur in the lungs." From these considerations he inferred tentatively that "it becomes very probable [from these experiments] that the inspired oxygen gas is absorbed in the lungs and is carried around in the body by the blood, so that it serves for oxidation in the so-called capillary vessels, and probably for the formation of carbonic acid."[27]

This was a restatement of the view that had been maintained ever since Lavoisier first offered it in his paper on respiration in 1790, as the main alternative to the view that the inspired oxygen forms carbonic acid in the lungs. Magnus thus implied that his investigation had "probably" resolved the controversy that had lasted for nearly 50 years. His belief that his results led to such a conclusion shows also how deeply embedded the question and its two potential answers had become in the collective thought of those who studied respiration. Knowledge of the circulation, the composition of the blood, and the respiratory gaseous exchanges had all changed substantially since the time of Lavoisier, yet the choices he had posed still defined the acceptable solutions, making it difficult to see that there might be a broader range of possibilities.

When Magnus wrote his paper on the blood gases, Berzelius was working on the ninth volume of the third edition of his authoritative *Lehrbuch der Chemie*, in which he treated animal chemistry. His discussion of "the lungs and respiration" featured a five-page summary that closely paraphrased Magnus's own description of his experiments and reproduced his entire table of results.[28] "As a consequence of these experiments," Berzelius commented, "we must change our conception of the operation of the respiratory process," and accept that oxygen is taken up and carbonic acid gas exhaled "as a result of the laws of diffusion of gases, rather than as a result of a chemical affinity combination." Berzelius perceived some difficulties in Magnus's conclusions, such as how to reconcile the displacement of carbonic acid gas by an equal volume of oxygen gas with the very unequal relative solubilities of the two gases in fluids. He also astutely identified a tension within Magnus's conception of the process:

Although oxygen gas, according to this view, is not chemically bound by the blood in the lungs, its chemical binding is nevertheless the final aim of this absorption, through which it is retained everywhere to be present and ready for the vital processes, where the metamorphoses of the materials required for life could not go on without it. We have grounds to suspect that the most essential of these metamorphoses take place in the capillary network, which completes the transition from arteries to veins, and that carbonic acid is a product of these metamorphoses.

Much remained to be investigated concerning these metamorphoses, Berzelius declared, but "much may remain forever uninvestigatible." Despite these few reservations, Berzelius clearly presented Magnus's investigation as a major event.[29]

Magnus had the good fortune that his paper was also ready just in time to be incorporated into the revisions that Johannes Müller was making for the third edition of his influential *Handbuch der Physiologie des Menschen*. In the previous edition he had discussed the difficulty of reconciling the observation that blood seemed to contain no carbonic acid with observations by Spallanzani, and experiments that he himself had carried out, which had shown that frogs kept in an atmosphere of pure hydrogen gas continue for a long time to breathe out carbonic acid. "Now," he wrote, "this riddle is solved to complete satisfaction. Both types of blood contain, according to Magnus's superb investigation, oxygen, nitrogen, and carbonic acid gas, the arterial blood more oxygen than the venous, the latter more carbonic acid gas than the arterial."[30] Müller too summarized Magnus's experiments in detail and included his table of results.[31]

Magnus's work thus quickly received from two of the most authoritative scientists of the time, with each of whom he maintained close personal connections, a strong approbation. The ready assent of Berzelius and Müller to his results and his interpretation of their significance undoubtedly provided some of the impetus that made his experiments appear soon afterward to have decided conclusively in favor of one of the "two principal theories that had been proposed concerning the chemical phenomena of respiration."

In 1844, however, Joseph Gay-Lussac, then in the twilight of a long and distinguished scientific career, challenged the consensus that seemed to him to have formed around "the recent work of Magnus on respiration." Rearranging and combining Magnus's data so as to produce average values for the oxygen and carbonic acid respectively in 100cc of arterial and of venous blood, Gay-Lussac asserted that the results actually showed less carbonic acid in the venous than in the arterial blood, contrary to what Magnus had sought to prove and to what he required for his theory. Gay-Lussac made several other criticisms, including a calculation that the quantity of oxygen that the arterial blood must contain to correspond with the quantity of carbonic acid exhaled in respiration would necessitate that oxygen be twenty-four times as soluble in blood as it is in water. Such a situation was not impossible, he thought, but implausible enough so that it was necessary to reconsider whether the union of the oxygen with the blood "occurs in virtue of an affinity which produces the combinations? Or is it simply in virtue of that which presides over dissolutions?"[32]

Magnus quickly rejected Gay-Lussac's objections. At the June 1844 meeting of the Berlin Academy of Sciences, he pointed out that in his original paper he had made clear that the conditions under which he had extracted the gases varied from experiment to experiment, yielding different proportions of the total gas in the blood. It was, therefore, not legitimate to combine the results into averages. In each *individual* experiment the proportion of oxygen to carbonic acid had been higher in arterial than in venous blood, which was all that was necessary to sustain his conclusions. Magnus appears to have been justified in his resistance to the way in which Gay-Lussac had manipulated his data. Nevertheless, he seems to have been tacitly influenced by Gay-Lussac's call to reopen the question of the nature of the union between oxygen and the blood, because he now enumerated four, rather than two, possible "theories of respiration." They were, (1) what he called the "older theory deriving from Lavoisier," that oxygen combines directly with carbon in the lungs; (2) a theory that oxygen enters a chemical combination in the lungs, but only in the capillaries enters "into other combinations by taking up carbon and hydrogen," these combinations then being decomposed by the further uptake of oxygen when the venous blood returns to the lungs, and released as carbonic acid and water. Magnus termed this the "chemical theory"; (3) the theory that the inspired oxygen is only absorbed into the blood and carried to the capillaries, where it seems to oxidize "certain substances, transforming them there into carbonic acid and water"; and (4) a theory that oxygen enters in the lungs into a chemical combination that is decomposed in the capillaries to yield carbonic acid and water.[33]

Magnus did not expressly choose between these theories at this time. Returning to his experiments, he searched for a way to extract from the blood a larger proportion of the oxygen he expected it to contain. By repeatedly shaking a sample of blood with carbonic acid gas in a closed vessel over mercury, and collecting the resulting gases again in tubes attached to the vessel, he was able to obtain a maximum quantity of oxygen of 16 percent of the volume of the blood. The fact that he could displace as much oxygen by shaking the blood with carbonic acid as the blood absorbed when he placed it in atmospheric air was, for Magnus, the "most convincing proof that the oxygen is not chemically bound to the blood, but only held in it by absorption." Now he firmly rejected the chemical theories also on grounds that they involved an internal contradiction:

One cannot grasp, if the arterial color of the blood is created by oxidation, how it becomes dark by shaking with carbonic acid, again becomes bright red through oxygen or atmospheric air and takes on again its arterial color. For carbonic acid cannot reduce the blood, and how can one imagine that blood once oxidized, without being reduced can become oxidized a second and a third time and as often as one wishes.

This contradiction did not arise in his theory, according to which the absorbed oxygen reaches the capillary vessels, where it is applied to the oxidation of particular substances, transforming these into carbonic acid and water.[34]

We may note the way in which the debate had shifted during the 8 years since Magnus had published his first experiments. Then the issue had been whether the respiratory oxidations take place in the lungs or in the course of the circulation. By 1844 Magnus could refer to the former as the "older" theory. The current question was how oxygen is connected with the cyclic changes between arterial and venous blood. The opposed chemical and physical absorption theories *both* shared the view that the processes of respiratory oxidation – whatever they might be – took place in the blood as it passed through the capillaries. This was a widely shared assumption during the 1840s. It was during this period that Jean-Baptiste Dumas and Justus Liebig proposed theories about the nature of the respiratory oxidations based on their knowledge of the chemical composition and properties of carbohydrates, fats, and proteins. Both of them presupposed that these oxidations occurred in the blood.

III

Shortly after Magnus's experiments had revealed the presence of the gases in arterial and venous blood, an equally significant development further elucidated the nature of the most distinctive constituent of the blood: its "red coloring matter." In 1838 René Lecanu, a physician and professor at the *École de Pharmacie* in Paris, published the results of research he had conducted on that substance for his thesis at the *Faculté de Médecine*. Because of the well-known difficulty in separating the clot of blood absolutely from the serum, Lecanu suspected that the coloring matter that Berzelius and other earlier chemists had described was mixed with some albumin. Searching for a way to confirm his surmise, Lecanu discovered a very simple way to separate them. When he boiled the coloring matter in alcohol and sulfuric acid, the acidified alcohol became charged with the "coloring part" and left the albumin as a residue. The close resemblance between coloring matter and albumin that his predecessors had observed was, according to Lecanu, actually due to the albumin present with it. "In a state of purity," he claimed, "the red coloring matter of blood differs essentially from albumin." He identified the distinctive chemical properties of red coloring matter by testing it with the usual solvents and reagents. Reviving a name suggested much earlier by Michel Chevreul, Lecanu proposed to name the coloring matter *hematosine*. He was also able to establish that his hematosine was contained in the globules of the blood.[35]

Lecanu's work too found a prominent place in the ninth volume of Berzelius's *Lehrbuch der Chemie*. Noting that Leopold Gmelin had come close to the discovery in 1826, when he had separated the coloring matter of the blood from its albumin in acid, but "had not pursued this important experiment further," Berzelius wrote that Gmelin's discovery had "remained unused until Lecanu . . . went a step further and separated the coloring matter from the acid bound with it." Thereby Lecanu had had the good fortune to prove that the blood corpuscles "are formed from two substances that had been confused with one another and regarded as one ... until he found a method to separate it into a red and a colorless substance." Following up the discovery himself, Berzelius worked out a "very simple" procedure for isolating the new substance. He changed its name to *hämatin*, on the grounds that this word represented the correct etymological derivation from the Greek word for blood. He also corrected Lecanu's identification of the colorless component as albumin. The colorless substance differed in its properties from both albumin and fibrin; from albumin, for example, in that it was insoluble in salt solutions in which albumin could be dissolved. Berzelius therefore made "the suggestion to give it its own name, *globulin*, from Globuli sanguinis." It is a mark of Berzelius's contemporary authority that both of his terms became permanently embedded into the nomenclature of this field. He did not take the obvious step of naming the combination of the colored and the colorless substance according to its components, but instead chose the more traditional, and less enduring name *Blutroth*.[36]

By the time Berzelius described these discoveries in 1840 in his *Lehrbuch*, the Dutch chemist Gerardus Mulder had subjected the well-known "albuminous bodies" to elementary combustion analysis, shown them all to have nearly identical combining proportions of carbon, hydrogen, oxygen, and nitrogen, and proposed that all of them contained the common radical that he called – following the suggestion of Berzelius – protein. Mulder also analyzed the newly defined globulin and hämatin. "According to Mulder's investigations," Berzelius stated in the *Lehrbuch*, "globulin obviously belongs to the protein-compounds." The elementary composition of hämatin did not resemble those of albumin, fibrin, or globulin, and Berzelius therefore concluded that "it does not belong to the animal matters that contain protein as their basis." The iron long associated with the coloring matter of blood was now shown to reside in its hämatin.[37]

As we have seen, Berzelius had already glimpsed 25 years earlier the potential significance of the iron in the coloring matter for the changes distinguishing arterial from venous blood. It did not escape his notice now, that hämatin could be central to understanding "the transformation of the blood," about which "nothing hitherto was known beyond what could be suspected from the changes that the air" undergoes in respiration. "Experiments have proven," he stated,

that in vertebrates the most important interaction between the air and the blood concerns the hämatin. Blood serum without hämatin transforms, to be sure, a portion of oxygen gas to carbonic acid gas, but extremely insignificantly in comparison to that [which occurs] when the serum is mixed with hämatin, whose brown color is thereby transformed into a very vivid red. It is believed that the oxidation state of the iron is thereby also changed, but [the colors involved are not of such a nature that the change] can be attributed to the presence of iron oxide. It is difficult to decide ... whether these color changes are accompanied by a chemical change, or consist only in a change of the physical constitution of hämatin ... Perhaps all of this will become clear, when one finally knows in what state the iron is contained in hämatin.[38]

Despite Berzelius's caution about how much remained unknown, it would appear that by the 1840s physiologists and chemists were in a position to begin to bring together what they knew about the gaseous exchanges of respiration, the gaseous contents of arterial and venous blood, and hämatin, to create a coherent outline of the respiratory processes that take place in the lungs and the circulation. Such a synthesis required two more decades to achieve. The decisive developments took place in relatively rapid succession between 1857 and 1866. We may ask whether this was a "normal" rate of progress, or whether those in the field between 1840 and 1857 were blocked by conceptual or technical obstacles that "delayed" the solution of a problem long viewed as central to the field. Talk of "delays" and "obstacles" implies some sort of reference rate of advance for which we have no real measure. Nevertheless, the rhythms of movement and apparent pauses that often attend the long-term investigations of problems in science invite attempts at explanation. I have not worked out the history of the intervening period in sufficient detail to provide a definitive answer to the question I have posed, but can offer some preliminary suggestions. We can first rule out the idea that there may have been a lack of serious attention to the problem. Intensive investigations took place during the period concerning each of the phenomena – the gaseous exchanges in the lungs, the composition of the blood and the differences between venous and arterial blood, and the physical and chemical nature of respiratory oxidations and their location. Close scrutiny of each of these developments would reveal, I believe, the complexity of the many sub-problems that required solution before one could move from provisional theories linking the properties of *Blutroth* – or the combination of globulin and hämatin – with these phenomena, toward a securely unified picture.

To take one example, the identification of hämatin was obviously a major event that invited chemists to investigate its properties and its relationships to other compounds. In 1844 Mulder showed that one could remove the iron from hämatin, producing an iron-free compound that retained the red color of the parent substance. He also prepared a chloro-derivative of hämatin.[39] The chemistry of hämatin, however, remained elusive. Nearly ten years later it was still not possible to relate its composition to any of the major classes of nutrient compounds regarded as the sources of the constituents of the body, or to any other constituents of the tissues and fluids of the body, except for the bile pigments, which were thought to be breakdown products. It was not even certain whether the hämatin obtained by chemical analysis was really a component of the coloring matter of the blood or a "decomposition product of the true blood pigment."[40]

As possible conceptual "obstacles" we may point to misfits between (1) the role that the iron in hämatin could be imagined to play in forming a chemical bond with oxygen that was reversible, (2) the contemporary idea that the substances with which oxygen combined chemically in the blood were "respiratory" nutrients oxidized in the blood to produce heat, and (3) the apparent demonstration by Magnus that oxygen and carbonic acid enter and leave the blood and are held there solely according to the laws of gaseous diffusion and the Henry–Dalton law of physical absorption. The vagueness of the alternative theories of respiration outlined by Magnus in 1844 and the contradictions in the chemical theory that he pointed out in 1845 can be seen as a mark of the incommensurability of these various sides of the problem.

Although no definitive resolution of any of these aspects of the problem emerged during the decade between 1845 and 1855, the cumulative effects of numerous investigations pertinent to it gradually altered the way in which the problem was viewed. These shifts are reflected in a lucid discussion of "theories of respiration" in another textbook that was authoritative in its time, the second edition, published in 1853, of the *Lehrbuch der physiologischen Chemie* of Carl Gottlob Lehmann. As Professor of physiological chemistry at the University of Leipzig, Lehmann was one of the first to hold a chair in a new discipline coalescing around a set of problems seen to lie permanently on the borderland between chemistry and physiology.

On the question of whether the gases were mechanically absorbed in, or chemically combined with the blood, Lehmann compromised. "Undoubtedly," he wrote, "it is both." It followed still from the experiments of Magnus, that oxygen must be taken up "in part" according to physical laws. More recent experiments by Carl Vierordt and by Gustav Valentin had undermined Magnus's earlier idea about how the uptake of oxygen in the lungs drives out the carbonic acid, but their measurements of the relations between the gaseous content of the blood and the partial pressures in the lungs indicated that the pulmonary exchanges could still be explained by the laws of gaseous diffusion. On the other hand, Lehmann thought, "it would be most astonishing" if the red blood cells, as sensitive as they are to chemical agents, were chemically unaffected by their contact with oxygen.[41] "One must, in fact, hold very stubbornly to a preconceived opinion if, in the face of the considerable differences between arterial and venous blood recently made known to us, one still insisted that the blood is completely unaltered by the oxygen with which it becomes laden in the lungs."[42] Lehmann thought it well proven that "a very large part of the oxygen absorbed in the lungs is genuinely bound to the blood corpuscles." Only a small portion of the oxygen found in blood can be accounted for by mechanical absorption in its "water." The rest "must be fixed through specific constituents of the blood; but this is only conceivable by means of a chemical attraction, no matter how weak."[43]

Lehmann's view of the way in which the oxygen produced carbonic acid represented a more fundamental revision of the opinions prevailing in the 1840s. Oxygen and carbonic acid were present not in the blood alone, he asserted, but in all the fluids of the body. "An investigation begun long ago in my laboratory," he wrote, "has given qualitatively positive proof" for what could "readily be expected a priori," that these gases penetrated into all animal tissues and the parenchymatic fluids of the organs. In the "search for the source of carbonic acid," he believed, more attention should be paid also to the respiratory mechanisms of insects, which have no blood, only "parenchymous fluids." In insects the gases pass through the trachea directly to the "elements of the organs." Moreover, the "superb" experiments of Georg Liebig (the son of the famous chemist) had shown that frog muscles isolated from the circulation absorb oxygen and expire carbonic acid. From such evidence Lehmann built up a new picture of the role of the blood: "The more probable view [is] that the oxygen is carried by the blood corpuscles in a loosely bound state to the capillaries, and transferred from there to the parenchymatous fluids, in order to bring about [in the tissues] oxidations, among the products of which carbonic acid and water appear."[44]

Here too, Lehmann left room for residues of the older view. Current investigators of nutrition, or the *Stoffwechsel* as it was called in the German literature, disagreed over whether all nutrient substances must become assimilated to the tissues before breaking down, or whether nutrients ingested beyond the nutritional needs of the animal were consumed directly in the blood. Siding with this latter conception, called *Luxusconsumption*, Lehmann allowed that not *all* of the oxidations took place in the parenchymous fluids of the organs.[45]

When Lehmann formulated these textbook overviews of problems as he saw them in 1853, he had already participated in the exploitation of a discovery that promised to make some of what he wrote obsolescent. We should be cautious, however, when we recount scientific events that appear by hindsight to be "leading toward" an outcome whose historical roots we are seeking to trace. We risk forcing into goal-directed seemingly unidirectional narratives, investigations originally undertaken for purposes unrelated to the contributions they may later turn out to have made toward that outcome. As the next episode in my story unfolds, its canonical standing in the history of the present subject will readily be recognized; but I will not describe it that way at the start, because the research in question began as an exploration of a different set of physiological problems.

In 1850 Lehmann published "some comparative analyses of the bloods of the portal vein and the hepatic vein." By comparing blood before and after its passage through the liver, he hoped to elucidate the function of that organ and the formation of the bile. He examined changes in the forms and quantities of the red and white corpuscles microscopically. Chemically he found, among other things, sugar in the hepatic vein and less fibrin there than in the portal vein. On this basis he was able to confirm Claude Bernard's recent discovery that the liver produces sugar, and to offer the explanation that the sugar arises from the breakdown of the fibrin.[46]

About the same time, a talented 22-year-old medical student named Otto Funke entered Lehmann's laboratory to do some research for his doctoral dissertation. Setting out to apply to the spleen the method his mentor had used to study the liver, Funke compared the composition of splenic vein blood with that of other venous and arterial blood. He began his research with blood samples taken from a horse at the veterinary school in Dresden and shipped in air-tight containers to him in Leipzig.[47]

The chemical aspects of his investigation provided little satisfaction for Funke. Diligently applying the best current methods to determine quantitatively the recognized constituents of the blood, he encountered mainly "difficulties and obstacles." In the end, he acknowledged, "the results of the analysis in most points were so variable as to render every conclusion futile." He devoted special attention to the microscopic examination, because a much disputed point about the function of the spleen was whether it produced or destroyed blood corpuscles. Despite his careful observations and thoughtful interpretations of their significance, he was unable here either to settle the question conclusively.[48] While conducting his microscopic studies, however, he accidentally discovered "a most remarkable phenomenon, one which," he wrote afterward, "one would sooner expect in any other animal fluid than in fresh blood." After spreading a drop of blood diluted with a little water on a slide and placing a cover slip over it, he was observing the edge of a clump of red corpuscles, where the fluid had already begun to dry, when he saw the corpuscles "suddenly change." Some of them disappeared, others became angular and elongated. Then

there formed an enormous number of embryonic crystals, too small for their exact form to be determined. They grew rapidly in length, while their cross-sectional width remained unchanged or increased only slightly, formed little prismatic balls which in part lay together like vertebra, and finally the entire visual field was covered with a thick network of needle-like crystals crossing each other in all directions.[49]

By following the development of the "intensely red-colored" crystals "with his eyes," Funke persuaded himself that they "originated from the blood corpuscles themselves." Energetically pursuing his surprising discovery of these elongated crystals, Funke ran into "the greatest difficulty in the determination of their crystallographic form." Part of the trouble was that the crystals were so thin that even with the strongest magnification he could barely make out their surfaces and edges. Nevertheless, with a microscopic goniometer he determined the approximate angles at their corners. In a second pleasant surprise, Funke found that he could readily obtain crystals also from fish blood.[50]

Lehmann's tireless student performed "a thousand experiments" to try to establish the chemical composition and properties of his crystals. All of them failed, because he could produce the crystals only under the cover slip of his microscope slide, where he could not treat them with the usual reagents. In all the solvents he tried the crystals quickly dissolved and disappeared. His many attempts to form the crystals in larger quantities also came to nothing.[51]

After publishing a first paper in which he included these observations within his broader investigation of the properties of splenic vein blood,

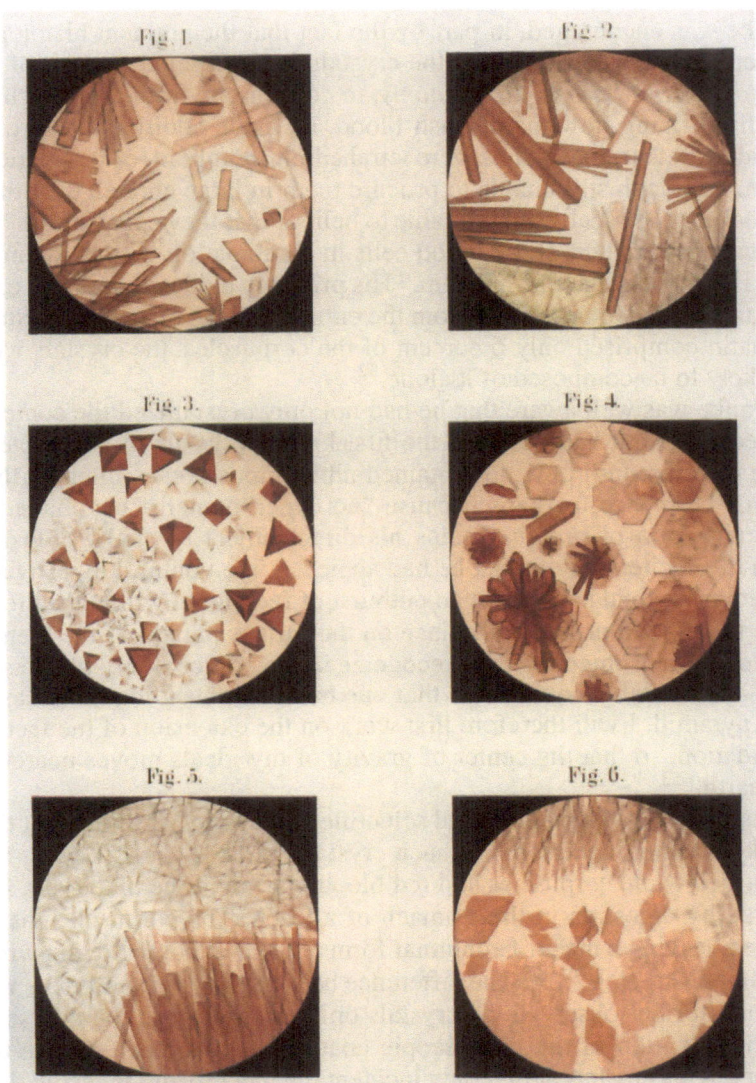

Fig. 2. Otto Funke's drawings of hemoglobin crystals: 1. from normal human blood; 2. from blood from the heart of a young cat; 3. from venous blood of a guinea pig; 4. from venous blood of a squirrel; 5. from blood from the heart of a fish; 6. from normal human splenic blood. Otto Funke, *Atlas der Physiologischen Chemie* (Leipzig: Engelmann, 1853). Table X.

Funke was encouraged, in part by the fact that the eminent histologist Albert von Kölliker included the crystals in the second volume of his *Handbook of Microscopic Anatomy*, to concentrate further research on his find. Using now mainly fish blood, he made additional efforts to decide whether his crystals were tetrahedral, rhomboid, or prismatic in form. Although still unable to procure them in large enough quantities to analyze chemically, Funke came to believe that they consisted of "the albuminous content of the blood cells in combination with hämatin" – that is, of "globulin plus hämatin." His principal evidence was that each crystal seemed to be formed from the entire contents of a red cell. Since hämatin comprised only 6 percent of the corpuscles, the crystals were unlikely to be composed of it alone.[52]

Funke was well aware that he had not only crystallized the contents of red cells, but had obtained the first known crystals of any protein. "We have up until now not obtained albuminous bodies in crystalline form," he mused, perhaps because "no experimentalist has made the crystallization of protein bodies his direct object of study." Flushed with the excitement of what he had achieved, the youthful investigator ended his second paper with an outburst of feeling: "I will not set forth here the bold hopes which I place on the further pursuit of the themes laid out here, someone might recognize in the edifice of my phantasies the same unstable equilibrium that one has while standing on the apex of a pyramid. I will therefore first work on the extension of the factual foundation, so that the center of gravity of my ideals moves nearer to the earth."[53]

As with so many unexpected scientific discoveries, it turned out that Funke was not the first to see such crystals. In 1849 Franz Leydig had described white corpuscles and red blood plasma of the blood of a fish altered by digestion in the stomach of a leech. The corpuscles disappeared, leaving a mass of columnar forms. In parentheses Leydig wrote "(Hämatin crystals?)."[54] The difference between Leydig and Funke was that the former observed the crystals only in passing, in the course of an investigation of the microscopic anatomy of the leech. Funke too observed his crystals at first only incidentally to a broader investigation. Every other aspect of his investigation being relatively unpromising, however, Funke diverted himself energetically to the further examination of what he had found.

We must again resist the temptation to identify from later developments what Funke had discovered. What his crystals appeared to be had not yet stabilized, even though the high hopes he had allowed himself while perched on the peak of his mental pyramid were, in one sense, "fulfilled in the most brilliant manner." His mentor Lehmann observed regular tetrahedral crystals in blood from a guinea pig. Funke himself made it his "first concern," in the spring of 1852, to determine whether normal human venous blood would yield the crystals. "To my greatest amazement," he reported at the beginning of June, "I succeeded on the first attempt." It was so easy that for Funke it now seemed "an incomprehensible riddle, how in spite of the innumerable times in which human blood has been treated with water, this phenomenon has never before appeared to the eyes of an observer." But not everything came easily to Funke. "To isolate the crystals and to obtain them pure for elementary analysis has up until now," he lamented, "always proven impossible. Perhaps Lehmann will succeed."[55]

Perhaps Lehmann also dissuaded Funke from his initial belief that the crystals consisted of a combination of globulin and hämatin, because Funke suggested in his final paper on the subject only that their formation was a characteristic of "globulin."[56] Lehmann did, however, succeed where Funke hoped that his teacher would. He was able to prepare crystallizable matter in large quantities from almost every type of blood. His first report to the Royal Saxon Academy of Sciences in 1852 on the nature of these crystals did not win "full belief everywhere," because it gave "some investigators the 'impression' that they were only mineral materials." Although he himself believed that the large quantities that the red cells yielded already precluded the possibility that the crystals were mineral, Lehmann intensified his investigation to convince his skeptics that the crystals were mainly organic in nature. He was able to determine their solubility properties, and to test the chemical properties in "aqueous solutions of the purest crystals." From these results he reported to the Academy, later in 1852, that "although the behavior of the crystal substance with regard to some reagents shows the most striking analogy to the albuminous bodies, it nevertheless is differentiated from every known protein compound by its indifference to other reagents."[57]

Continuing his study into 1853, Lehmann examined exhaustively the conditions under which the crystals formed, established that different types of blood yielded respectively tetrahedral, rhomboidal, prismatic, or tabular crystals, and recrystallized his crystals repeatedly in order to obtain them in purer form. Variabilities in the quantities of the iron oxide, mineral ash, and "extractive substances" that could be obtained from the material, as well as in their solubilities, persuaded him, however, that despite his efforts, the crystalline matter he had isolated "is not a chemically pure substance." Confident that it was some form of protein body, he thought it nevertheless differed from globulin, and he gave it the provisional name "haemato crystallin" in order to avoid identifying its "true constitution" until it could be isolated in "chemically pure condition."[58]

IV

For nearly two decades Magnus's investigation of the blood gases dominated the field, shaping the way in which others examined related phenomena and posing a formidable obstacle to the inferences that could be drawn from the tangible evidence that the exchanges of oxygen and carbonic acid must exert some chemical effect on the blood.[59] In 1855 the situation suddenly changed.

Magnus's contention that the gases are absorbed in the blood "according to the laws that Dalton has given for the absorption of different types of gas in a fluid"[60] had been hard to challenge, in large part because of the absence of rigorous quantitative methods for testing the Henry-Dalton law that is, the claim that the quantity of each gas absorbed in a fluid is directly proportional to the partial pressure of that gas in the atmosphere with which the fluid is in contact. In 1855 Robert Bunsen published a major paper on "The Laws of Gas Absorption," which supplied both the theoretical framework and the experimental method necessary to determine the accuracy of the law and the conditions under which it held. He defined the coefficient of absorption, α, as the quantity of gas at standard pressure (76 mmHg) absorbed in a unit volume of fluid. To measure the coefficients, he derived the formula

$$\alpha = \frac{1}{H_1} \left(\frac{VP}{P_1} - V_1 \right),$$

where H_1 is the fluid volume, V and P the volume and pressure of the gas before absorption, V_1 and P_1 the reduced volumes after the gas is absorbed into the fluid. He designed an apparatus, appropriately called an absorptiometer, that enabled him to measure these pressures and volumes with great accuracy (see Figure 3). Testing various gases with

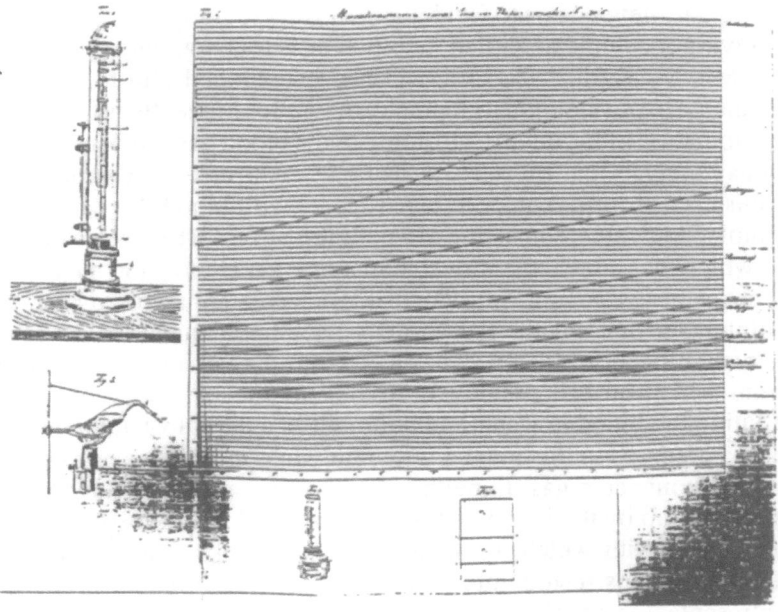

Fig. 3. Robert Bunsen's Absorptiometer. Upper left, apparatus; upper right, absorption curves for various gases in water. *Annalen der Chemie und Pharmacie* **93** (1859): Plate I.

water and other solvents, he found that, so long as there was no chemical interaction between gas and fluid, the absorption coefficient was, at a given temperature, constant within well-defined experimental limits. He discussed various applications of his method, including the analysis of the composition of mixed gases whose coefficients of absorption were known.[61]

Through most of his distinguished career Bunsen was interested primarily in inorganic chemistry, geology, and industrial applications. He seldom ventured into physiological questions. He did not, however, overlook the physiological implications of these results. At the end of his paper, he remarked, "I reserve for myself to return later to the theory ... of the processes of absorption of air in the blood that can be developed from the law of absorption."[62] Very quickly two students in his laboratory at Heidelberg began to work on that problem.

Lothar Meyer came to Heidelberg in 1854 to study with Bunsen, after studying medicine at Zürich and then at Würzberg, where he received his MD. At Zürich, Carl Ludwig had encouraged him to move from medicine to physiological chemistry. It was for Meyer therefore a natural combination of his interests to apply to the blood gases the new methods of gas analysis that Bunsen was using when he arrived. Unable to adapt Bunsens's absorptiometer to experiments with blood, Meyer devised a simpler apparatus, consisting of a cylindrical vessel to receive the blood, to which was attached a movable manometer. He lost the advantage of a constant temperature provided in Bunsen's apparatus by the water column surrounding the measuring vessel, but he nevertheless attained very significant results. Meyer placed into the vessel samples of blood from which he had previously driven off the gas they contained by diluting the blood in water and boiling it in a partial vacuum. He tested blood prepared in this manner separately in atmospheres of oxygen gas and carbonic acid gas. In neither case was the absolute quantity of the gas taken up by the blood directly proportional to the absolute pressure of the gas with which it was in contact. Meyer found, however, that he could fit his results into an equation in which the absorbed gas is represented as consisting of "two parts, one of which is independent of the pressure, whereas the other obeys the Henry-Dalton absorption law."

$$A = kh + \alpha h P,$$

where A = the total quantity of absorbed gas, α = absorption coefficient, h = volume of blood, P = gas pressure, k = a constant independent of the pressure. In the case of carbonic acid the absorption coefficient was close to that which Bunsen had established for the absorption of that gas in water (α = 1.20 at 12°C). For oxygen α was much smaller, barely outside the limits of error (the largest value he obtained was α = 0.4 at 18°C). The constant k was 0.166. The implications of these results were clear to Meyer. The exchanges of carbonic acid could be "viewed very probably as a pure absorption phenomenon." For oxygen, on the other hand, "the uptake of oxygen in the blood is essentially independent of the pressure of the free gas." It must be due, therefore, to a chemical attraction by one or more constituents of the blood.[63] "Through what part of the blood this attraction is exerted," Meyer commented,

and whether a combination in fixed atomic proportions, therefore a chemical compound in the true and strict sense is produced, I have not undertaken to investigate. But in any case, the combination is a very loose one. If the pressure of free oxygen gas is very small, or entirely removed, it decomposes. In the air pump the blood gives off its entire content of oxygen.[64]

What Meyer did not undertake, someone else working at the same time in Bunsen's laboratory was willing to try. George Harley, a "teacher of practical physiology and histology in University College, London," had already spent several years learning research methods in French and German laboratories before he came to Heidelberg. There, on Bunsen's advice, he developed a new method to test whether "the blood is able to bind the inspired oxygen chemically to itself." After shaking a quantity of blood with a portion of the atmosphere, he transferred the mixture into a closed graduated cylinder also containing atmospheric air. When the air and the blood had been thoroughly mixed, he left the cylinder on its side for several hours, displaced the gas into a tube inverted over mercury, and analyzed it by Bunsen's eudiometric methods. In each of the many such experiments he tried, Harley found that the proportion of oxygen had been reduced, compared to that of the atmosphere, whereas the proportion of carbonic acid increased. The procedures that he had followed having ruled out the possibility that these changes could have resulted from the mechanical absorption or release of gases, he concluded that the contact between the gases and the blood produced a chemical change in which oxygen was bound and carbonic acid produced.[65]

Harley's "next objective" was to determine the "number and identities of the substances through which these changes were brought about." To do so he subjected each of the "organic constituents of the blood" – fibrin, albumin, coagulum, serum, and hämatin – to the same processes that he had previously carried out with the blood itself. Each substance altered the atmosphere in the same sense that whole blood did, but to varying degrees. His conclusion, that "a part of the oxygen enters chemical combination with the various constituents of the blood," did little to narrow down the possibilities. The enthusiastic young physician recognized that his investigation of the respiratory mechanism was incomplete, and he hoped for a new opportunity to study further these "complicated changes."[66]

To have described Harley's experiments on blood gases at nearly as great length as those of Meyer may appear disproportionate to their relative importance. The comparison between these two investigations, directed at the same general problem, and carried out by two experimenters under the supervision of the same eminent chemist and teacher, provides, however, an instructive test of the extent to which we rely on hindsight to distinguish major from minor contributions in an advancing field. Both young men helped to end the twenty-year reign of the physical interpretation of the absorption and release of the blood gases. Both were able to apply rigorous methods of gas analysis that they owed mainly to the outstanding technical skill of Bunsen. Both presumably were influenced by his sense for what might prove to be effective approaches to current questions of central concern. Harley ventured more boldly than Meyer to the question of what constituent of the blood combines chemically with oxygen. Yet we judge, perhaps too easily, that Harley grasped too rapidly for the prize and failed to reach it, whereas Meyer established a fundamental prerequisite for the future identification of that constituent. Moreover, with respect to the broader implications, Harley seems to have reached backward with his view that "the circulatory system [is] a great laboratory, in which the most important combinations, conversions, and decompositions of the animal body are carried out."[67] Meyer, in contrast, appears to have reached forward with his view that "the oxidizing actions of the oxygen do not take place in the blood itself, but essentially in the ... tissues."[68] Was Meyer more sensitive than Harley to the direction of movement, or is it only in the light of subsequent trends that he appears in these ideas more progressive? Are we apt to be influenced in our appreciation of these two parallel but unequal achievements by the fact that, of the two, only Meyer afterward became a prominent scientist? What role did Bunsen play in determining which of the two would receive the more powerful project? Did he apportion the tasks according to his estimate of their respective talents? Did he anticipate in advance which approach would lead to the more significant conclusion?

The promise that Bunsen's new methods of gas analysis held for the problem of blood gases and respiration was pursued almost as quickly outside his laboratory as within it. In Paris, Émile Fernet, working in the chemical laboratory of Henri Sainte-Claire Deville at the *École normale*, saw immediately that these methods offered the precision necessary to illuminate "the influence that the diverse constituent parts of the blood exert on the absorption and the disengagement of the respiratory gases." Very soon after Bunsen published his paper on the laws of gaseous absorption, Fernet had already designed an apparatus that enabled him to apply Bunsen's basic absorption methods to experiments with oxygen and blood, experiments for which Bunsen's own absorptiometer was

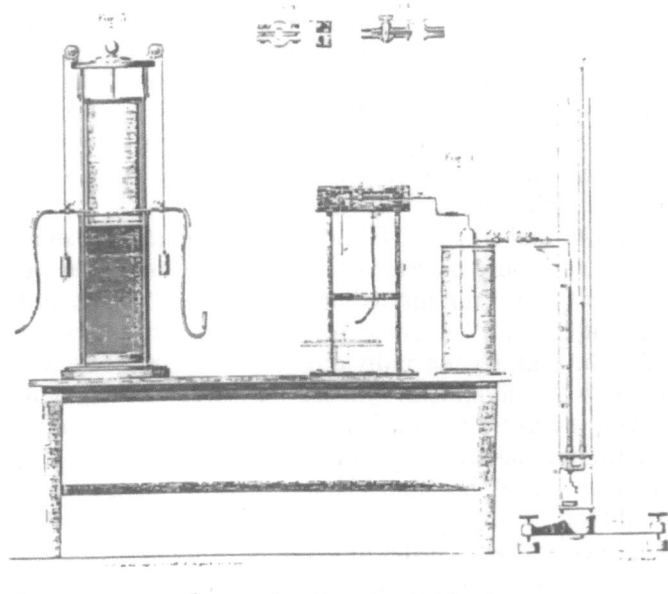

Fig. 4. Emile Fernet's apparatus for disengagement and absorption of gases by saline solutions and blood. *Annales des Sciences Naturelles (Zoologie)* 8 (1857): Plate 4.

unsuited. Fernet's apparatus separated the mixing cylinder from the mercury in an attachable manometer, so that the mercury would not be oxidized (see Figure 4). Before Fernet had time to complete his study, Lothar Meyer's paper on the same subject appeared, and Fernet lost his priority. He compensated by extending his experiments further and by pointing out that his methods were more precise than those of Meyer.[69]

Fernet studied the absorption of carbonic acid and oxygen in carbonic acid, phosphoric acid, and hydrochloric acid solutions, as well as in blood serum and whole blood. Like Meyer, Fernet found his results to fit the interpretation that one portion of each gas was absorbed according to Dalton's law, and the other portion fixed in the fluid in a manner independent of the pressure. Like Meyer, Fernet assumed that the latter portion must be chemically bound, "in a very loose combination," because the gases can be removed in a vacuum. Where Fernet went beyond Meyer was in comparing the results in whole blood and in serum. The serum absorbed only a very small quantity of oxygen, in proportion to the pressure. Whole blood fixed a large quantity of oxygen, independent of pressure. For carbonic acid, on the other hand, the results in whole blood and in serum were similar. From these contrasts Fernet inferred (1) that with respect to carbonic acid, blood behaves as the solution of mineral salts that it contains, and (2) that the oxygen is "fixed by the globules" of the blood.[70]

Mirko Grmek regards Fernet's experiments as "superior to those of Meyer, not only because of his perfected apparatus, but especially because they were not limited to whole blood." He nevertheless credits Meyer with having "executed a delicate and fastidious work with minute care."[71] In ranking the individual merits of the two achievements, I agree with this judgment. For the collective development of the field, however, it was more significant that both sets of experiments, carried out in different laboratories with different elaborations of the same basic method, arrived at the same conclusion regarding the fixation of oxygen in the blood. Both investigators were led by their results to infer that a certain constituent of the blood must combine with the oxygen in a "loose" chemical combination, from which the oxygen can be removed by sufficiently lowering the pressure of the oxygen gas with which the blood is in contact. The publication in 1857 of these papers by Meyer and Fernet moved the investigative problem into a new, more coherent phase.

To contemporary scientists attentive to current developments pertinent to the role of the blood gases in respiration, it must have been clear by now what lay ahead: to search for a chemical substance in the blood that bore the functional properties that the long series of investigations culminating in those of Meyer and Fernet specified that it must possess. Furthermore, according to Fernet's results, that substance must be found in the red corpuscles. Nor were the candidate substances difficult to identify. It was already evident that, aside from their limiting membranes, red blood cells were made up almost entirely of globulin and hämatin. Investigators of the problem were in the position of having assembled all of the pieces of a puzzle, seeing in a general way how they ought to be placed, but not yet knowing just how closely they could be made to fit together.

J.Fuchs, Strassburg, Phot.   Meisenbach Riffarth & Co. grav.

Felix Hoppe-Seyler

Fig. 5.

V

In 1846 Felix Hoppe matriculated in medicine at the University of Halle. A year later, however, he accidentally met Ernst Heinrich Weber and Eduard Weber during a mountain hike, and transferred to Leipzig to study with the two distinguished brothers, who were both anatomists and physically oriented physiologists. Hoppe also attended the lectures of the third Weber brother, the even more distinguished physicist Wilhelm, as well as the organic chemistry lectures of Otto Erdmann, and the lectures of Lehmann in physiological chemistry and pharmacology. Hoppe worked for a time in Lehmann's laboratory, but his closest associations were with the Webers. In 1850, Hoppe transferred to Berlin to complete the clinical stages of his medical education, and he entered private practice there in 1852. Preferring scientific investigation to medicine, however, he sought an academic position, and became professor of anatomy in Greifswald in 1854. Not entirely happy in this setting, he was happy to accept an invitation from Rudolf Virchow, in 1856, to become prosector of anatomy in Virchow's new Pathological Institute in Berlin and leader of the chemical laboratory that Virchow established within that institution.[72]

It was probably sometime before he came back to Berlin that Hoppe took up the problem of the quantitative determination of the "moist or dry corpuscles in the blood." Perhaps reflecting a physicalist viewpoint derived from the Webers, Hoppe attempted to measure the mass of the blood cells by means of the velocity with which they sank in defibrinated blood. His efforts were foiled by the difficulty of measuring the viscosity of the fluid.[73] After he joined Virchow's Institute, a communication sent to him by a physician in Silesia prompted Hoppe to examine red blood cells from a different point of view. A Dr. Wolff informed him that the blood of miners who had died of carbon monoxide poisoning was bright red in color. Wolff also killed rabbits with carbon monoxide and drew bright red blood from their hearts. Dispensing with the animal, Hoppe showed that ox blood placed directly in contact with carbon monoxide also becomes bright red, and that the result was the same whether he used arterial or venous blood. The color could not be further altered by shaking the blood with carbonic acid or oxygen gas.[74]

Going further, Hoppe showed that the same color change occurred when, instead of whole blood cells, he subjected an aqueous solution of their dissolved contents, *Blutroth*, to carbon monoxide gas. The effect did not occur, however, with solutions of hämatin isolated from the red cells.[75]

In a "preliminary communication," published in *Virchow's Archiv* in March 1857, Hoppe drew from these experiments a tentative conclusion "that carbon monoxide is not only absorbed as a gas, but brings about a significant alteration of the *Blutroth*, which cannot be [further] changed by introducing oxygen, or by putrefaction." This statement describes his result in straightforward, empirical terms, but as he completed his brief discussion, Hoppe shifted his language and attached to the immediate conclusion a deeper level of interpretation:

The behavior toward carbonic acid, atmospheric air, a vacuum, and heating, of *Blutroth* modified by carbon monoxide showed that this change introduced into the hämato-globulin is a very definite one, not merely conditioned by an absorbed gas, and that hämatoglobulin treated with carbon monoxide has lost its property of existing in a venous and an arterial condition. It is, accordingly, in the highest degree probable, that hämatoglobulin changed in this way is no longer capable of fulfilling its very important function as the carrier of oxygen for the blood and the entire organism.[76]

There is much to ponder in this passage, stated so succinctly that neither its sources nor its implications are fully visible. How deliberate and how significant was Hoppe's shift from the term *Blutroth* to "hämato-globulin"? Did he think of these as synonyms, hämatoglobulin merely expressing directly the view already stated 17 years earlier by Berzelius that *Blutroth* is a combination of hämatin and globulin? Or did Hoppe introduce the term hämatoglobulin to represent a more theoretical conception of a chemical compound underlying the tangible *Blutroth*?

More striking is Hoppe's reference to hämatoglobulin as the "carrier of oxygen." As we have seen, the current state of the problem was that investigators were still trying to close in on the substance in the blood that combines with oxygen. Had Hoppe reasoned, perhaps on the basis of his own experiments, that hämatoglobulin was the substance for which the others were looking? If so, why did he not present his conclusion as a novelty? Perhaps he was merely giving expression to a shared expectation in the field that had not yet coalesced in the published literature as the accepted view. Finally, his phrase "carrier of oxygen" has a modern ring that may tempt us to believe that he understood hä-matoglobulin to transport oxygen from the lungs to the tissues. Yet the rest of his sentence, "for the blood and the organism," reminds us that his words were used in a context in which the question of where the respiratory oxidations occur was still open.

Hoppe remarked that he intended to continue his study of the chemical action of carbon monoxide. In January 1858, he did publish a report of his examination of the blood of 5 persons who had suffered carbon monoxide poisoning (4 of them fatal), but he did not substantially advance his conception of the underlying process.[77] As is well known, Claude Bernard had already arrived independently at a similar explanation of the mechanism of carbon monoxide poisoning. I shall not pursue further here the history of this subject, about which Mirko Grmek has given a beautiful, detailed account.[78]

In October 1857, in a paper summarizing his continued efforts to determine the weight of the red blood cells, Hoppe referred to hämatoglobulin as "the chief constituent of the cells," and as the agent that exerts the "condensing effect" of the cells on oxygen. "The quantity of hämatoglobulin contained in a blood can suffice," he asserted, "to evaluate the vital capacity of the blood."[79] That Hoppe nevertheless had at the time no better knowledge of the nature of hämatoglobulin than anyone else then had, is indicated by a description that he provided in a handbook for practical laboratory instruction in pathological-chemical analysis that he published in 1858:

Haematoglobulin. Blutroth. This combination of haematin with globulin, that is supposed to be contained in the red blood cells, has not yet been isolated, its properties as well as its existence are therefore still too doubtful to say anything here about it. If it does exist, then to it alone must be attributed the capacity to condense oxygen and thereby become bright red, whereas the true pigment of the body would be dark red brown, in thin layers yellow green, just as are alkaline haematine solutions.[80]

Hoppe expressed similar skepticism about a substance that one of his former mentors had defined, Lehmann's haematocrystallin. "This substance," Hoppe's textbook stated, "is supposed to be contained in all vertebrates and to possess the capacity to separate out in beautifully formed crystals"; but the different crystal forms obtained from the blood of different animals, the complexity of albuminous substances, and other factors "speak against this capacity to crystallize" and suggested that the question of "the existence of this matter be left to future investigation." Lehmann responded in the edition of his *Lehrbuch* that appeared shortly afterward, with several implied criticisms of Hoppe's views, drawing from the combative younger man an angry rebuttal in *Virchow's Archiv* in 1859.[81]

In 1861 Hoppe was appointed extraordinary Professor of applied chemistry at the University of Tübingen, where he inherited a primitive laboratory that he quickly turned into an active center of teaching and research.[82] There he began to investigate intensively the colored substances in the blood whose very existence he had two years earlier questioned. Because "fine chemical methods for the recognition of colouring matters and their changes were very lacking,"[83] he turned to a mode of investigation that had until recently been pursued mainly by physicists. The principle was based on the observation made early in the century by Joseph von Fraunhofer that sunlight passed through a slit and a prism produced a series of fine dark lines, visible through a telescope, and characteristically spaced along its color spectrum. During the next four decades physicists had great trouble interpreting these lines and their relation to the distinctive emission spectra of certain incandescent substances. The landmark experiments published by Bunsen with the physicist Robert Gustav Kirchhoff in 1860 opened a new era by demonstrating that the emission spectra obtained by burning very pure substances were simple and characteristic, and that they coincided with the absorption lines of the same substances. Besides advancing the physical understanding of the phenomena, Kirchhoff and Bunsen's publication transformed spectroscopy into a delicate instrument of chemical analysis.[84]

It is not evident whether or not the "sensation that these observations of Kirchhoff and Bunsen aroused"[85] among physiologists as well as chemists induced Hoppe to apply spectral analysis to his study of the coloring matter of the blood. It was the emission spectra that had caused most difficulty before Kirchhoff and Bunsen. Hoppe needed to rely only on the absorption spectra of colored solutions through which sunlight passed, and analyses of this type had already been carried out successfully before 1860 by physicists such as Johann Müller on other organic coloring matters, such as indigo.[86] Hoppe referred to such prior studies, but not to Kirchhoff and Bunsen. He employed what he described as

the known combination of apparatus: a heliostat throws light through a slit into a darkened room ... onto an achromatic lense, in whose focal point the slit is placed, from there onto a prism ... One allows the spectrum so created to pass through the solution to be investigated, which is contained in a narrow vessel with parallel sides, and observes it either with a telescope, or with the naked eye on a strip of white paper.[87]

The instrument Hoppe employed may have looked like the contemporary illustration shown in Figure 6. Following customary practice, he used alphabetically labelled Fraunhofer lines as a reference scale on which to locate the absorption bands he saw. Here is a contemporary

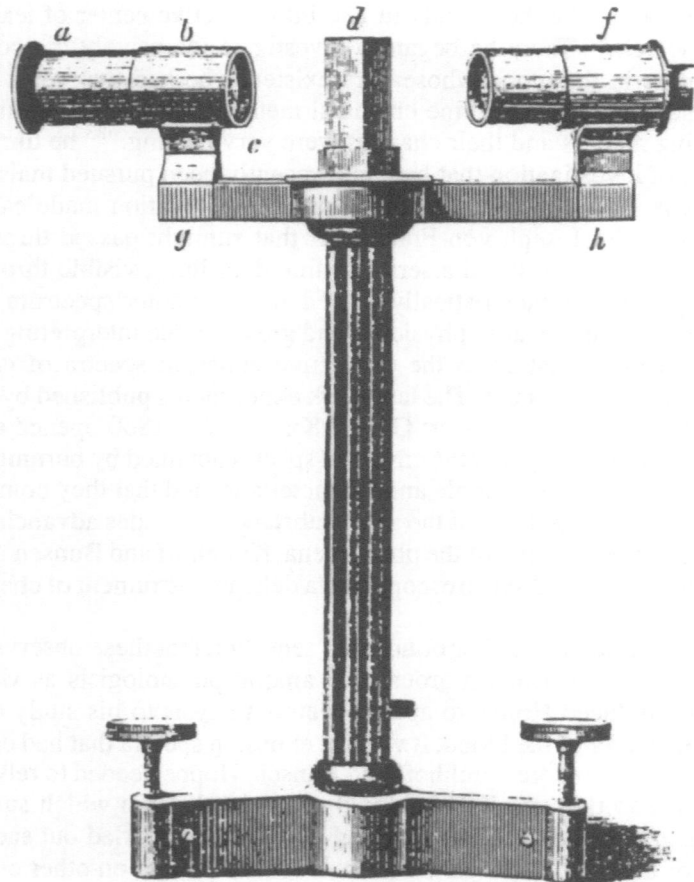

Fig. 6.   Spectroscope, a, b. tube for incoming rays; c. lense for incoming rays; d. prism; e, f. telescope on rotatable arm h. From G. Valentin, *Der Gebrauch des Spektroskopes* (Leipzig: Winter, 1863), p. 18.

representation of them.[88]

When Hoppe placed a dilute solution of blood in water in the spectroscope, he saw, "two distinct lines in the yellow and green. Both lines lay between the Fraunhofer lines D and E." He could observe the lines also with "undissolved blood cells" by passing the rays through a thin layer of blood on a microscope slide. Testing the blood in solution with various gases, and with the reagents ordinarily used to examine the chemical properties of albuminous substances (proteins), he found that only those agents that coagulate or destroy such substances caused the dark lines to disappear. Globulin separated from the coloring matter did not produce the lines. From all these observations Hoppe concluded that "one would have to accept that the same substance that gives the contents of blood cells their red color also produces that absorption [spectrum];" and that the substance must be decomposable into an albuminous substance and hämatin. "Without doubt," he asserted, "this substance is the same one that forms Funke's crystals."[89]

His new observations had evidently stilled Hoppe's earlier doubt that these crystals, with their great variety of forms, could represent a real, well-defined chemical substance (although the fact that he did not in this paper refer to the substance as haematoglobulin may suggest some embarrassment about the reversal that had obviously taken place in his attitude since writing the entry for that term in his analytical textbook). His reference to Funke's crystals without mention of haematocrystallin was an intended further rebuff to Lehmann's view that the crystals did not represent a chemically pure substance, and to unsuccessful efforts Lehmann had later made to purify the crystals until they would become colorless. Their color was, from Hoppe's current perspective, integral to the pure crystallizable substance that he could now identify within the blood cells by its distinct spectral lines.

There was one statement in his report of his observations that Hoppe passed by without further comment. "Arterial as well as venous blood shows both lines. Sustained treatment of the blood solution with carbonic acid does not alter them."[90] These observations ought to have disturbed him. Having identified the substance that gave rise to the dark lines with the substance that gave blood its red color, how could he reconcile the fact that his spectrally defined substance was unaltered by the change from arterial to venous blood with the long-standing expectation that a chemical change underlies the difference in the color of blood in these two conditions? His silence on the question may indicate that he had no answer for it. One of the pieces of the puzzle was now more sharply outlined than ever before, but Hoppe gave no hint that its shape might not fit the space reserved for it.

Thirty years ago Robert Merton stated the hypotheses that the dominant pattern in science is that discoveries are multiple – that is, are made independently by more than one investigator – and that even those which appear on the surface to be single are in principle multiple. The latter effect Merton attributed to the fact that normally, when an investigator who has made a discovery independently finds out that someone else has already done so, he reports his own work merely as a confirmation.[91] Merton's generalization applies aptly to the discovery of the spectral lines of the coloring matter of blood. In Bern, Switzerland, the physiologist Gustav Valentin enthusiastically took up spectroscopy in the wake of Kirchhoff and Bunsen's publication. The blood, Valentin quickly saw, "belongs among the most fruitful fluids for spectral investigations." He made many observations on the spectral bands of arterial and venous blood, and identified in dilute solutions the same two sharp lines that Hoppe found, he too observing them in both types of blood. In more concentrated solutions Valentin observed, when he viewed the spectrum through an enlarging spectroscope, some differences between arterial and venous blood in the extent of a weaker band that appeared under those conditions in the red. Despite these distinctions, the main conclusion of Valentin, like that of Hoppe, was that the bands "appear in bright red as in dark red blood."[92]

Valentin claimed that he had "made these observations" and written down an account of his results when it came to his attention that Hoppe had already seen the spectral lines and investigated them "from a chemical point of view" more fully than he himself had done. Consequently, "what I believed to have found as new" Valentin did not publish in a journal, but included in 1863 in a handbook on the use of the spectroscope that he put out, and "only as confirmation of the results that the predecessor unknown to me had already obtained."[93] Hoppe later wrote, rather ungenerously, that "my statements about the optical properties of the coloring matter of blood ... were confirmed by Valentin,"[94] without mentioning the independence of the investigation that Valentin had conducted a little too late to share in the credit for a major discovery.

By the time Hoppe published a "second communication" on "the chemical and optical properties of the coloring matter of the blood," in the first number of *Virchow's Archiv* for 1864, he had advanced considerably in his study of this substance. In particular he had investigated very thoroughly what he had earlier called "Funke's crystals." To obtain them in a state of great purity, he recrystallized them as often as 5–

6 times, a condition in which, contrary to Lehmann's expectation, they retained the red color of the blood corpuscles. Now placing solutions made from this crystalline material in the spectroscope, he found that it too absorbed "with particular intensity in green and yellow light." In dilute solutions the two sharp bands appeared. There was now, however, a significant distinction to report:

If a solution is freed from O_2 by means of CO_2 or putrefaction, it exerts the least effect on the least refracted light of the solar spectrum, as far as the Fraunhofer C line. If the solution is shaken with air, the absorption of light from line C until line D is very much diminished. The solution allows the light to pass through in rather significant intensity. As this part of the spectrum already possess great light intensity, whereas the part of the spectrum from A to C is very weak, these differences allow us to explain the brightness of color and transparency of arterial blood in contrast to the darkness of venous blood.[95]

This description Hoppe introduced with the misleading phrase, "as had already been stated in the first communication," thus masking the critical differences between his observations in 1862 and his present ones. Then he had produced his absorption spectrum with "blood dissolved in water," now with solutions made from the crystallized coloring matter of blood. Then he had found no difference between the spectra of venous and arterial blood; now he had found a distinction between the spectra of the coloring matter containing and that free of oxygen, which offered him an explanation for the difference between venous and arterial blood. Hoppe had surmounted the main obstacle that he had encountered in his quest to identify the long-sought "oxygen carrier," yet he did not even acknowledge that that obstacle had ever come in his way.

Hoppe showed also that the crystalline coloring matter can be decomposed, by means of caustic alkalis, into hämatin and globulin, thereby confirming the prediction about its composition that had been made by Berzelius a quarter century earlier. Finally, Hoppe gave the substance a new name. "To avoid confusion," he wrote, "I name the coloring matter of blood hämatoglobulin or hämoglobin."[96] Henceforth dropping the more cumbersome choice, he always afterward used the name which has been attached to the substance ever since. Hoppe also gave himself a new name. Adding to his surname the name of the relative who had raised him after the death of his own parents,[97] he published this paper under the name Felix Hoppe-Seyler.

Later in 1864, Hoppe-Seyler published a "third communication," which included the latest refinements in his investigation of hemoglobin. These included new elementary analyses of haematin and globulin, and further descriptions of the crystals, which clarified the differences between their properties in a dry state and with water of crystallization. For our purposes the most significant addition he made was to show that:

The bright red color of the blood crystals of dry dog or goose blood depends on their content of loosely bound oxygen. To be sure, the oxygen content of the crystals, which can be removed by warming in a vacuum, is small, and the smaller the drier the crystals are, but they do contain loosely bound oxygen as long as they remain undecomposed.

The amount was too small to account for "the quantity of oxygen absorbed by the circulating blood," but was consistent with the observation that the "uptake of chemically bound oxygen through hemoglobin is to some extent a function of its concentration or its water content."[98] As these passages suggest, Hoppe-Seyler was in 1864 close to being able to correlate the chemical properties of hemoglobin with its expected physiological functions, but had not yet achieved a complete fit. By then, moreover, he no longer had the field to himself.

On June 16, 1864, a communication by the British physicist George Stokes, "On the Reduction and Oxidation of the Colouring Matter of the Blood," was read at a meeting of the Royal Society of London. It began:

Some time ago my attention was called to a paper by Professor Hoppe, in which he has pointed out the remarkable spectrum produced by the absorption of light by a very dilute solution of blood, and applied the observation to elucidate the chemical nature of the colouring matter. I had no sooner looked at the spectrum, than the extreme sharpness and beauty of the absorption-bands of blood excited a lively interest in my mind, and I proceeded to try the effect of various reagents.[99]

Stokes found it "easy to verify" Hoppe's observations with a solution of the colouring matter obtained by merely allowing sheep or ox blood from a butcher to clot, cutting the clot into small pieces, and extracting the pieces in water. Then he simply placed the solution in a test tube behind a slit, and viewed it through a prism. But Stokes did far more than to confirm in so playful a style what Hoppe had reported in his first paper on the subject. Instead of comparing the absorption of arterial and venous blood as Hoppe (and Valentin, of whom Stokes was not aware), had done, he treated his coloring matter with "reducing agents" (such as protosulphate of iron), treated so as to maintain a slightly alkaline solution and avoid changes that acids produced in the substance.[100] The color changed almost instantly to a purple red. "The change of colour, is striking enough," he wrote,

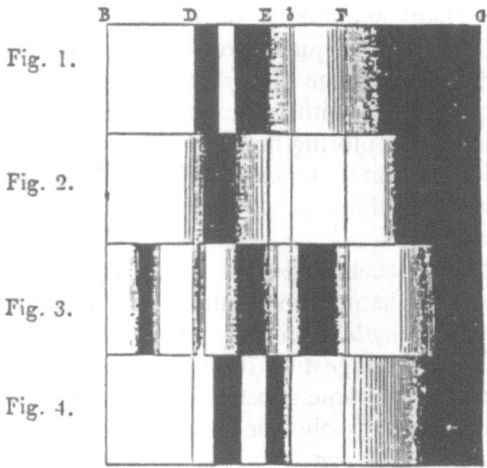

Fig. 7. G.G. Stokes's Absorption Spectra: 1. Aqueous extract of ox blood clot; 2. same in alkaline solution (reduced condition); 3. in acidic solution of blood (decomposed coloring matter); 4. haematin in alkaline solution. *Philosophical Magazine* (1864): 393.

but the change in the absorption spectrum is far more decisive. The two highly characteristic dark bands seen before are now replaced by a *single* band, somewhat broader and less sharply defined at its edges than either of the former and occupying nearly the position of the bright band separating the dark bands of the original solution . . .

If the purple solution be exposed to the air in a shallow vessel, it quickly returns to its original condition, showing the two characteristic bands the same as before . . . If an additional quantity of the reagent be now added, the same effect is produced as at first, and the solution may thus be made to go through its changes any number of times.[101]

With remarkable ease Stokes was able to achieve another demonstration that had eluded Hoppe, showing that hämatin too yields a different characteristic absorption spectrum when reduced than when "oxidized by shaking up its solution with air." Stokes was able to explain the fact that Hoppe had found the same two sharp absorption lines in venous and arterial blood by showing that the former had probably contained enough unreduced coloring matter to produce them.[102] The conclusions Stokes drew seemed crystal clear:

the colouring matter of blood, like indigo, is capable of existing in two states of oxidation, distinguishable by a difference of colour and a fundamental difference in the action on the spectrum. It may be made to pass from the more to the less oxidized state by the action of suitable reducing agents, and recovers its oxygen by absorption from the air.[103]

Moreover, the "facts which have been adduced" seemed to him sufficient to settle the long "disputed point whether the oxygen introduced
into the blood in its passage through the lungs is simply dissolved or
is chemically combined with some constituent of the blood." Having
shown that there is a coloring matter in blood "capable of undergoing
reduction and oxidation," Stokes maintained, "we have all that is necessary to account for the absorption and chemical combination of the
inspired blood."[104]

Unaware of the recent paper in which Hoppe-Seyler had named
the coloring matter hemoglobin, Stokes proposed the names "*scar-
let cruorine* and *purple cruorine* respectively" for its two states
of oxidation.[105] His suggestion obviously failed to dislodge Hoppe-
Seyler's choice for the same substance. Stokes also did not know that
Hoppe-Seyler had already observed a distinction between the spectra of
hemoglobin containing oxygen and that from which the oxygen had been
dislodged. The two spectra observed by Stokes were, however, so much
more decisive that he is often given credit for the discovery of the spectroscopic changes undergone by hemoglobin.[106] Hoppe-Seyler found it
difficult to concede that the British physicist who had followed so blithe-
ly in his footsteps had so easily surpassed his own more painstakingly
acquired observations. In the same passage, partially quoted above with
respect to Valentin, Hoppe-Seyler attributed to Stokes only the confirmation of his own statements "later in a more detailed way," and barely
mentioned a little further on Stokes's work "on the reduction of the
coloring matter of the blood."[107]

That Stokes was able to upstage Hoppe-Seyler at a crucial point in the
demonstration of the functionally critical properties of hemoglobin does
not mean that he had wrested the leadership in this field from the tena-
cious German physiological chemist. If the vision of the British physicist
was piercing on this point, it was also the narrowed gaze of a scientist
seizing on a problem far from his own discipline. The quoted passages suggest that Stokes was unfamiliar with the distinction, painfully
acquired among physiologists and chemists, between oxygen "loosely
bound" to the coloring matter and the oxidations supposed to occur
in the capillaries or tissues. It was Hoppe-Seyler who finally reached
an overview that integrated the chemical properties of hemoglobin, the
constitution of the blood, and the problem of respiratory oxidations as
it had developed since the time of Lavoisier.

In an article probably written about 1866, Hoppe-Seyler reviewed the ideas about "the oxidations of organic matter in the living animal body" that had been pursued since "the famous experiments of Lavoisier." The place where these oxidations occur had been sought in the lungs, until Magnus and others proved that the arterial blood contains much oxygen, the venous blood less oxygen. Afterward the open question became "whether the oxygen lost [during the change from arterial to venous blood] is given off into the organs and consumed there, or whether instead easily oxidizable substances from the organs pass into the blood and are here oxidized." It was, Hoppe-Seyler asserted, the new knowledge of the properties of hemoglobin and of hemoglobin containing oxygen, which he now called oxyhemoglobin, that enabled one also to gain a clearer understanding of the oxidative processes. Having juxtaposed the alternative views of these processes in unusually lucid language (although not with perfect historical accuracy), he set out to determine experimentally which view was correct.[108]

Hoppe-Seyler first tested whether oxyhemoglobin was able to oxidize substances such as sugar or uric acid that would be most likely to be oxidized in the blood, and found that it could not. Oxyhemoglobin was not to be regarded as a strong oxidizing agent. To ascertain whether "oxidations take place in the blood itself," he tried a variety of experiments in which he tied off portions of an artery full of blood in a living or dead animal, sometimes replacing the section of artery with a glass or rubber tube, then opened it after 2 hours and looked to see whether the blood had turned dark. From these experiments he inferred that arterial blood does lose a portion of its oxygen during its passage through the arteries, but that this process is "dependent on the contact of the blood with the living vessel walls." "The loss of oxygen from the oxyhaemoglobin of arterial blood is not caused by the oxidation of substances that diffuse from the vessel walls into the blood, but the oxygen itself is given off into the walls." Consequently,

There is now no basis left for accepting that under normal conditions oxidative processes take place in the blood of vertebrates.

On the contrary, the properties of hämoglobin, as well as the preceding experiments, prove definitively that oxyhaemoglobin, and through it the arterial blood, are only oxygen carriers, that they give off oxygen to the vessel walls, that in the arterial walls as well as in the muscles oxidative processes go on that maintain the organs free of oxygen. Only in this way is it conceivable that the oxygen is given off from oxyhaemoglobin into the organs.[109]

In this statement Hoppe-Seyler had formulated clearly the basic functional definition of hemoglobin that has, with numerous refinements, endured to our own time. We have, therefore, reached a point of closure in the story we have been following. No story, however, scientific or otherwise, ends quite so cleanly. Hoppe-Seyler's simple experiments were not compelling enough to settle the controversy over whether oxidations take place in the tissues or in the blood. Studies of the diffusion of gases between the blood and the tissues in Carl Ludwig's laboratory during the 1860s revived the idea that the oxidations must occur in the blood. The final proof that they are tissue phenomena is generally held to have come only during the 1870s with experiments by Edward Pflüger, particularly one in which a student in his laboratory showed that the respiratory activity of a frog continues unabated after its blood is replaced by saline solution.

I believe that no single set of experiments decided this question, but that the gradually deepening influence of the cell theory on physiological thought eventually made it appear self-evident that all such fundamental vital activities occur within the cells. When Magnus's experiments led him and others to think that the oxidations occur in the course of the circulation, Theodor Schwann had not yet formulated the broad generalization that the cells are the vital units in which respiration and nutrition are centered. During the 1840s and 1850s the cell theory had already spawned the sub-field of histology and transformed embryological investigations, but had not widely influenced the way in which experimental physiologists and chemists viewed processes such as respiration and the *Stoffwechsel*. By the 1860s a few leading physiologists, such as Ernst Brücke and Claude Bernard, were reorienting their approach to vital processes around the idea that cells are the "elementary organisms." By the 1880s this was a nearly universal viewpoint. It was well summarized by Hoppe-Seyler himself in his textbook of physiological chemistry in 1881:

Only the completely subordinate, preparatory chemical processes of life are completed in fluids, all of the more important ones happen on and within soft imbibed masses that are neither really solid bodies nor truly fluids. The cells are the chemical instruments and workshops, they are the chemical organs of the organism.[110]

If this perspective relegated hemoglobin to a subordinate, preparatory role, it was nevertheless, for vertebrates, an extremely important one. It was a role which the chemical properties that had been established about it by the 1860s fitted it to play exceedingly well, and a role whose understanding has been further enhanced by everything that has been learned since then about this remarkable molecule.

VI

I want to end by reflecting briefly on some of the more general patterns of scientific change manifested in the developments that I have traced. We may, if we choose, fix on innovations in experimental technique as the driving forces that both initiated and closed the several epochs into which the early history of hemoglobin can be divided. The method that Magnus invented to produce a stronger vacuum enabled him to produce the blood gases that had eluded others. The precise gasometric methods devised by Bunsen were capable of demonstrating for the first time that oxygen could not be held in the blood, in the quantities that were present, by physical absorption. The application of spectral analysis afforded a more positive way to identify the two physiologically critical forms of hemoglobin than had previously been possible. These inventions, each brought to the physiological problem by individuals, or from fields, that had stood outside its prior development, reshaped the problem and ultimately enabled its resolution.

That is one side of the picture. Each of these transitions occurred, however, within the boundaries of mental formulations of the range of possible solutions that preceded and shaped the experimental ventures themselves. The experimental departures did not so much create new theoretical perspectives as refine and adjudicate between the previously conceived alternatives. The changes they wrought were superimposed on a gradually evolving problem whose continuity over the long era connecting Lavoisier in the 1790s to Hoppe-Seyler in the 1860s is as evident as are the mutations. Even the choices between apparently discrete alternatives were less abrupt than logic would lead us to expect. Between the time in which the prevailing view was that the respiratory oxidations took place in the blood and the time when it was recognized that they took place in the cells of the tissues, there was a prolonged interval in which it was acceptable to believe that they occurred in both places. Between these two alternatives themselves lay the intermediate opinion that they occur in the parenchymatous fluids of the tissues. The choice between physical dissolution of the gases in the blood and their chemical fixation turned out not to be absolute, because it was possible, and in the case of carbonic acid turned out to be important, that both processes were involved.

Another theme illuminated in this story is the various forms of interdisciplinary interaction that can impinge on the investigation of a multidimensional problem such as the functional properties of hemoglobin. The interplay between physiology and chemistry, particularly organic chemistry, was sustained, as the problem resided permanently on the borderland linking these two fields. In general the problem did not lie on the intersection between physiology and physics in the nineteenth century, but opportunities did arise for occasional critical interventions by physicists such as Stokes – or by inorganic chemists such as Bunsen – who ordinarily worked in areas remote from physiology.

Finally, I would like to ponder the meaning of the foregoing account as a story. Behind every scientific event that we count as a landmark discovery, or other memorable change lies such a story. As I constructed this narrative I was acutely conscious that it is both natural and artificial. I clearly did not "invent" the story, which has been told many times before. The significance of the events included within it is obviously determined in part by the outcome; but these were not events unrelated in their own time, their apparent unity only imposed on them by retrospective association. The participants in the earlier stages of the story were acting with purpose, heading deliberately toward a solution whose details they could not foresee, but with expectations that bore sufficient resemblance to the eventual resolution to assure us that the successive generations who took up the quest do belong in the same story. The teleological character of the narrative is, at least in part, an extension of the intrinsically teleological character of individual and collective human action.

On the other hand, I have been equally aware that the coherence and transparency of the story has been maintained by ruthlessly shearing away intimate connections between developments that fit into it and developments that belong to other stories. For example, most of the investigations I have summarized gave as much attention to the place of carbonic acid in the blood and its role in respiration as to oxygen. In doing so, they examined the resemblances between the properties of the blood and the interactions of carbonic acid, bicarbonates, and carbonates in aqueous solution. I have given only very limited treatment to the problem of carbonic acid, because a more balanced account of the relation of the two gases, both in the blood and in the respiratory process, would be a diversion from my focus on the history of hemoglobin during this period. For later periods, of course, when the story becomes more complex, carbonic acid has to be brought back more fully into the picture. I have totally eliminated the prominent part that nitrogen gas played in investigations of the blood gases in the 1830s and 1840s, because the problem that concerned physiologists then – whether nitrogen gas is absorbed or released in respiration – had disappeared before the end of the story and contributed little to the characterization of hemoglobin.

Similarly, it has been necessary to overlook aspects of Hoppe-Seyler's investigations that were integral to the work that I have discussed, in order to bring the story to an end. The same experimental methods with which he was resolving the long-standing problem of the functional properties of hemoglobin were simultaneously opening up new problems, such as the nature of the compounds that arose from the decomposition and other modifications of hemoglobin and its two major components. The same spectral methods that allowed him to characterize hemoglobin and oxyhemoglobin were also leading him to examine the relations between hemoglobin and the bile pigments. Stories must end, but scientific investigation finds no true resting places.

To divide the ongoing stream of scientific investigation into manageable stories, therefore, inevitably sunders much that belongs together. As in all forms of narrative, the process creates fictions as well as meanings – limitations of vision along with its vistas. But how else can we understand the enormous complexity of scientific change than by picking out, from the network of endless reticulations that bind together the numberless thoughts and operations of daily scientific life, a few avenues of continuity narrow enough for us to follow? A more difficult question is: how can we integrate such stories into the larger scale movements of a subfield or field of science without losing the texture of individual human activity that allows the history of science to come alive? That is a question for which I have no present answer, and I suspect that as historians of science we have no collective answer.

Department of History of Medicine
Yale University
U.S.A.

NOTES

[1] See, especially, John T. Edsall, "Blood and Hemoglobin: The Evolution of Knowledge of Functional Adaptation in a Biochemical System," *Journal of the History of Biology* 5 (1972): 205–257; Joseph S. Fruton, *Molecules and Life: Historical Essays on the Interplay of Chemistry and Biology* (New York: Wiley-Interscience, 1972), pp. 277–286; Joseph S. Fruton, *Contrasts in Scientific Style: Research Groups in the Chemical and Biochemical Sciences* (Philadelphia: American Philosophical Society, 1990), pp. 77–78; Mirko D. Grmek, *Raisonnement expérimental et recherches toxicologiques chez Claude Bernard* (Genève: Librairie Droz, 1973), pp. 83–207; Claude Debru, *L'esprit des protéines* (Paris: Hermann, 1983), pp. 131–188.

[2] Felix Hoppe-Seyler, "Beiträge zur Kenntniss der Constitution des Blutes," in *Medicinisch-chemische Untersuchungen* (Berlin: August Hirschwald, 1866–71), pp. 133–140.

[3] Frederic Lawrence Holmes, *Lavoisier and the Chemistry of Life* (Madison: University of Wisconsin Press, 1985), p. 462.

[4] Everett Mendelsohn, *Heat and Life: The Development of the Theory of Animal Heat* (Cambridge, MA: Harvard University Press, 1964), pp. 166–176.

[5] A.F. Fourcroy, *Elements of Chemistry*, 5th ed. (Edinburgh: Mundell, 1800), vol. 3, pp. 262–270.

[6] Holmes, *Lavoisier*, pp. 462, 475–478.

[7] J. Berzelius, "Sur la composition des fluides animaux," *Annales de chimie* 88 (1813): 26–72.

[8] *Ibid.* 39–56.

[9] *Ibid.* 55.

[10] *Ibid.* 71.

[11] Gustav Magnus, "Ueber die im Blute erhaltenen Gase, Sauerstoff, Stickstoff und Kohlensäure," *Annalen der Physik und Chemie* 40 (1837): 583.

[12] L. Gmelin and F. Tiedemann, "Versuche über das Blut," *Ibid.* 31 (1834): 289–290.

[13] Frederic Lawrence Holmes, *Claude Bernard and Animal Chemistry* (Cambridge, MA, 1974), pp. 149–159.

[14] Gmelin and Tiedemann, "Versuche," 289–303.

[15] Magnus, "Blute enthaltene Gase," 584–585; George B. Kauffman, "Magnus, Heinrich Gustav," *Dictionary of Scientific Biography*, C.C. Gillispie, ed. (New York: Charles Scribner's Sons, 1970–1980), vol. 9, pp. 18–19.

[16] *Aus Jac. Berzelius and Gustav Magnus' Briefwechsel*, Edvard Hjelt, ed. (Braunschweig: Vieweg, 1900), p. 96.

[17] Magnus, "Blute enthaltene Gase," 589.

[18] *Berzelius und Magnus' Briefwechsel*, p. 103

[19] Magnus, "Blute enthaltene Gase," 592–593.

[20] *Berzelius und Magnus' Briefwechsel*, p. 123.

[21] *Ibid.*; Magnus, "Blute enthaltene Gase," 593–598.

[22] *Berzelius und Magnus' Briefwechsel*, p. 123.

[23] Magnus, "Blute enthaltene Gase," 599–600.

[24] *Ibid.* 591; *Berzelius und Magnus' Briefwechsel*, pp. 123–125.

[25] *Berzelius und Magnus' Briefwechsel*, pp. 125–126.

[26] Magnus, "Blute enthaltene Gase," 600–601. Magnus actually stated 'less' where I have placed "more" in brackets, but this must have been a slip.

[27] *Ibid.*, 602.

[28] J.J. Berzelius, *Lehrbuch der Chemie*, trans. F. Woehler, 3rd ed. rev. (Dresden: Arnold, 1840), vol. 9, pp. 127–132.

[29] *Ibid.*, pp. 131–132.

[30] Johannes Müller, *Handbuch der Physiologie des Menschen*, 3rd ed. rev. (Coblenz: Hölscher, 1838), vol. 1, p. 334.

[31] *Ibid.*, pp. 321–323, 334–335.

[32] [Joseph] Gay-Lussac, "Observations critiques sur la théorie des phénomènes chimiques de la respiration," *Annales de Chimie et de Physique* 3rd. ser. 11 (1844): 5–16.

[33] [Gustav] Magnus, "Bemerkungen über den Vorgang bei der Respiration," *Bericht über die zur Bekanntmachung geeigneten Verhandlungen der Königlich Preussischen Akademie der Wissenschaften Berlin* 51 (1844): 234–242.

[34] G. Magnus, "Über die Respiration," *Ibid.* 52 (1845): 115–119.

[35] L.-R. Lecanu, "Études chimiques sur le sang humain," *Annales de Chimie et de Physique* 67 (1838): 54–63.

[36] Berzelius, *Lehrbuch der Chemie*, pp. 60–62.

[37] *Ibid.*, pp. 66–67, 71.

[38] *Ibid.*, pp. 123–124.

[39] Theodor Bischoff, "Bericht über die Fortschritte der Physiologie im Jahre 1844," *Archiv für Anatomie Physiologie und wissenschaftliche Medicin* (1846): 40–41.

[40] C.G. Lehmann, *Lehrbuch der Physiologischen Chemie*, 2nd ed. rev. (Leipzig: Engelmann, 1853), vol. 1, pp. 283–295.

[41] *Ibid.*, vol. 2, pp. 240–241.

[42] *Ibid.*, vol. 3, p. 339.

[43] *Ibid.*, vol. 2, pp. 161, 241.

[44] *Ibid.*, vol. 3, pp. 335–338.

[45] *Ibid.*, vol. 3, p. 339.

[46] C.G. Lehmann, "Einige vergleichende Analysen des Blutes der Pfortader under der Lebervenen," *Berichte über die Verhandlungen der Königlich Sächsischen Gesellschaft der Wissenschaften zu Leipzig* (1850): 131–163.

[47] "Funke, Otto F.," in *Biographische Lexikon der hervorragenden Ärtzte*, August Hirsch, ed. (Berlin: Urban and Schwarzenberg, 1930), vol. 2, p. 647; Otto Funke, "Ueber das Milzvenenblut," *Zeitschrift für rationelle Medizin* 1 (1851): 172–177.

[48] Funke, "Milzvenenblut," 177–218.

[49] *Ibid.* 185.

[50] *Ibid.* 185–191.

[51] *Ibid.* 191–192.

[52] O. Funke, "Neue Beobachtungen über die Krystalle des Milzvenen- und Fisch-Blutes," *ibid.* 2 (1852): 198–217.

[53] *Ibid.* 215–217.

[54] Franz Leydig, "Zur Anatomie von Piscola geometrica mit theilweiser Vergleichung anderer einheimischer Hirudineen," *Zeitschrift für Wissenschaftliche Zoologie* 1 (1849): 115–116.

[55] O. Funke, "Ueber Blutkrystallisation," *Zeitschrift für rationelle Medizin* 2 (1852): 288–292.

[56] *Ibid.* 288.

[57] C.G. Lehmann, "Ueber den krystallisirbaren Stoff des Blutes," *Gesellschaft der*

Wissenschaften zu Leipzig (1852): 78–84.

[58] Lehmann, "Weitere Mittheilungen über die Krystallisirbare Proteïnsubstanz des Blutes," *ibid.* (1853): 101–133.

[59] Emile Fernet, "Du rôle des principaux éléments du sang dans l'absorption ou le dégagement des gaz de la respiration," *Annales des Sciences Naturelles (Zoologie)* **8** (1857): 127–129.

[60] Magnus, "Blute enthaltene Gase," 589.

[61] R. Bunsen, "Ueber das Gesetz der Gasabsorption," *Annalen der Chemie und Pharmacie* **93** (1855): 1–50.

[62] *Ibid.* 50; Susan G. Schaher, "Bunsen, Robert Wilhelm Eberhard," *Dictionary of Scientific Biography*, C.C. Gillispie, ed. (New York: Charles Scribner's Sons, 1970–1980), vol. 2, pp. 586–590.

[63] Lothar Meyer, "Ueber die Gase des Blutes," *Annalen der Physik und Chemie* **178** (1857): 299–307.

[64] *Ibid.* 305.

[65] Georg Harley, "Ueber die chemischen Veränderungen des Blutes bei der Respiration," *Archiv für pathologischen Anatomie* **11** (1857): 107–111.

[66] *Ibid.* 112–115.

[67] *Ibid.* 111.

[68] Meyer, "Gase des Blutes," 305.

[69] Fernet, "Éléments du sang," 125–153.

[70] *Ibid.* 153–220.

[71] M.D. Grmek, *Raisonnement expérimental* (Paris: Droz, 1973), pp. 119, 122.

[72] E. Baumann and A. Kossel, "Zur Erinnerung an Felix Hoppe-Seyler," *Hoppe-Seyler's Zeitschrift für Physiologische Chemie* **21** (1895): VI–XI.

[73] Felix Hoppe, "Zur Blutanalyse," *Archiv für pathologischen Anatomie* **12** (1857): 483–486.

[74] Felix Hoppe, "Ueber die Einwirkung des Kohlenoxydgases auf das Hämatoglobulin," *ibid.* **11** (1857): 288–289.

[75] *Ibid.*

[76] *Ibid.* 289.

[77] *Ibid.*; Hoppe, "Ueber die Einwirkung des Kohlenoxydgases auf das Blut," *Archiv für pathologischen Anatomie* **13** (1858): 104–105.

[78] Grmek, *Raisonnement expérimental*, pp. 71–207.

[79] Hoppe, "Zur Blutanalyse," 484.

[80] Felix Hoppe, *Anleitung zur Pathologisch-Chemischen Analyse für Aertze und Studirende* (Berlin: Hirschwald, 1858), p. 139.

[81] *Ibid.*; F. Hoppe, "Ueber Hämatokrystallin und Hämatin: Erwiderung an Herrn Prof. C.G. Lehmann," *Archiv für pathologischen Anatomie* **17** (1859): 488–491.

[82] Baumann and Kossel, "Erinnerung," XI.

[83] Felix Hoppe, "Ueber das Verhalten des Blutfarbstoffs im Spectrum des Sonnenlichtes," *Archiv für pathologischen Anatomie* **23** (1862): 446.

[84] William McGucken, *Nineteenth-Century Spectroscopy* (Baltimore: The Johns Hopkins Press, 1969), pp. 1–34; G. Kirchhoff and R. Bunsen, *Chemische Analyse durch Spectralbeobachtungen* (Vienna: Lenoir, 1860).

[85] G. Valentin, *Der Gebrauch des Spektroskopes* (Leipzig: Winter, 1863), p. 11.

[86] Johann Müller, "Prismatische Zerlegung der Farben verschiedener Flüssigkeiten,"

Annalen der Physik und Chemie **72** (1847): 76–82.

[87] Hoppe, "Verhalten des Blutfarbstoffs," 446.

[88] Valentin, *Gebrauch*, pp. 17–21.

[89] Hoppe, "Verhalten des Blutfarbstoffs," 446–449.

[90] *Ibid.* 447.

[91] Robert K. Merton, "Singletons and Multiples in Science," in *The Sociology of Science: Theoretical and Empirical Investigations*, Norman W. Storer, ed. (Chicago: University of Chicago Press, 1973), pp. 343–370.

[92] Valentin, *Gebrauch*, pp. 73–86.

[93] *Ibid.*, pp. 73–74.

[94] F. Hoppe-Seyler, "Beiträge zur Kenntniss des Blutes des Menschen und die Wirbelthiere," in *Medicinisch-Chemische Untersuchungen aus dem Laboratorium für angewandte Chemie in Tübingen*, Felix Hoppe-Seyler, ed. (Berlin: Hirschwald, 1866–71), p. 180.

[95] Felix Hoppe-Seyler, "Ueber die chemischen und optischen Eigenschaften des Blutfarbstoffs," *Archiv für pathologischen Anatomie* **19** (1864): 233–235.

[96] *Ibid.* 233.

[97] Baumann and Kossel, "Erinnerung," VI.

[98] Felix Hoppe-Seyler, "Ueber die chemischen und optischen Eigenschaften des Blutfarbstoffs," *Archiv für pathologischen Anatomie* **29** (1864): 597–600.

[99] G.G. Stokes, "On the Reduction and Oxidation of the Colouring Matter of the Blood," *Philosophical Magazine* (1864): 391.

[100] *Ibid.*

[101] *Ibid.*

[102] *Ibid.* 397.

[103] *Ibid.* 394.

[104] *Ibid.* 397.

[105] *Ibid.* 394.

[106] Joseph S. Fruton, *Contrasts in Scientific Style: Research Groups in the Chemical and Biochemical Sciences* (Philadelphia: American Philosophical Society, 1990), pp. 77–78.

[107] Hoppe-Seyler, "Kenntniss des Blutes," 180.

[108] F. Hoppe-Seyler, "Ueber die Oxydation im lebenden Blute," in *Medicinisch-chemische Untersuchungen*, pp. 132–133.

[109] *Ibid.*, pp. 134–139.

[110] Felix Hoppe-Seyler, *Physiologische Chemie* (Berlin: Hirschwald, 1881), p. 73.

A. J. KOX

EINSTEIN, SPECIFIC HEATS, AND RESIDUAL RAYS:
THE HISTORY OF A RETRACTED PAPER*

I. INTRODUCTION

Towards the end of 1911 Albert Einstein submitted a paper to the *Annalen der Physik* with the title "Zur Theorie der Reststrahlen" ("On the Theory of Residual Rays"). It was inspired by the experimental work of the Berlin experimentalist Heinrich Rubens and his collaborators on the optical properties of solids in the far infrared. Some substances, for instance NaCl, exhibit selective reflection of infrared radiation: they strongly reflect radiation of certain characteristic wavelengths, whereas they are transparent for other infrared wavelengths. The reflected rays are known as residual rays, and Rubens had determined the wavelengths of the residual rays for a number of substances. Because the frequencies of the residual rays were thought to be connected with characteristic frequencies of the substances that produce them, their determination was important for theories in which proper frequencies of solids played a role, in particular the quantum theory of specific heats first developed by Einstein.

In his paper Einstein tried to explain an unexpected feature of Rubens's experimental results. Several weeks after the paper was submitted, however, Einstein retracted it. These facts are known from correspondence published in Volume 5 of *The Collected Papers of Albert Einstein*.[1] But Einstein's letters from late 1911 and early 1912 provide more information: they also allow a partial reconstruction of the contents of the paper and of the reasons why Einstein retracted it. In this paper I will give such a reconstruction. In addition, I will put the paper in the context of the contemporary theoretical and experimental work on specific heats, both by Einstein and by others.

245

A.J. Kox and D.M. Siegel (eds.), No Truth Except in the Details, 245–257.

II. EINSTEIN'S THEORY OF SPECIFIC HEATS

Einstein was the first to apply the quantum hypothesis to the problem of specific heats. In his first paper on this topic, published in late December 1906,[2] he introduced a model in which a solid consists of a collection of three-dimensional monochromatic harmonic oscillators, the energy of which is quantized in units of $\varepsilon = \frac{R}{N}\beta\nu$, with R the gas constant, N Avogadro's number, and $\beta = h/k$ (in modern notation). Writing the mean energy of a one-dimensional oscillator as

$$
(1) \qquad \bar{E} = \frac{\int E e^{-(NE/RT)}\omega(E)\,\mathrm{d}E}{\int e^{-(NE/RT)}\omega(E)\,\mathrm{d}E} \, ,
$$

where $\omega(E)$ is a function that is sharply peaked around the values ε, 2ε, $3\varepsilon, \ldots$, it is easily found that the mean energy U of N three-dimensional oscillators is

$$
(2) \qquad U = 3N\bar{E} = 3R\frac{\beta\nu}{e^{\beta\nu/T} - 1} \, .
$$

Differentiation of this expression gives the specific heat c as a function of the temperature and the frequency of the oscillators:

$$
(3) \qquad c = 3R\frac{(\beta\nu/T)^2 e^{\beta\nu/T}}{(e^{\beta\nu/T} - 1)^2} \, .
$$

Although in retrospect this derivation seems a straightforward application of the quantum hypothesis to solids, it was in fact a bold extension of the quantum hypothesis from radiation theory to another field. As Martin Klein has emphasized, Einstein's paper addresses a fundamental problem concerning the description of the properties of matter.[3]

The problem is the following. In radiation theory, for example, the validity of the equipartition theorem is disputed, because it leads inescapably to the Rayleigh–Jeans radiation law, which only holds for the low frequency region. From the success of Planck's radiation theory it appears as if we have to adopt a new view of the mechanism of energy exchange between matter and radiation, a new view that transcends the usual molecular-kinetic theory. If that is the case, we have to modify our theory not only for the case of radiating oscillators, but for all cases where oscillating objects play a role. Or, in Einstein's own words:

Wenn die Plancksche Theorie der Strahlung den Kern der Sache trifft, so mussen wir erwarten, auch auf anderen Gebieten der Wärmetheorie Widersprüche zwischen der gegenwartigen molekular-kinetischen Theorie und der Erfahrung zu finden, die sich auf dem eingeschlagenen Wege heben lassen.[4]

In Einstein's paper this insight is worked out further and applied to a situation where a discrepancy between molecular-kinetic theory and experience does indeed exist: the theory of specific heats. For a one-dimensional harmonic oscillator the mean energy according to the equipartition theorem is kT. For a collection of N oscillators in three dimensions – the system considered above – the total energy becomes $3RT$, so that the specific heat takes the constant value $3R$ (approximately 6 cal/degree). This is the well-known rule of Dulong and Petit. For high temperatures this rule was confirmed in general through experimental results, but for low temperatures deviations occurred. Moreover, there were substances, such as carbon and silicium, that already showed anomalous low values for the specific heat at room temperature.

Einstein also drew attention to another, even more serious problem in the theory of specific heats. In two recent papers,[5] Paul Drude had developed a theory of dispersion, based on the assumption that dispersion phenomena are caused by the interaction of electromagnetic waves with charged microscopic oscillators. He had shown that the ultraviolet proper frequencies observed in solids are due to vibrations of particles with masses comparable to the electron mass, whereas the infrared proper frequencies could be associated with vibrations of larger masses, i.e., the atoms themselves. That result caused problems for the theory of specific heats: if all electronic vibrations would contribute equally to the specific heat – as the equipartion theorem demanded – specific heats should have much greater values than was observed.

Einstein's theory of specific heats went a long way toward solving these problems, as becomes clear from a closer inspection of Equation (3). For values of $T/\beta\nu$ smaller than 0.1 the specific heat is practically zero; for increasing values of $T/\beta\nu$ it first increases, then the curve flattens off and approaches the Dulong–Petit value $3R$. In fact, already for $T/\beta\nu > 0.9$ the specific heat lies close to the Dulong–Petit value. This behavior has several important implications. In the first place it means that for all substances the specific heat approaches zero as the temperature goes to zero. Furthermore, it implies that if the temperature is kept constant, the specific heat decreases with increasing oscillator frequency (or decreasing wavelength). At room temperature the specific heat turns out to have a negligible value for wavelengths smaller than 4.8μ. In other words, the very rapid oscillations of small oscillator masses – Drude's electronic vibrations – do not contribute to the specific heat.

In order to obtain numerical results for individual substances from Einstein's formula, their characteristic frequencies needed to be known and this would remain one of the central problems in Einstein's theory. One possible way to solve this problem was to take the experimental value of the specific heat for a certain temperature, calculate the corresponding frequency from Equation (3), and use this value to determine the specific heats for other temperatures. Another possibility was to identify the oscillator frequencies with observed residual ray frequencies. In that case Einstein's formula predicts that substances with residual-ray wavelengths smaller than 4.8μ will show significant deviations from Dulong and Petit's rule at room temperature. Einstein found this behavior confirmed qualitatively in experimental data. He found a correlation between small specific heats on the one hand, and small atomic mass and small infrared wavelengths on the other hand. Particularly impressive was the agreement between experimental data and the theoretical prediction for diamond. Because it turned out that its characteristic wavelength (calculated from Einstein's formula) was quite small (11μ), the range of values of $\beta\nu/T$ corresponding to temperatures for which the specific heat is known runs from 0.17 to 0.95. All measurements lie close to the theoretical curve.

In spite of these successes, Einstein was aware of the theory's shortcomings. He listed several problematic points. One of those was the assumption that the frequency of the oscillators was independent of their energy. Also, it was possible that the thermal proper frequencies were different from the observed optical ones. A problematic point not explicitly mentioned by Einstein, but one he was undoubtedly aware of, was his assumption that all oscillators had the same proper frequency. In any case, for a further test of the formula new data were needed, especially at low temperatures, where the deviations from the classical theory were most striking.

The first such measurements became available in 1910 and were made by Walther Nernst. They were part of a series of investigations of the low-temperature behavior of the properties of solids. This work was meant to give experimental support for Nernst's new heat theorem, later known as the Third Law of Thermodynamics. From Nernst's measurements it followed that for low temperatures the specific heat goes to zero, in qualitative agreement with Einstein's formula.[6] Einstein was delighted when he heard about Nernst's results. In a letter to Jakob Laub he wrote: "I am certain about quantum theory. My predictions concerning specific heats appear to be brilliantly confirmed. Nernst, who just visited me, and Rubens are busily working on the experimental confirmation."[7] After their meeting Nernst showed himself much impressed with Einstein, calling him a "Boltzmann redivivus."[8] That does not mean that Nernst fully accepted Einstein's premises: in an often-quoted remark in a lecture given at the Prussian Academy of Sciences in January 1911 he called quantum theory a "calculational rule, indeed one may say a very odd, even grotesque one."[9]

The success of Einstein's theory did not just earn praise from Nernst, but was also rewarded in a more concrete way. In the fall of 1910 Einstein received a letter from Emil Fischer, a professor of chemistry at the University of Berlin, in which he was offered a grant of 15,000 marks, to be spent at his own discretion.[10] The money was made available by a "gentleman from the chemical industry" who wished to remain anonymous. From a draft of the letter, however, we know that the donor was Franz Oppenheim, director of the Aktiengesellschaft für Anilinfabrikation (Agfa). In his letter Fischer explicitly mentioned Einstein's work on the theory of specific heats and Nernst's experimental confirmation of it. There is also evidence that Nernst had discussed Einstein's work with Fischer at an earlier time. In his reply to Fischer Einstein gratefully accepted the offer, assuring him that he would use the money in the most conscientious way.[11] He also added some comments that throw more light on his own views: he warned against too much confidence in his work, calling his theory of specific heat very unsatisfactory, and pointing out that all efforts to revise molecular mechanics to conform to experience in this field had been without result. He clearly referred to the difficulty that remained central in his thinking in those years: how can quantum theory be reconciled with existing classical theory, or, alternatively, how must classical mechanics and electrodynamics be modified in order to incorporate quantum phenomena?

III. RUBENS'S MEASUREMENTS

In January 1910 Rubens and his co-worker Hollnagel published a paper in which they reported on their measurements of the wavelengths of the residual rays of various crystals.[12] The residual rays were produced by successively reflecting light off four slabs of the substance in question; their wavelengths were determined with the help of an interference method. The interferometer they used consisted of two parallel quartz plates, placed perpendicularly in the incident beam, the distance of which could be varied. As the distance of the plates was gradually increased, the intensity of the residual rays was observed to fluctuate. Not in the expected periodic, sine-like manner, however, but in a pattern that resembled frequency beats. Rubens and Hollnagel concluded that the residual rays were not strictly monochromatic, but had two characteristic frequencies that lay close together. They found this behavior confirmed for three of the four substances they studied and succeeded in calculating the values of the two frequencies in those cases.[13]

A little more than a year later Nernst and his co-worker Frederick Lindemann commented on Rubens's measurements in two joint papers and tried to draw some consequences for the theory of specific heats, not only from the residual-ray experiments, but from their own work as well.[14] Earlier that year, Nernst had published measurements that showed that the specific heat converged more slowly to zero than Einstein's formula predicted.[15] But, as Nernst and Lindemann pointed out, the discrepancies did not only show for low temperatures: for KCl, for instance, they also found deviations for higher temperatures, although qualitatively the agreement remained. In their second paper, Nernst and Lindemann discussed Rubens's results and tentatively suggested that the two measured frequencies could be associated with the two types of atoms present in the substances investigated by Rubens. In any case, they claimed that Rubens's results excluded a possible explanation of the KCl results, namely the presence of a wide resonance curve, and came to the conclusion that Einstein's theory had to be modified. Their modification was a revised formula for the specific heat that gave a better description of its low-temperature behavior:

$$(4) \qquad c = \frac{3}{2} R \left[\frac{(\beta\nu/T)^2 e^{\beta\nu/T}}{(e^{\beta\nu/T} - 1)^2} + \frac{(\beta\nu/2T)^2 e^{\beta\nu/T}}{(e^{(\beta\nu/2T)} - 1)^2} \right].$$

This formula corresponds with the following expression for the energy of a system of harmonic oscillators, which looks very much like Einstein's formula:

$$(5) \qquad U = \frac{3}{2} R \left[\frac{\beta\nu}{e^{\beta\nu/T} - 1} + \frac{\beta\nu/2}{e^{\beta\nu/2T} - 1} \right].$$

Instead of one, two proper frequencies appear, one exactly one half of the other one. This is what Einstein later called Nernst's "double-quantum theory." As Nernst and Lindemann admitted, the formula had been found by trial and error. But once they had it, they made an attempt to give it a theoretical foundation. The best they could come up with was that the two terms represented respectively the kinetic and potential energy of the oscillators. These were thus no longer equal on the average and moreover the quantum of action for the potential energy was half that of the kinetic energy. No reasonable explanation could be found for this. But it was the only way in which this formula allowed the derivation of Planck's radiation law.

Einstein commented on this curious result in a paper from May 1911.[16] Nernst had sent him the proofs of his first paper with Lindemann and Einstein showed himself much impressed with the usefulness of the formula, calling it "surprisingly useful."[17] He realized that this development touched on the central problem of the frequency spectrum of solids. As he pointed out, the Nernst–Lindemann formula could be found by assuming that each atom oscillates half of the time with frequency ν, and the other half with frequency $\nu/2$. In this way, the non-monochromatic character of the atomic vibrations found its "most primitive expression."[18] Of course, the quantity that really mattered was the frequency at which an oscillator would be in equilibrium with a thermal radiation field.

In a letter to Nernst[19] Einstein further analyzed Nernst and Lindemann's formula in terms of a frequency spectrum for the solid under consideration. While he had represented the spectrum by a function that was sharply peaked at one frequency, Nernst and Lindemann's spectrum showed two peaks. In his letter Einstein pointed out that of course the real spectrum was a continuous one.

In their second paper, Nernst and Lindemann explicitly rejected the implication of Einstein's analysis of their formula, namely that the substances they studied exhibited two kinds of oscillations. They argued that there was no experimental evidence for a second frequency at half the value of the first one, because residual ray measurements always showed only one frequency. The possibility that for one of the frequencies charged oscillators were responsible, while neutral ones oscillated at the other frequency (so that there would be only one observable infrared frequency) was rejected by them, also on the grounds that it was extremely unlikely that there were equal numbers of charged and

neutral oscillators. They did not discuss the possibility that each oscillator might have two equally likely proper frequencies – which is what
Einstein seems to imply.

IV. EINSTEIN'S PAPER

The problem of the frequency spectrum of solids remained on Einstein's
mind. It had also been discussed during the first Solvay Congress, held
in Brussels from 30 October to 3 November 1911, in particular during the discussion following Nernst's contribution.[20] In December 1911
Einstein wrote to Heinrich Zangger that he had just finished a paper on
the properties of bodies in the infrared region, adding that the experimental results had been misinterpreted.[21] Later that month, in a letter to
Michele Besso, he gave more information on the contents of the paper.[22]
On the basis of this letter, the following reconstruction can be made of
the paper.

Einstein's attempted explanation of Rubens's results has two important features. The first one is based on some fundamental considerations
on dispersion, similar to the ones given by Drude in his theory of
dispersion.[23] From a model in which a solid consists of charged harmonic oscillators it is straightforward to derive the following expression
for the index of refraction:[24]

$$(6) \qquad n^2 = 1 + \sum_i \frac{\theta_i}{1 - (\nu/\nu_i)^2} \ .$$

The quantities ν_i are the resonance frequencies of the oscillators; the θ_i
are related to their harmonic force constants and their masses. For simplicity's sake we shall assume in the following that only one resonance
frequency ν_0 exists. Equation (6) can be combined with the expression
for the reflective power

$$(7) \qquad R = \left| \frac{n-1}{n+1} \right|^2 \ .$$

For increasing frequency R increases and reaches a value of 1 for $\nu = \nu_0$.
For greater ν, n^2 becomes negative and n purely imaginary, so that the
substance reflects totally. The region of negative n^2 runs from $\nu = \nu_0$
to $\nu = \nu_1$, where ν_1 is the root of the equation

$$(8) \qquad 0 = 1 + \frac{\theta_0}{1 - (\nu/\nu_0)^2} \ .$$

Thus, for ν between the values of ν_0 and ν_1 the substance reflects totally. How sharp the reflection peak is depends on how close those two values lie together. In his letter to Besso, Einstein drew a diagram of n^2 as a function of the wavelength that corresponds to the above argument. On the basis of this argument Einstein predicted a broad resonance curve instead of a narrow peak. Furthermore, Einstein argued – without giving any details – that a sharply defined wide region of total reflectivity would cause beats like the ones Rubens had observed.

A second aspect of Einstein's modified theory, but one we know much less about, is the inclusion of damping in the motion of his oscillators. In the spring of 1911, in the same paper in which he commented on Nernst's double quantum theory, Einstein had already pointed out that the vibrations of his oscillators were influenced by their interaction with neighboring ones.[25] Because of this influence they behaved as strongly damped, non-monochromatic oscillators. He had also tried to calculate the proper frequencies of those coupled oscillators through a dimensional consideration and found that his calculation supported earlier results by Lindemann, who had related a substance's proper frequency with its melting temperature.[26] Later that year, in a letter to H. A. Lorentz of November 1911, Einstein mentioned further work on damped oscillators, referring to it as "quite a bit of calculation."[27]

Although in his letter to Besso of December 1911[28] Einstein made no mention of damping, it becomes clear from other sources that his paper did deal with this subject as well. Although there are no indications as to how Einstein tried to incorporate this feature, it appears from a later letter[29] that he did more than include a simple standard damping term proportional to the speed in the equation of motion of the oscillator. Such a term would only result in a smoothing of the resonance curve and would neither provide an explanation of Rubens's results nor modify the low-temperature behavior of the specific heat. It is much more likely that Einstein tried to improve the qualitative treatment of his earlier paper.

On 27 January 1912, Einstein wrote to Wilhelm Wien, the editor of the *Annalen der Physik*, asking him not to publish his paper, or, if that was not possible, to allow him to add a postscript to it.[30] Rubens had convinced him, he wrote, that his paper contained errors and needed revisions. Although Einstein spoke of submitting a revised version in the future, nothing came of this. More details of what was wrong with the paper can be found in later letters. To Zangger Einstein wrote that

Rubens had informed him of experimental results that showed the reality of the two maxima,[31] and in a letter to Besso he repeated this.[32] But, as Einstein pointed out, this did not mean that there were really two proper frequencies: "It is absolutely impossible that two proper frequencies exist."[33] Absorption could have the effect of splitting the theoretically predicted wide peak into two narrower ones. This absorption is obviously connected with the energy dissipation involved in the damping of the oscillators, but, as Einstein pointed out in his letter to Besso, a simple term proportional to the speed was not sufficient to explain the observations. Einstein did not succeed in solving the problem. At the end of February he was still trying: "Furthermore, I struggle with dispersion in the infrared. Friction term all messed up."[34] That seems to be the end of Einstein's efforts to refine his theory.[35]

The observations that convinced Einstein of the reality of the two peaks were presumably measurements by Hollnagel. That can be inferred from a passage in a paper by Rubens from 1913, which mentions Hollnagel's investigation of the possible influence of absorption by water vapor on the intensity distribution of residual rays and its negative outcome.[36] Rubens referred to this investigation because it contradicted the conclusion he presented in his paper that the two resonance peaks are in fact caused by selective absorption by water vapor in the air. Hollnagel's negative result was caused by the fact that in his experimental setup the effect was too small to be measured.

A final remark on Einstein's attitude toward experiments is in order. It is sometimes claimed that Einstein had a habit of ignoring or dismissing experimental evidence that contradicted his own work. A striking example is his rejection of Walter Kaufmann's experimental results on the specific charge of electrons, which seemed to favor Max Abraham's and Alfred Bucherer's theories of the electron over special relativity, and which worried Lorentz so much that he was prepared to abandon his theory.[37] Perhaps this is true in matters of such fundamental importance as a theory that Einstein fully believed in and that he found superior to all alternatives – in this case the special theory of relativity.[38] It is also true that Einstein never accepted experimental results at face value. But, as becomes clear from Einstein's work on specific heats, and, in particular, from the episode of the paper on residual rays, Einstein took experimental results very seriously, never doubting Nernst's data on the low temperature behavior of specific heats and eventually accepting the correctness of Rubens's measurements.[39] It is ironic that precisely these measurements turned out to be wrong.

Instituut voor Theoretische Fysica
Universiteit van Amsterdam
The Netherlands, and
The Collected Papers of Albert Einstein
Boston University
U.S.A.

NOTES

* I am grateful to the Albert Einstein Archives of the Hebrew University of Jerusalem for permission to quote from Einstein's letters.

[1] *The Collected Papers of Albert Einstein*, Volume 1, *The Early Years, 1897–1902*, John Stachel *et al.*, eds.; Volume 2, *The Swiss Years: Writings, 1900–1909*, John Stachel *et al.*, eds.; Volume 3, *The Swiss Years: Writings, 1909–1911*, Martin J. Klein *et al.*, eds.; Volume 4, *The Swiss Years: Writings, 1912–1914*, Martin J. Klein *et al.*, eds.; Volume 5, *The Swiss Years: Correspondence, 1902–1914*, Martin J. Klein *et al.*, eds. (Princeton: Princeton University Press, 1987, 1989, 1993, 1995, 1993). In the following this edition will be abbreviated as *CPAE*, followed by a volume number.

[2] Albert Einstein, "Die Plancksche Theorie der Strahlung und die Theorie der spezifischen Wärme," *Annalen der Physik* **22** (1907): 180–190. Reprinted in *CPAE*, vol. 2, pp. 379–389.

[3] See Martin Klein, "Einstein, Specific Heats and the Early Quantum Theory," *Science* **148** (1965): 173–180.

[4] "If Planck's radiation theory strikes to the heart of the matter, contradictions between the current molecular-kinetic theory and experience must be expected in other areas of the theory of heat as well and could be resolved in the same manner." Einstein, "Plancksche Theorie," 184.

[5] Paul Drude, "Optische Eigenschaften und Elektronentheorie. I, II," *Annalen der Physik* **14** (1904): 677–725, 936–961.

[6] See Walther Nernst, "Untersuchungen über die spezifische Wärme bei tiefen Temperaturen. II, *Königlich Preußische Akademie der Wissenschaften* (Berlin), *Sitzungsberichte* (1910): 262–282.

[7] "Die Quantentheorie steht mir fest. Meine Voraussagungen inbetreff der spezifischen Wärmen scheinen sich glänzend zu bestätigen. Nernst, der eben bei mir war und Rubens sind eifrig mit der experimentellen Prüfung beschäftigt." Einstein to Jakob Laub, 16 March 1910, in *CPAE*, vol. 5, Doc. 199.

[8] *Ibid.*, note 9.

[9] "Zur Zeit ist die Quantentheorie wesentlich eine Rechnungsregel, und zwar eine solche, wie man wohl sagen kann, sehr seltsamer, ja grotesker Beschaffenheit." Walther Nernst, "Über neuere Probleme der Wärmetheorie," *Königlich Preußische Akademie der Wissenschaften* (Berlin), *Sitzungsberichte* (1911): 65–90, on 86.

[10] See Emil Fischer to Einstein, 1 November 1910, in *CPAE*, vol. 5, Doc. 230.

[11] Einstein to Emil Fischer, 5 November 1910, in *CPAE*, vol. 5, Doc. 232. It is not clear what Einstein did with the money.

[12] Heinrich Rubens and H. Hollnagel, "Messungen im langwelligen Spektrum," *Königlich Preußische Akademie der Wissenschaften* (Berlin), *Sitzungsberichte* (1910): 26–52.

[13] The behavior of NaCl, KCl, and KBr was clear-cut; for KI there were only indications of two maxima.

[14] See Walther Nernst and Frederick A. Lindemann, "Untersuchungen über die spezifische Wärme bei tiefen Temperaturen. V," *Königlich Preußische Akademie der Wissenschaften* (Berlin), *Sitzungsberichte* (1911): 494–501 and their more detailed paper of a few months later: "Spezifische Wärme und Quantentheorie," *Zeitschrift für Elektrochemie* 17 (1911): 817–827.

[15] See Walther Nernst, "Untersuchungen über die spezifische Wärme bei tiefen Temperaturen. III," *Königlich Preußische Akademie der Wissenschaften* (Berlin), *Sitzungsberichte* (1911): 306–315. In spite of the discrepancies, in this paper Nernst characterized his experimental results as "a brilliant confirmation of the quantum theory of Planck and Einstein." ("eine glänzende Bestätigung der Quantentheorie von Planck und Einstein," p. 310). Later that year he used a similar expression, but with the word "qualitatively" ("qualitativ") added to it. See Walther Nernst, "Die Energieeinhalt fester Stoffe," *Annalen der Physik* 36 (1911): 395–439, on 423.

[16] Albert Einstein, "Elementare Betrachtungen über die thermische Molekularbewegung in festen Körpern," *Annalen der Physik* 35 (1911): 679–694. Reprinted in *CPAE*, vol. 3, pp. 460–475.

[17] "überraschend brauchbar." *Ibid.* 685–686.

[18] "findet ihren primitivsten Ausdruck." *Ibid.* 686.

[19] Einstein to Walther Nernst, 20 June 1911, in *CPAE*, vol. 5, Doc. 270.

[20] The Proceedings of the Solvay Congress were first published in a French and later in a German edition. French: Paul Langevin and Maurice de Broglie, eds., *La théorie du rayonnement et les quanta. Rapports et discussions de la réunion tenue à Bruxelles, du 30 octobre au 3 novembre 1911* (Paris: Gauthier-Villars, 1912); German: Arnold Eucken, ed., *Die Theorie der Strahlung und der Quanten. Verhandlungen auf einer von E. Solvay einberufenen Zusammenkunft (30. Oktober bis 3. November 1911)* (Halle a.S.: Knapp, 1914). Nernst's lecture: "Application de la théorie des quanta à divers problèmes physico-chimiques," pp. 254–290, 291–302 (discussion); "Anwendung der Quantentheorie auf eine Reihe physikalisch-chemischer Probleme," pp. 208–233, 234–244 (discussion). The German version renders Nernst's lecture and the discussion remarks by the German participants in the form and language in which they were actually delivered.

[21] Einstein to Heinrich Zangger, 13–16 December 1911, in *CPAE*, vol. 5, Doc. 325.

[22] Einstein to Michele Besso, 26 December 1911, in *CPAE*, vol. 5, Doc. 331.

[23] See note 5.

[24] See, e.g., Paul Drude, *Lehrbuch der Optik* (Leipzig: Hirzel, 1990). Einstein owned a copy of this book.

[25] See Einstein, "Elementare Betrachtungen."

[26] Frederick A. Lindemann, "Über die Berechnung molekularer Eigenfrequenzen," *Physikalische Zeitschrift* 11 (1910): 609–612.

[27] "eine ziemliche Rechnerei." Einstein to H. A. Lorentz, 23 November 1911, in *CPAE*, vol. 5, Doc. 313.

[28] See note 22.

[29] Einstein to Michele Besso, 4 February 1912, in *CPAE*, vol. 5, Doc. 354.

[30] Einstein to Wilhelm Wien, 27 January 1912, in *CPAE*, vol. 5, Doc. 343. It is from this letter that we also know the title of the paper.

[31] Einstein to Heinrich Zangger, 27 January 1912, in *CPAE*, vol. 5, Doc. 344.

[32] See note 29.

[33] "Es ist gar nicht möglich, dass zwei Eigenfrequenzen existieren." *Ibid.*

[34] "Ausserdem schlage ich mich mit der Dispersion im Ultrarot herum. Reibungsglied ganz faul." Einstein to Ludwig Hopf, after 20 February 1912, in *CPAE*, vol. 5, Doc. 364.

[35] A major step towards a solution of the problem of molecular proper frequencies was taken in April 1912 by Born and Von Kármán in their theory of lattice vibrations. That development might have prompted Einstein to abandon the problem. See Max Born and Theodor von Kármán, "Über Schwingungen in Raumgittern," *Physikalische Zeitschrift* 13 (1912): 297–309. For a modern quantum-mechanical treatment of lattice vibrations, see Max Born and Kun Huang, *Dynamical Theory of Lattice Vibrations* (Oxford: Clarendon Press, 1954).

[36] Heinrich Rubens, "Über die Absorption des Wasserdampfs und über neue Reststrahlengruppen im Gebiete der großen Wellenlängen," *Königlich Preußische Akademie der Wissenschaften* (Berlin), *Sitzungsberichte* (1913): 513–549.

[37] As Lorentz put it in the first edition of his *Theory of Electrons* (Leipzig: Teubner, 1909, p. 212): "But, so far as we can judge it at present, the facts are against our hypothesis." The reference is to the hypothesis of the deformable electron. On p. 213, speaking about the same hypothesis, he says: "[...] it is very likely that we shall have to relinquish this idea altogether."

[38] In his 1907 review paper on relativity Einstein discussed Kaufmann's experiments and found them inconclusive. Furthermore, he commented on the rival theories of Abraham and Bucherer in the following way: "Jenen Theorien kommt aber nach meiner Meinung eine ziemlich geringe Wahrscheinlichkeit zu, weil ihre die Maße des bewegten Elektrons betreffenden Grundannahmen nicht nahe gelegt werden durch theoretische Systeme, welche größere Komplexe von Erscheinungen umfassen." Albert Einstein, "Über das Relativitätsprinzip und die aus demselben gezogenen Folgerungen," *Jahrbuch für Radioaktivität und Elektronik* 4 (1907): 411–462, on 439. Reprinted in *CPAE*, vol. 2, pp. 433–484.

[39] There are two, more or less contemporary episodes that show similarities with the one described in this paper and that further illustrate Einstein's attitude towards experiments. The first one is Einstein's work on photochemical decomposition from late 1911, which was stimulated by his conviction that experimental results by Emil Warburg were incorrect. In this case Einstein was right. The second one occurred in 1912, when Einstein and Otto Stern wrote a paper to explain Arnold Eucken's measurements of the specific heat of molecular hydrogen. The key feature of their explanation was the introduction of a zero-point energy; several months later, however, Einstein abandoned this idea, stating that he no longer believed in the existence of zero-point energy. Eucken's measurements were later shown to be problematic. See *CPAE*, vol. 4, the editorial notes, "Einstein on the Law of Photochemical Equivalence" and "Einstein and Stern on Zero-Point Energy," for more detailed discussions.

ROBERT SCHULMANN

FROM PERIPHERY TO CENTER: EINSTEIN'S PATH FROM BERN TO BERLIN (1902–1914)

In 1916, two years after moving to Berlin, Albert Einstein wrote to one of his closest friends in Zurich that he had accepted membership in the Prussian Academy of Sciences and a position at the University of Berlin without teaching obligations because of "a cousin, who drew me to Berlin in the first place."[1] Recently recovered letters reveal that Einstein had been corresponding with this cousin, a divorcée named Elsa Löwenthal née Einstein, since at least 1912.[2] Relying in part on this correspondence, I would like to suggest here that Einstein's reasons for coming to Berlin were far more complex than he indicated to his friend.

In what follows I will trace Einstein's career in the period from 1900 (when he graduated from the Swiss Polytechnic in Zurich) to April 1914 (when he assumed a place as Van 't Hoff's successor in the Prussian Royal Academy of Sciences) and follow this trajectory against the background of some developments in theoretical physics at the beginning of the twentieth century in Switzerland and Germany. My main objective is to place the outlines of Einstein's early professional career into the context of two sets of institutional constraints within which he operated.

To begin with, I will give the bare outlines of Einstein's career from 1900 until spring 1914; I then turn my attention to two critical phases of that development. The first phase, extending from 1902 until 1909, and coinciding with Einstein's work at the Swiss Patent Office, I call the Tactical Retreat;[3] the second, from 1910 until 1914, when Einstein already held university appointments, I refer to as the Call to Olympus. Discussion of both these phases rests on new material which the editors of the Einstein Papers project have gathered in recent years. For the first phase, a close examination of the administrative records of the University of Zurich have proved invaluable; for the second, the discovery of letters from Einstein to Elsa Löwenthal – his cousin, lover, and future second wife – give us new perspectives on the Berlin call. In the first phase (1902–1909), I will argue, Einstein went to the Patent

259

A.J. Kox and D.M. Siegel (eds.), No Truth Except in the Details, 259–271.
© 1995 *Kluwer Academic Publishers.*

Office in Bern with a strong expectation of returning to the academic fold, and a central figure in the Zurich establishment groomed him for the return. I will also claim that there was a strong professional motivation for Einstein to pursue physics while employed at the Swiss Patent Office, though the road back to Zurich was littered with obstacles, and he had no guarantees in advance that his return ticket to academia would prove valid. In the second phase (1910–early 1914), Einstein had already achieved legitimacy among his colleagues. Here I will argue that the lack of a stable funding model served to delay a summons to Berlin in which Einstein, in contrast to his activity during the Swiss phase played a primarily passive role. I want to show how a long-range plan was developed by key figures in the Prussian bureaucracy and by M. Planck, W. Nernst, E. Warburg, H. Rubens, and F. Haber to ensure Berlin's primacy as a center for research in the physical sciences and to secure a position for Einstein as the capstone to this achievement.

Let me now outline Einstein's career. In early 1902, one and a half years after graduating with a teacher's certificate in physics and mathematics from the Swiss Federal Polytechnic (ETH) and offering intermittent private instruction, Einstein received a provisional appointment at the Patent Office in Bern as a technical expert third class. By 1904 he had received a permanent position there, and in 1906 he was advanced to expert second class. In 1908 he succeeded in becoming a Privatdozent at the University of Bern, while retaining his position at the Patent Office. The following year Einstein resigned his position as patent officer and his Privatdozentur, and from the fall of 1909 until spring 1911 he taught physics at the University of Zurich. He then moved to the Charles University of Prague (the oldest German university), where he assumed a full professorship with an institute renamed especially for him as the "Institute of Theoretical Physics"; he was called back to Switzerland and to his alma mater, the Federal Polytechnic, as full professor, in fall 1912, after turning down a competing offer from the University of Utrecht and declining to become H.A. Lorentz's successor in Leyden. Finally, it is from Zurich that Einstein was summoned to Berlin to assume his place among the illuminati in spring 1914.

When Einstein graduated from the ETH in 1900 there were no theoretical physics chairs in all of Switzerland. Establishing a second position in physics to complement the experimental chair at Zurich was first broached in September 1901, in a memorandum from Alfred Kleiner, Professor of Physics, to his dean at the University of Zurich.[4] The argument was straightforward, based on Kleiner's need to lighten his teaching load as the only physics professor, who was required to teach in the areas of "experimental physics, theoretical physics and subjects from the border regions of chemical and mathematical physics." Kleiner agreed that it might not yet be time to create a second chair, but mentioned that he was encouraging a number of students of physics to obtain the Habilitation, and insofar as one individual might meet expectations, that he might be granted a position, after obtaining some teaching experience.

Quite clearly it was not a chair of theoretical physics that was being proposed. Kleiner's cautious memo reflected a general practice in German-speaking countries, wherein second professorships of physics were initially designed to absorb increasing numbers of students pursuing higher-level, if traditional, physics curricula and were only later converted into definitive theoretical physics positions.[5]

What, then, were the possibilities surrounding this second chair in 1901 which proved attractive to Einstein and afforded him the prospect of becoming an academic physicist? First, it must be understood that the definition of the Zurich position was a very vague one. Einstein and other candidates were not confronted with a precise set of criteria for eligibility. At the turn of the century, physics instruction at Swiss universities and the ETH offered a traditional fare of mechanics, thermodynamics, and electrotechnical and electromechanical applications to its students.[6] Even in Germany, there were only four chairs of theoretical physics.[7] It is instructive to recall that the theoretician Arnold Sommerfeld could, in 1907, still associate unalloyed theoretical work, ungrounded in the empirical tradition of the laboratory, with "an unhealthy dogmatism" reflecting "the abstract-conceptual manner of the Jew."[8]

There was, however, a physicist at the University of Bern – in the same town where Einstein was employed as a patent clerk – whose research interests lay in the area of theoretical physics and whose professional career provided a model for Einstein. A Privatdozent, Paul Gruner was the only one of nineteen academic physicists in Switzerland at this time who, in his research, was engaged primarily in the area of theoretical physics.[9] The attractiveness of the Gruner model became evident to Einstein no later than early 1903, when, already at the Patent Office, he made his acquaintance at a meeting of the Bern Natural Science Society, thereafter continuing the friendship throughout the Bern years.[10] The lustre of Gruner's example, however, was dimmed somewhat by the fact that he had, in 1903, not advanced beyond the status of Privatdozent in nine years.

We know that the 16-year-old Einstein, on entering the ETH in 1895, contemplated a career as an engineer and that he modified his professional goal to that of secondary-school teacher of physics and mathematics by the time he reapplied a year later. On graduating in 1900 from the ETH, he in turn abandoned the idea of becoming a physics instructor and attempted to obtain a doctorate in physics. Einstein chose his ETH physics mentor, H.F. Weber, as Doktorvater for a dissertation that would presumably deal with an experimental topic in Weber's primary fields of research: heat conduction and the anomalous temperature dependence of certain specific heats. This choice had another very practical feature: a position as Weber's *Assistent* became available in the summer of 1900, and I assume that Einstein presented himself as a candidate. He proved unsuccessful.

Einstein probably continued his doctoral work with Weber into early 1901, at which point he continued his attempt for the doctorate under the supervision of Kleiner at the University of Zurich – the same Kleiner who, later in the year, drew up the memorandum calling for a second chair in physics.

Though Einstein withdrew his dissertation under somewhat mysterious circumstances in early 1902, I think that the Kleiner phase of the dissertation effort (summer 1901 to February 1902) bore a markedly theoretical cast. Though the dissertation itself is not available, we know that Einstein completed a paper in late 1901 "on the electromagnetic light theory of moving bodies," which Kleiner thought so highly of that he urged Einstein to publish it.[11] I speculate that Kleiner encouraged Einstein's theoretical bent, drew the consequences of his failure in the more conventional realm of experimental physics, and urged him to position himself for a second chair of physics with a generously imprecise set of constraints.[12]

A number of discussions with Kleiner in late 1901 developed in the failed doctoral candidate a long-term strategy for winning his professional spurs. The first step, and a striking illustration of this strategy, is visible in Einstein's boast to his friend Michele Besso in January 1903 that he "has only recently decided to join the ranks of Privatdozenten, provided that I can pull it off. I won't even try to obtain the doctorate since it doesn't help me much ... Toward this end I will now set about working on the molecular forces in gases and then turn to comprehensive studies in electron theory."[13] Einstein based his hopes for becoming a Privatdozent – a position which to this day is a prerequisite for a teaching career in German-speaking universities – on a clause in the regulations of the University of Bern, stating that "in exceptional cases, when outstanding publications exist, the candidate can dispense with submitting a doctoral diploma."[14] The relevance to Einstein, who had already completed four papers (of which three had already seen publication in the *Annalen der Physik*)[15] was obvious. On the other hand, his failure was signaled two months later, when he pungently described "the university here [as] a pigsty" in a follow-up letter to Besso.[16]

A reversal of fortunes came two years later by a more conventional path. The dissertation completed by Einstein in 1905 presents further evidence of the effect of Kleiner's advice, on both the intellectual and institutional levels. This dissertation was a successful combination of a single, powerful theoretical claim – the existence of molecules – with a description of the law governing their behavior, buttressed by experimental data gleaned from Landolt and Börnstein.[17] It gave free play to the speculative, within the constraints imposed by an experimentally-oriented academic physics environment. The other hallmark papers of 1905 similarly reveal a combination of theoretical impulses with a concern for experimental verification, and in their content, and as means to a professional goal, bear witness to the encouragement and support of the Zurich physicist.[18]

The next step in the Kleiner–Einstein strategy was the preparation of a Habilitationsschrift, the traditional requirement for the conferral of what the Swiss call the *venia docendi* (teaching permission). Einstein's first concerted effort to obtain it at the University of Bern came in 1907, when he submitted seventeen publications in theoretical physics, almost all of which had been prepared while he was at the Patent Office.[19] His failure to obtain the *venia docendi* on the first attempt was probably due to the fact that he did not submit a tailor-made paper, but rather tried to overwhelm the authorities with the sheer number of his publications.[20] Einstein's attempts to secure secondary-school teaching positions at the Technikum in Winterthur and at the Kantonsschule in Zurich in late 1907 and early 1908 must be seen as parallel attempts to legitimize himself for a call to Zurich. They should not be regarded as desperate attempts to keep himself financially above water[21] or to return to his earlier goal of becoming a secondary-school teacher in physics.

At the beginning of 1908 Einstein submitted a Habilitationsschrift on black-body radiation and the *venia docendi* was conferred at the end of February.[22] In the next two semesters, summer 1908 and winter 1908/09, Einstein gave courses on the molecular theory of heat and on radiation theory, which together with the Habilitation topic illustrate his theoretical interests. Kleiner, in the meantime, was appointed rector of the University of Zurich for two years, beginning winter semester 1908/09, and was able to move swiftly to consummate the lengthy process initiated seven years earlier. He travelled to Bern even before his rectorship officially began, in order to attend an Einstein lecture, from which he returned dissatisfied.[23] Undeterred, he requested that Einstein address the Physical Society of Zurich in February 1909, an "examination" which Einstein passed handily.[24]

The final step in the Einstein appointment was a recommendation that Kleiner prepared for the Zurich authorities.[25] It is interesting to compare the language of this document with the one Kleiner had prepared eight years earlier. He began by describing the candidate "as one of the most important theoretical physicists" and went on to characterize Einstein's publications as possessing "an unusual keenness in the conception and implementation of ideas and a profundity which penetrates to the most fundamental level." Whereas the document of 1901 never touched on the theme of theoretical physics, the 1909 recommendation was built around it. The confidence of Kleiner's language reflected the fact that the imagery of the physicist as theorist had gained an established legitimacy, even linguistic currency in Switzerland. To a considerable degree, this was due to the international resonance that Einstein's work in physics had found while he was employed at the Patent Office.

With Einstein finally ensconced in Swiss academic life as extraordinary professor of theoretical physics in May 1909, I want next to ask how he made the move from the periphery to the center, where both his legitimacy and the acceptance of theoretical physics were unquestioned, but where a new set of obstacles barred the way. As we examine the background to the Berlin appointment, let me remind you of Friedrich Adler's prophetic statement made a year before Einstein's appointment in 1909, that he "will be coming now to Zurich, but only as to a transit station, as he will soon be called to Germany."[26] Why, then, did it take five years?

By 1910, it is very clear that the eyes of Berlin, if not of the world, were already on Einstein. In October of that year Emil Fischer of the

University of Berlin arranged a three-year stipend of 5,000 marks per year for Einstein, contributed by Franz Oppenheim, the director of Agfa, a major German anilin-dye producer.[27] Fischer singled out Einstein, together with Planck and Nernst, as examples of scientists who had given Germany the lead in Europe in fundamental research on thermodynamics.[28] Earlier in the same year, Nernst, Fischer's colleague in Berlin visited Einstein in Zurich, praised his quantum hypothesis, and called him a "Boltzmann redivivus."[29] On drawing up a list of participants, in July 1910, for the first of the Solvay Conferences the following year, Nernst gave special prominence to Planck and Einstein as seminal contributors to "the new development of principles which serve as the basis of classical molecular theory and the kinetic theory of matter."[30]

Einstein's special theory of relativity was not neglected in these paeans of praise. When Einstein's name was put forward as a candidate for a full professorship in Prague in 1910, Planck's stirring phrase from his Columbia lectures of the previous year was incorporated into the memorandum of recommendation: "In its breadth and profundity, this principle [Einstein's special theory] is comparable only to the revolution in the physical world-view occasioned by the introduction of the Copernican world-system."[31]

A number of studies – most importantly that of Forman, Heilbron, and Weart[32] – have shown how the Big Four (Germany, France, England, and the United States) had about the same level of expenditures on scientific research at the beginning of the century, and how by 1914, the United States had assumed the lead, with England and Germany about tied for second place, and with France bringing up the rear, in parallel with these countries' respective positions in aggregate industrial output. The beginnings of big science and national support for scientific research before the Great War were based on, and accelerated by, increasing pressures on these states to compete in industrial terms and match each others' military growth.

Bureaucratic mills grind slowly however, and the difficulty of coming to terms with these new realities are typified by the attitude of Adolf Harnack, Prussian court theologian and a major ideologue of the Prussian state, who had the ear of the Emperor. Harnack rejected the purely private organizational and financial model that the Americans had developed and which assured them primacy. Harnack had a pre-modern distrust of the motives of private capital, fearing that it would be difficult to harness its interests to those of the state.

What the German state did develop toward the end of the first decade of the century was a cautious policy of Mischfinanzierung (complementary funding), drawing on a combination of industrial and state resources to further commonly agreed-upon research programs.[33] Thus, for example, the Koppel Stiftung (set up in 1905 with an endowment by one of the richest men in Prussia, Leopold Koppel) financed Fritz Haber's Institute for Physical Chemistry and Electrochemistry in conjunction with the Prussian state, while the newly-founded Kaiser Wilhelm Society and the Verein Chemische Reichsanstalt simultaneously supported its twin, Ernst Beckmann's Institute for Chemistry.[34] Still, the concept of Mischfinanzierung was in its infancy and three earlier attempts to create a physics institute outside the Prussian university system had already failed by this time. The first was a proposal made by Philipp Lenard in December 1906, calling for an Institute of Physics Research – obviously a thinly veiled attempt to obtain a beachhead for Lenard in Berlin.[35] The second was a memorandum by Nernst in April 1908, which suggested the creation of an Institute for Radioactivity and Electron Research.[36] Finally, in November 1909 Harnack had proposed setting up a physics institute that would serve to "strengthen experimental research."[37] But all three proposals fell upon deaf ears. Having given some indication of the unwillingness of the Prussian bureaucracy to embrace these suggestions let us pick up on the Einstein chronology in April 1912 and observe the maturation of plans to bring him to Berlin.

From a newly-discovered letter to Elsa Löwenthal of 30 April 1912,[38] we know that Einstein visited Berlin in early April of that year to hold discussions with Planck, Nernst, Warburg, Haber and Rubens. At first I had assumed that talk centered not only on mutual research interests, which Einstein mentioned in a contemporary letter to Besso,[39] but also on prospects for a call to Berlin. Apparently, though, only a position at the Physikalisch-Technische Reichsanstalt was discussed with Warburg, its President. How do we know this? Because we have also come into the possession of a letter written by Haber on 4 January 1913 to a Referent in the Prussian Ministry of Education, Hugo Andres Krüss,[40] in which Haber made it very clear that he had not breathed a word to Einstein about the possibility of bringing him to Berlin. More importantly, Haber went on to forge a plan, which he had apparently discussed with Krüss some weeks earlier, for bringing "this extraordinary man" to Berlin and giving Einstein rooms and a salary within his Institute for Physical Chemistry of the Kaiser Wilhelm Society. Haber proposed that the Prussian government avail itself of the Richard Willstätter model for attracting Einstein – that is, dangle the idea of a co-directorship in front of Einstein, offer him 15,000 marks, and completely renovate the second floor of his institute for Einstein's use. Haber was convinced that Leopold Koppel would shoulder the cost of 50,000 marks for renovation and also pick up Einstein's salary. At the same time Haber recapitulated and accepted an earlier objection from Krüss's superior, Friedrich Schmidt-Ott, that the creation of a new institute for Einstein would be out of the question "because Einstein is not an experimentalist."[41]

I assume that Nernst and Planck were privy to Haber's deliberations, although I cannot prove it. In any case, in late May of 1913, these two together with Emil Warburg and Heinrich Rubens made an announcement in the Prussian Academy of Science that they would be proposing Einstein as a member of the Academy.[42] The Haber model had apparently been modified in such a way as to combine membership in the Academy with miraculously revived prospects of a directorship of a Kaiser Wilhelm Institute of Physics. We know the first part of the modification from the famous Planck proposal to the Academy of 12 June 1913[43] and the second from three letters to Löwenthal in late 1913 and early 1914,[44] the last one written two months before Einstein's arrival in Berlin. In these letters, besides talking about a wealth of other topics, Einstein blew hot and cold about the prospects of obtaining an institute. In the one from mid-October, he wrote "I haven't heard anything about the question of an institute; I've pushed it out of my mind. I'm sure that it will fall well-deservedly into the water."[45] Three weeks later he announced to Löwenthal that "the matter of my purported institute has been postponed until I arrive in Berlin. It would be good, were I to obtain some kind of institute; then I could collaborate with others, which I much prefer to working alone."[46] In the last of the letters dealing with the question of an institute, that of early 1914, he told Löwenthal that "there is apparently nothing to be done with the institute. Thank God, I will be free as a bird."[47]

It was wise of Einstein to express skepticism about the fortunes of his institute. His last comment was written shortly after a meeting between Planck, Nernst, and representatives of the Prussian Ministry of Education, where all sides decided to postpone the creation of a Kaiser Wilhelm Institute for theoretical physics and to bring Einstein to Berlin solely as successor to Van 't Hoff at the Academy. Unfortunately, the minutes of the early January meeting are very sketchy,[48] but again it seems that a research institute given over to purely theoretical matters was considered unacceptable. It was, in fact, not until 1917 that Einstein finally got his institute after many twists and turns.[49]

In conclusion, let me touch on some themes suggested by the discussion above. It is clear that Germany (i.e. Prussia) had as much trouble in 1913 calling a theoretician of international repute to Berlin as the Swiss had had in 1909 in calling a Bern patent office clerk to the University of Zurich. Moreover, there was a rivalry between the Prussian Academy of Sciences and the Kaiser Wilhelm Society, which Günter Wendel has pointedly touched on, but which should be re-examined in light of the Einstein call. Can a case be made that the Academy was rooted in a more conservative research tradition, still strongly influenced by the historical-literary class, while the Kaiser Wilhelm Society adopted a more modern, aggressive, basic research program?

The development of big science was thus a much more tentative, halting process than is commonly thought: new forms of financial support only emerged gradually in Germany and elsewhere. A commitment to a program of scientific research, including theoretical work that only yields fruit far down the line, was problematic at the beginning of the century, just as it is today. Then, new forms of state-organized science were crystallizing, just as the discipline of theoretical physics began to come into its own. Let us hope that the end of the Cold War will loosen the bonds that bind science to the state in our own time.

The Collected Papers of Albert Einstein
Boston University
U.S.A.

NOTES

[1] "eine[r] Cousine, die mich ja überhaupt nach Berlin zog." Einstein to Heinrich Zangger, 7 July 1915, Estate of Heinrich Zangger, Zurich.
[2] See the twenty-four letters from Einstein to Elsa Löwenthal in *The Collected Papers of Albert Einstein*, vol. 5 (Princeton: Princeton University Press, 1993). In the following this edition will be abbreviated as *CPAE*, followed by a volume number.
[3] Discussed first in Robert Schulmann, "Einstein at the Patent Office: Exile, Salvation, or Tactical Retreat?," *Science in Context* 6 (1993): 18–25.
[4] Dean Hans Schinz to Erziehungsdirektor Albert Locher, 10 September 1901, Zurich Cantonal Archive, U 110 b.1 (25).
[5] One must be careful not to include Germany, where theoretical physics was in flower a

full decade earlier. See, e.g., Christa Jungnickel and Russell McCormmach, *Intellectual Mastery of Nature: Theoretical Physics from Ohm to Einstein*, vol. 2, *The Now Mighty Theoretical Physics, 1870–1925* (Chicago: University of Chicago Press, 1986), p. 123.
[6] Einstein's ETH Record and Grade Transcript in *CPAE*, vol. 1, Doc. 28.
[7] Max Planck (Berlin), Woldemar Voigt (Göttingen), Ludwig Boltzmann/Arnold Sommerfeld (Munich), and Theodor Des Coudres (Leipzig).
[8] "ein ungesunder Dogmatismus ... die abstrakt-begriffliche Art des Semiten." The reference is to Einstein specifically and is from a letter of 26 December 1907 to H.A. Lorentz, Archief H.A. Lorentz, Rijksarchief Noord-Holland, Haarlem, The Netherlands.
[9] For the purpose of this paper I do not consider academic practitioners of electrochemistry, physical chemistry, etc., as physicists, although they had overlapping interests. Only two Swiss universities – Geneva & Lausanne – and the ETH had second chairs of physics, while another three had Privatdozenten in this field (Schinz to Locher, cited in note 4).
[10] On 2 May 1903 Gruner lectured to a meeting of the Society at which Einstein was inducted (Minutes of the Society, SzBe, Archiv der Naturforschenden Gesellschaft Bern). Between 1903 and 1907 when Gruner supported his efforts toward Habilitation, Einstein gave numerous informal talks on scientific topics at Gruner's home in Bern (oral communication from Gruner's son).
[11] "über elektromagnetische Lichttheorie bewegter Körper." Einstein to Mileva Maric, 19 December 1901, in *CPAE*, vol. 1, Doc. 130.
[12] Indirect evidence of who was under consideration is provided in Kleiner's memorandum of 27 February 1909, Zurich Cantonal Archive, U 110 b.2 (44).
[13] "hab mich nun neuerdings entschlossen, unter die Privatdozenten zu gehen, vorausgesetzt nämlich, daß ichs durchsetzen kann. Den Doktor werde ich hingegen nicht machen, da mir das doch wenig hilft ... In der nächsten Zeit will ich mich mit den Molekularkräften in Gasen abgeben, und dann umfassende Studien in Elektronentheorie machen." Einstein to Michele Besso, January 1903, in *Albert Einstein/Michele Besso Correspondance 1903–1955*, Pierre Speziali, ed. (Paris: Hermann, 1972), no. 01, p. 4, and in *CPAE*, vol. 5, Doc. 5.
[14] See the Reglement über die Habilitation an der philosophischen Fakultät der Hochschule Bern, para. 2.
[15] *CPAE*, vol. 1, Docs. 1–4.
[16] "Die hiesige Universität ist ein Schweinestall." Einstein to Michele Besso, 17 March 1903, in *Einstein/Besso Correspondance*, no. 03, p. 14, and in *CPAE*, vol. 5, Doc. 7.
[17] Hans Landolt and Richard Börnstein, *Physikalisch-chemische Tabellen*, 2nd ed. (Berlin: Julius Springer, 1894).
[18] Notes by a confidant of Einstein's, Heinrich Zangger, suggest that Zangger consulted on a Zurich appointment for Einstein with Kleiner in 1905 (notes in the Estate of Zangger, Zurich). By 1908, the evidence is more direct: Alfred Kleiner to Einstein, 28 January and 8 February 1908, in *CPAE*, vol. 5, Docs. 78 and 80.
[19] Einstein to Director of Education, 17 June 1907, *CPAE*, vol. 5, Doc. 46.
[20] Bern University regulations required "a specialized investigation of a scientific topic." See Reglement über die Habilitation an der philosophischen Fakultät der Hochschule Bern of 1891, para. 2.
[21] Einstein's salary at the Patent Office – 4,500 Swiss francs – was the same as his beginning salary at the University of Zurich (Minutes of the Zurich Governing Council

1909, no. 888, 7 May 1909 in: Zurich Cantonal Archive, U 110 b.2 (44)).

[22] Minutes of the Faculty, 24 February 1908, Philosophical Faculty II, Bern Cantonal Archive.

[23] Einstein to Michele Besso, 6 March 1952, in *Einstein/Besso Correspondance*, no. 182, p. 464.

[24] Alfred Schweitzer to Einstein, 19 January 1909, in *CPAE*, vol. 5, Doc. 135.

[25] Memorandum of 27 February 1909, Zurich Cantonal Archive, U 110 b.2 (44).

[26] "wird wohl jetzt nach Zürich kommen, allerdings wohl nur *als Durchgangsstation*, denn er wird wohl bald nach Deutschland weiterberufen werden." Friedrich Adler to Victor Adler, 19 June 1908, Archive for the History of the Workers' Movement, Vienna, file 77 in the Adler Archive.

[27] Emil Fischer to Einstein, 1 November 1910, in *CPAE*, vol. 5, Doc. 230.

[28] One may note here how Einstein is being appropriated as a German, though he gave up Württemberg citizenship fourteen years earlier and was teaching at a Swiss university.

[29] Walther Nernst to Arthur Schuster, 17 March 1910, Archive of the Royal Society, London, Sc. 130.

[30] "die neue Entwicklung der Prinzipien, die als die Grundlage der klassischen Molekulartheorie und der kinetischen Theorie der Materie dienen." Draft enclosure in Walther Nernst to Ernest Solvay, 26 July 1910, Archive of the Free University of Brussels, 11Z, dossier 1er Conseil de Physique, no. 1688.

[31] "Mit der durch dies Prinzip im Bereich der physikalischen Weltanschauung hervorgerufenen Umwälzung ist an Ausdehnung und Tiefe wohl nur noch die durch Einführung des Copernikanischen Weltsystems bedingte zu vergleichen." The citation is from the Columbia lectures he delivered in 1909 (Max Planck, *Acht Vorlesungen über Theoretische Physik, gehalten an der Columbia University in the City of New York im Frühjahr 1909*. Leipzig: Teubner, 1910) and is restated in the memorandum An das Professorenkollegium der philosophischen Fakultät der k. k. deutschen Universität in Prag, before 21 April 1910, Czech State Archive, MKV/R, no. 101, Einstein Dossier.

[32] Paul Forman, John L. Heilbron, and Spencer Weart, "Physics *circa* 1900. Personnel, Funding, and Productivity of the Academic Establishments," *Historical Studies in the Physical Sciences* 5 (1975): 1–185.

[33] Cf. Günter Wendel, *Die Kaiser-Wilhelm-Gesellschaft 1911–1914. Zur Anatomie einer imperialistischen Forschungsgesellschaft*. Studien zur Geschichte der Akademie der Wissenschaften der DDR, vol. 4 (Berlin: Akademie-Verlag, 1975), and Lothar Burchardt, *Wissenschaftspolitik im Wilhelminischen Deutschland. Vorgeschichte, Gründung und Aufbau der Kaiser-Wilhelm-Gesellschaft zur Förderung der Wissenschaften* (Göttingen: Vandenhoeck & Ruprecht, 1975).

[34] The Kaiser Wilhelm Society was founded in 1911 as were its two flagship institutes, that of Haber and of Beckmann. See *Forschung im Spannungsfeld von Politik und Gesellschaft. Geschichte und Struktur der Kaiser-Wilhelm-/Max-Planck-Gesellschaft*, Rudolf Vierhaus and Bernhard vom Brocke, eds. (Stuttgart: Deutsche Verlags-Anstalt, 1990), pp. 98–101, on Haber's institute; pp. 144–151 on Beckmann's.

[35] Burchardt, *Wissenschaftspolitik*, p. 22 and Vierhaus and vom Brocke, *Forschung*, pp. 132–133.

[36] Burchardt, *Wissenschaftspolitik*, p. 27.

[37] Burchardt, *Wissenschaftspolitik*, pp. 31–34, and Vierhaus and vom Brocke, *For-*

schung, pp. 138–140.

[38] *CPAE,* vol. 5, Doc. 389.

[39] Einstein to Michele Besso, 26 March 1912, in *CPAE,* vol. 5, Doc. 377.

[40] GyB, Acta Preußische Staatsbibliothek, Kaiser-Wilhelm-Institute XXVII.

[41] "weil Einstein kein Experimentator ist."

[42] Memorandum of Planck *et al.,* 29 May 1913, Archive of the former Academy of Sciences of the DDR, II–III, vol. 36, p. 35.

[43] C. Kirsten and H.-J. Treder, *Albert Einstein in Berlin, 1913–1933* (Berlin: Akademie-Verlag, 1979), vol. 1, pp. 95–97.

[44] Einstein to Löwenthal, 16 October 1913, 7 November 1913, and February 1914, in *CPAE,* vol. 5, Docs. 478, 482, and 509.

[45] "Über die Institutsfrage hörte ich nichts; ich denke nicht mehr daran. Das wird sicher in das wohlverdiente Wasser fallen."

[46] "Die Angelegenheit meines ev. Institutes wurde hinausgeschoben, bis ich nach Berlin komme. Es wäre doch gut, wenn ich eine Art Institut bekäme; ich könnte dann mit andern zusammen arbeiten statt nur allein. Das entspricht sehr meiner Vorliebe."

[47] "Mit dem Institut ist es vorerst nichts, Gott sei Dank. Ich werde frei sein wie ein Vogel."

[48] Minutes of a meeting at Prussian Ministry of Education, 9 January 1914, German State Archive Potsdam, Rep 76 V c, Sekt. 2, Tit. 23, Litt. A, no. 116, p. 18.

[49] A good account of these twists and turns is given in Wendel, *Kaiser-Wilhelm-Gesellschaft,* pp. 198–199, Burchardt, *Wissenschaftspolitik,* pp. 118–119, and Vierhaus and vom Brocke, *Forschung,* pp. 101 and 177.

GERALD HOLTON

EINSTEIN AND BOOKS

Albert Einstein loved books. A photograph taken of him in his study at Princeton toward the end of his life (Figure 1) shows him at a desk placed in such a way that, sitting in his comfortable chair, he would be practically surrounded by books of all kinds. His was clearly not a library devoted only to physics books. One can make out the spine of copy of Ghandi's *Autobiography* and a copy of the Bible, on shelves that are deep enough to have one row of books behind the other. As is clear from Einstein's published writings and his correspondence, his interests went far beyond science and the philosophy of science; he also thought and wrote knowledgeably about religion and politics, literature and music, education and human rights, pacifism and anti-Semitism, and the plight of the oppressed.

From the very beginnings, Einstein was fond of books, and often deeply affected by them. Born into a cultured, closely-knit German-Jewish family that valued *belles lettres* and music and earned its bread through an engineering enterprise, he came early into contact with a great variety of humanistic and scientific books, and this was reinforced in his school years in the exacting German *humanistische Gymnasium* system. According to the biography, *Albert Einstein* (London: Thornton, Butterworth Ltd., 1931) by Anton Reiser – the pseudonym for Rudolf Kayser, who in fact was Einstein's son-in-law, and knew him well – it was a family habit on evenings to read aloud from such works as those of Friedrich Schiller and Heinrich Heine. Through a talented school instructor, Einstein later came also under the spell of works by Shakespeare and Goethe.

Einstein tells in his own Autobiography (Paul A. Schilpp, *Albert Einstein: Philosopher-Scientist*, 1949, pp. 3–5) that he had been so caught up by his reading in the "stories of the Bible" that at the age of 11 he underwent a period of "deep religiosity," followed, however, soon by an equally powerful opposite reaction after reading popular scientific books; he identifies Aaron Bernstein's *People's Books on Natural Science* [*Naturwissenschaftliche Volksbücher*, in many volumes]. From other sources we know also of the influence on young Einstein of Ludwig Büchner's *Kraft und Stoff*, Immanuel Kant's *Kritik der reinen*

273

A.J. Kox and D.M. Siegel (eds.), No Truth Except in the Details, 273–279.
© 1995 *Kluwer Academic Publishers.*

Fig. 1.

Vernunft (given to Einstein at age 13), and Alexander von Humboldt's *Kosmos.* At age 16, he encountered the multi-volume physics text by Jules Violle, *Lehrbuch der Physik* (1892–93), which he annotated and kept in his library.

But perhaps most important for the young boy's identity formation and his growing feeling of self-confidence was what he called in his Autobiography the experience of the "wonder" of discovering, at age 12, a little book on Euclidean plane geometry (probably the *Planime-trie,* Part 2 of Adolf Sickenberger's (1888) *Leitfaden der elementaren Mathematik*), whose lucidity and certainty of results, he said, "made an indescribable impression upon me." The effect of the "holy geometry booklet," as he termed it, on Einstein's later, typically axiomatizing method of theory formation has even now not been sufficiently appreciated. It is appropriate that the earliest example of writings from Einstein's pen, as given in vol. I of his *Collected Papers*, is a sharp marginal comment on a geometrical position, entered in his copy of Eduard Heis and Thomas Joseph Eschweiler's *Lehrbuch der Geometrie* (1881), during the period 1891–95. Among the other mathematics books the youth owned were three by Heinrich Borchert Lübsen; they remained part of his Library, one of them carrying annotations by Einstein (cf. p. 4, vol. I, *Collected Papers*).

Einstein's Autobiography further records that during his years as a student at the Polytechnic Institute in Zürich, he read the works of "Kirchhoff, Helmholtz, Hertz, etc." at home, even at the cost of neglecting a sound mathematical instruction through the lectures of his professors. It was more evidence of his preference for self-study out of the books. Through letters written between 1899 and 1902 to Mileva Marić, we can conclude they studied together books by Boltzmann, Drude, Helmholz, Hertz, Kirchhoff and Mach. From Einstein's letter to Michele Besso we know that while still a student at the Polytechnic he was introduced by Besso to Mach's *Mechanics* and the *Wärmelehre*, both making "big impressions" (letter of 6 January 1948); similarly, his Autobiography makes clear that Ernst Mach's *Mechanics* (1883), in his words, "exercised a profound influence upon me while I was a student."

The influence of these and other books on his subsequent work has only recently begun to be carefully traced. For example, we now know that his reading of the text by August Föppl, *Einführung in die Maxwell'sche Theorie der Elektricität* (1894), provided some of the ideas and structure of his 1905 paper on relativity. And perhaps most important, Einstein confessed that especially the crucial step toward the special theory of relativity, namely freeing himself from the axiom of the absolute character of time and of simultaneity, required critical reasoning which "was decisively furthered, in my case, especially by the reading of David Hume's and Ernst Mach's philosophical writings" (Autobiography, p. 53). In a letter to Besso of 6 March 1952, Einstein added a third author, read during the Bern years, who also had "influence" – namely, Henri Poincaré. And from his contemporaries at that time as well as his biographer Carl Seelig we know that the list of authors, read and debated by the little band of friends about Einstein in their "Olympia Akademie" meetings, included Plato, Spinoza, Hume, J. S. Mill, Ampère, Kirchhoff, Helmholtz, Poincaré, and Karl Pearson – as well as Sophocles and Racine.

We shall come back to a further examination of some of the authors and books that impinged on Einstein's work. Suffice it to say now that, to a higher degree than for other scientists I have studied, the book has a major role in Einstein's thought and life. Nowadays, we have become used to the scientist typically finding professional guidance through the study of published journal articles, and more and more frequently even through prepublication reprints. Books now play a secondary role in the list of publications, and in citations in scientific articles. For example, a study made a few years ago of citation habits of physicists showed that only 6.5% of the references were to books. But in this respect Einstein was even more conservative and book-bound than his contemporaries; it was not accidental that his own Ph.D. dissertation made references in its text or footnotes only to a few books, the exception being a citation to one of his own papers.

Even though Einstein received, and kept, a great number of offprints by other scientists and scholars (now preserved at the Weizmann Institute at Rehovot, Israel), it is perhaps not too much to say that this relatively unusual interest in books, at least in his early, most productive years, was merely another indication of the fact that he tended to believe more firmly in older, well-established findings that had stood the test of time (and hence came to be incorporated in books), rather than allowing himself to be carried away by the latest news from the laboratories. Thus Einstein could confess from time to time without evident embarrassment that he had not kept up with the journal literature on points where one would have expected him to do so. (See for example the letters of Einstein to Stark and to Seelig, respectively on p. 272 and p. 307, vol. 2 of the *Collected Papers*.)

Yet another evidence of Einstein's fondness for books is simply the unusually large number of Prefaces or Introductions which Einstein furnished for books on a great variety of subjects that interested him. These include volumes by or about such authors as Lucretius, Galileo, Newton, Spinoza, Planck, Erwin Freundlich, David Reichinstein, Leopold Infeld, Peter G. Bergmann, Philipp Frank, and Upton Sinclair. One should also not forget that he was evidently proud of his own two books on relativity, and allowed his essays to be collected in such volumes as *Mein Weltbild* and *Out of My Later Years*.

Here, someone might interrupt to ask: Why is it important to know about the books that may have been read by a scientist, and why should a catalogue be made of those that remained in his library? To answer, one would begin by saying that in assessing the intellectual work of a major figure, and his debt to others, the historian desires to know as much as possible about the personal holdings. Such a library can tell much about the owner's range of interest, habits of work, acknowledged or unacknowledged but possible transmission of ideas from predecessors, objections to these ideas (as seen through marginal notes), and the like. It is, for example, of interest to know that Niels Bohr from reading Paul Møller, Søren Kierkegaard, and quite possibly William James's *Principles of Psychology*; or that, on the way to the discovery of the double-helix structure of DNA by Francis Crick and James Watson, Erwin Schrödinger's book *What Is Life* played an important role.

Therefore it is not surprising that enormous efforts have been put into the reconstruction and cataloguing of the libraries that were once owned, for example, by John Winthrop, Jr., Samuel Pepys, Robert Boyle, Robert Hooke, John Locke, Charles Darwin, Ernst Mach, and William James. With respect to Newton, it took about two and a half centuries to reestablish with authority the composition of Newton's library. (See John Harrison, *Library of Isaac Newton*, Cambridge University Press, 1978). We must count ourselves very lucky that in the case of Einstein, through the mighty labors of the librarians and curators of the Jewish National and University Library at Hebrew University, and with the welcome financial support of NHK (Japan), a catalogue of the collection of Einstein's books is now being organized.

To be sure, all such inventories give only a snapshot of the state of the holdings at a given moment. Furthermore, we do not always know which books in Einstein's study or house were his own, and which were those of members of the family who lived or had lived there or of his devoted secretary and housekeeper, Helen Dukas – who joined the family in 1928 and who, by Einstein's will, was heir to the library, with the stipulation that upon her death all the remaining books were to be transferred to their final destination in Jerusalem, where they are now kept.[1] Nor do we know for certain whether any books originally in Einstein's possession in Europe were left behind when his library and papers in Berlin were removed (by Diplomatic Pouch, through the efforts of the French Embassy) after the ascent of the Nazis in 1933, and sent on to Einstein's new home in America.

But in any case, most if not all the surviving books had been accessible to Einstein himself at some point. Indeed, many may well have come to him from his parents or their close relatives. In the present inventory of the collection, one notes the considerable number of books with publication dates prior to Einstein's own birth in 1879. These, reminders of the wide range of reading that may have been possible for him during his youth, include volumes by Dante Alighieri, Julius Caesar, Charles Dickens, J. W. Goethe, Alexander von Humboldt, Immanuel Kant, Hokusai, Gotthold Lessing – to list only a few in the early part of the alphabet. Another interesting point is the considerable number of books on religion, and particularly on Jewish religious themes, indicative of his growing interest in this subject after the early years.

Finally, we can illustrate the fruitful way books interacted with Einstein's own writings, by referring to some examples from his early publications, as documented in vols. 1 and 2 of his *Collected Papers*. In his articles on kinetic theory and statistical mechanics, Einstein referred variously to his prior familiarity with the work of Rudolf Clausius, *Die mechanische Wärmetheorie* (3 volumes, 1879–1891), Mach's *Wärmelehre* (1886) (read in 1897 or soon thereafter), Ludwig Boltzmann's powerful *Vorlesungen über Gastheorie* (2 volumes, 1896, 1898, still in Einstein's library, with some annotations), and Gustav Kirchhoff's *Vorlesungen über die Theorie der Wärme* (1897).

Wilhelm Ostwald's *Lehrbuch der allgemeinen Chemie* (1891) and the tables of Landolt and Börnstein (1894) are the only references Einstein makes in the text of his very first published paper (on Capillarity, 1901). Evidence that he had read Heinrich Hertz's book, *Untersuchungen über die Ausbreitung der elektrischen Kraft* (1892) can be found in his essay on the state of the ether in a magnetic field, sent to his uncle Caesar Koch in 1895. Another book still in Einstein's library, Drude's *Lehrbuch der Optik* (1900), was a possible source of ideas in his grappling with the puzzle of black-body radiation. Helpful concepts on what he later recognized to be Brownian Motion may have come from Poincaré's *Science et Hypothèse* (1902), read between 1902 and 1905, which contains a discussion of that phenomenon (as it does also of the difficulty of intuiting the simultaneity of two events at different localities). It is suggestive of Einstein's penchant for books that prior to writing his 1905 relativity paper he had read only H. A. Lorentz's book of 1895, *Versuch einer Theorie der electrischen und optischen Erscheinungen in bewegten Körpern*, but not Lorentz's more recent papers in journals.

These examples are merely meant to signal that historians of science will find much to guide them in a coordinate study of Einstein's *Nachlass* of over 45,000 letters, manuscripts, and publications. Indeed, the total collection is not only a treasure in its own right, but also a magnifying glass with which to study important aspects of the cultural history of this century.

Jefferson Physical Laboratory
Harvard University
U.S.A.

NOTES

[1] Miss Dukas told me that after Einstein's death she distributed a few books from the library to close friends who had been helpful during the last illness; however, she could not remember either the titles or the recipients.

DANIEL M. SIEGEL*

TEXT AND CONTEXT IN MAXWELL'S
ELECTROMAGNETIC THEORY

It is a common observation that the considerable effort invested in historical and philosophical study of the work of James Clerk Maxwell in electromagnetic theory has not been handsomely repaid: the effort seems not to have been cumulative, generating rather more questions than answers. Characteristically, trenchantly – and perhaps a bit extravagantly – Paul Forman has "described the [Maxwell] enterprise as a vortex in a draining sink that continually sweeps in new intellects and new ideas, which just as continually vanish."[1] (The allusion is to Maxwell's vortex theory of electricity and magnetism;[2] I shall have more to say both about allusions in various kinds of discourse and about Maxwell's vortex theory.)

Given the general agreement concerning the problems of the secondary literature on Maxwell's electromagnetic theory, the next question is, what is to be done? How shall we proceed so as to make better progress in the analysis and understanding of Maxwell's work in this area? Among the various answers that can and have been given to this question, two are interesting for their polar opposition: one side counsels a broader and more synthetic approach to Maxwell studies, with greater attention to the intellectual, social, academic, and national contexts of Maxwell's work; the other suggests more careful analysis, in textual and mathematical detail, of Maxwell's writings in electromagnetic theory.[3] These two approaches are, of course, not mutually exclusive, in that we would all agree that we must give attention to both text and context: on the one hand, we must read the individual texts, the documents on which we ground our historical enterprise, carefully and thoughtfully; on the other hand, we must never lose sight of the task of broader synthesis. Nevertheless, life is short, and we need to set priorities in our scholarly work: we need to allocate our time as between burrowing into texts and coming up to contemplate broad contextual horizons.

Ultimately, of course, the proof of the pudding will be in the eating: each scholar will make his or her own decisions with regard to the balance between text and context, and the community of scholars will

A.J. Kox and D.M. Siegel (eds.), No Truth Except in the Details, 281–297.
© 1995 *Kluwer Academic Publishers.*

judge and make use of the resulting contributions according to their merit and utility. Nevertheless, judgment and assimilation of individual contributions by the community will take place in the context of some broader vision of the needs and directions of the history of science discipline in general, and – in the present instance – Maxwell studies in particular. Thus, my purpose here is to suggest that, in Maxwell studies, we need more in the way of proper reading of texts. In order to do this, however, it is necessary to overcome certain hindrances to the proper reading of texts – certain trends and outlooks, certain practices, and certain *shibboleths* current in the history of science discipline – which tend to deter people from reading texts properly and reporting on that reading. In particular, I will be concerned with the reading of texts in mathematical physics, and how that enterprise is hindered by recent trends in the history of science discipline.

How should one read a text in mathematical physics? Let me begin a bit far afield, with a Maxwell text of a different sort. In 1852, while an undergraduate at Cambridge, Maxwell wrote a poem, entitled "A Vision," a brief analysis of which will call attention to some modalities in reading texts that will carry over to the case of Maxwell's mathematical physics. The poem, as presented in the biography of Maxwell by Campbell and Garnett, begins as follows:

A VISION

Of a Wrangler, of a University, of Pedantry, and of Philosophy
10th November 1852

Deep St. Mary's bell had sounded,
And the twelve notes gently rounded
Endless chimneys that surrounded
    My abode in Trinity.
(Letter G, Old Court, South Attics),
I shut up my mathematics,
That confounded hydrostatics –
    Sink it in the deepest sea!

In the grate the flickering embers
Served to show how dull November's
Fogs had stamped my torpid members,
    Like a plucked and skinny goose.[4]

One could begin to read this through as if it were prose, but one knows this is not the appropriate way to read poetry; one must, instead, read it aloud, or at least imagine it read aloud. When read in this way, the poem begins to disclose something that was otherwise not evident: The meter and rhyme scheme, and the rhyming syllables themselves, are somehow familiar, reminding one of another nineteenth-century poem in English:

> Once upon a midnight dreary, while I pondered, weak and
> weary
> . . . .
> Ah, distinctly I remember it was in the bleak December
> And each separate dying ember wrought its ghost upon the
> floor.
> Eagerly I wished the morrow; vainly had I sought to borrow
> From my books surcease of sorrow – sorrow for the lost
> Lenore.[5]

From this point on, the parallels to Edgar Allen Poe's "The Raven" tumble into place: the *a a a b c c c b* rhyme scheme, for Maxwell's lines or Poe's half-lines; the characteristic trochaic tetrameter, with the final foot shortened on the *b* lines; the solitary narrator in his room at midnight, poring over his arcane books – which do not satisfy his needs; the embers in the grate, flickering or dying, but nevertheless throwing a light that shows something; the dull November or bleak December season. Then – if we were to go on – a sound impinging, and an unexpected, grim, diabolical, and domineering visitor appearing, bringing a dread message whose grave portent drives the rest of the poem. (Poe's "The Raven" was first published in 1845 and appeared in a British edition in 1846; Maxwell, an omnivorous reader, wrote his piece in 1852.)[6]

Maxwell's poem, as poems will be, is rich in allusion. First and foremost, there is the overarching allusion to Poe's "The Raven." Next, there are Cambridge allusions, as to St. Mary's, the University Church, and to the layout of Trinity College, in which Maxwell was enrolled. Finally, there are physics allusions, the first of these involving the deep notes of St. Mary's bell gently rounding the chimneys, this alluding to the diffraction of sound waves around obstacles that are small compared to the wavelength. (The reader will have to know, or be able to calculate, that the wavelengths of "deep ... notes" will be of the order of tens of meters.)[7]

What maxims or procedures for the reading of texts does our reading of Maxwell's poem suggest? First, the text must be read in the appropriate manner, must be rendered appropriately – here out loud – if we are to get the full message. This in turn may involve some input from the reader – here, the reader must pronounce the words – and there are problems associated with this: Is the reader to affect a New England, or a Scottish – Gallovidian – or a characteristic Cantabridgian accent? Such questions must be addressed, but these kinds of concerns must not deter us from reading aloud, from rendering the text appropriately, for then we will have little chance of getting the full message. Second, in reading certain kinds of texts, we must be alert for allusions, and, once again, this means that we must bring something to the text, we must quite literally read between the lines. There are dangers associated with this, and we may make mistakes, for it is in the nature of allusion to be subtle – a word to the wise, a joke for those in the know. Nevertheless, we must not be deterred in attempting to puzzle out the allusions, for they are central to the message.[8]

What does all of this have to do with reading texts in mathematical physics? A lot, I would suggest. First, there is the question of reading the text appropriately, of rendering it appropriately. Texts in mathematical physics (and other kinds of mathematics or physics as well) are generally written in a compact and elliptical form; they are meant to be rendered in fuller form by readers who want to apprehend their full meaning. Textbooks will leave the explicit rendering of the mathematical steps that intervene between given equations A and B, or the working out of certain details or applications, as exercises for the reader: "The reader should verify ... " is a characteristic call for some reader involvement; "solving [the] Eq[uation] ... we obtain ... " is an invitation for the reader to carry out the steps of the solution, as only the final result is presented in the text.[9] The reader of a research-front paper in physics – in the nineteenth century, or before, or after – knows that many intermediate steps have been left out, and that he or she will have to reconstruct these, in order to get the full meaning; "the length of the other path [along an arm of the interferometer] is evidently $2D\sqrt{1 - v^2/V^2} \ldots$ " write Michelson and Morely, in typical fashion, while leaving it to the reader to both draw the diagram and carry out the calculation in order to see why it is evident.[10] The reader is thus meant to – required to – have pen or pencil in hand while reading, so as to fill in as necessary.[11]

This reconstruction or filling in is, of course, never completely unproblematic; it requires, in the most literal and direct way, reading and filling in between the lines, and this is in the nature of a hypothetical enterprise. Beyond this, there is another reason for the elliptical nature of mathematical writing. The writer knows that the reader will not make the argument his own until he translates the mathematical formalism into his own dialect – or indeed his own idiolect. Christian Huygens, the seventeenth-century mathematician and natural philosopher, on seeing an argument leading to a new and interesting result, would reconstruct the argument in his own way, using his own favored procedures and nota- tion, in order to "verify [the] conclusion"; in this way, he would "attain [a] deeper familiarity [than] comes from ... merely reading another's proof."[12] The situation has not changed in the intervening 300 years. Mathematics and mathematical physics are demanding disciplines, in which an individual trying to follow an argument must be prepared to deploy all of the intellectual tools at his disposal, in order to make head- way; and translation into a favored notation is one of those tools. The favored notation may be one taken over from a teacher or a textbook, but individuals who are functioning well as producers or consumers of mathematical physics will usually have developed an individual style in rendering the mathematics, even as one who writes will develop a writing style. The difference is, that while the reader of prose will not usually rewrite whole paragraphs in order to assimilate an argument, the reader of mathematics will often rewrite whole chains of equations. The elliptical form of mathematical writing facilitates, encourages, and all but requires this process of rendering into an idiolect; thus, this process is an integral part of reading a mathematical paper as it was meant to be read.[13]

One way in which the rendering of an equation in familiar or favored notation is extremely useful is in catching allusions. Mathematical physics, like poetry, is compact because it is highly allusive. If the lecturer says, "We begin by writing down $\mathbf{F} = m\mathbf{a}$," one has to rec- ognize that this is an allusion to an element of the physics tradition – namely, Newton's second law, where $\mathbf{F}$ is the force, m the mass, and $\mathbf{a}$ the acceleration.[14] Maxwell's exposition of his vortex model of the elec- tromagnetic field provides a somewhat more complex – and also more significant and instructive – example. Maxwell represented the velocity

of a point on the surface of a vortex in terms of the three components of
the vectorial velocity, as follows:

$$n\beta - m\gamma \quad \text{parallel to } x$$
$$l\gamma - n\alpha \quad \text{parallel to } y$$
$$m\alpha - l\beta \quad \text{parallel to } z$$

where l, m, n were identified as the direction cosines of the normal to the
surface, and α, β, γ were taken proportional to the direction cosines of
the rotation axis of the vortex (with the proportionality constant related
in a somewhat complex way to the rotational velocity of the vortex).
Looking at this set of symbols – constituting an implied equation for the
components of the velocity – I found myself unable to make sense of
it: Why should this combination of symbols give the appropriate velocity; what allusions, what words to the wise, was I missing? (Maxwell,
apparently, thought that it should have been obvious to the reader. Giv-
en the definitions of the symbols, he simply asserted that, "Then the
components of the velocity of the particles of the vortex at this part of
its surface will be ...")[15]

   If I render the implied equation in the vector notation with which I
am comfortable; use the definitions and notations for angular velocity
ω, linear velocity $\mathbf{v}$, and unit normal $\mathbf{n}$ that are part of my mathematical
dialect; and define a to be the distance from the center of the vortex to
its "circumference"; then the whole equation becomes

$$\mathbf{v} = \omega \times a\mathbf{n}$$

where $\times$ represents the vector product, and in this form I am able to catch
the allusion: this is the equation for the velocity of a point on the surface
of a rotating rigid body, where ω is the angular velocity of rotation,
and $a\mathbf{n}$ corresponds to the displacement of the moving point from the
origin or fixed point; this in turn requires that the surface be spherical.
What I learn from all of this is that, at this point in the argument,
Maxwell was treating the fluid vortices as rigid bodies – that is, he
was assuming uniform angular velocity – and he was approximating
the shapes of the vortices as spherical. (This conclusion concerning
spherical vortices turns out to have bearing on larger issues in the
interpretation of Maxwell's electromagnetic theory, as we shall see in
due course.)[16]

   Of course, I can be no more certain of my mathematical rendering of
this text, and my identification of the allusions it makes, than I can be of

my reading of Maxwell's poem, and my identification of it as a parody of "The Raven." Indeed, my situation is more difficult with respect to the vortex rotations than with respect to "The Raven." Catching allusions depends not only on rendering the text appropriately, but also on being well versed in the universe of discourse to which the allusions refer. An educated American (or English-reading) audience will know enough of poetry, and "The Raven," to recognize that this is not something brought up artificially and *ad hoc*, to construct a forced interpretation of Maxwell's poem. An audience of historians of science, however, will not necessarily know enough of classical mechanics, and the equations of rigid body rotation, to be able to make a similar favorable judgment concerning the relevance of those equations to Maxwell's vortex theory. Allusions that have to be belabored work no better than jokes that have to be explained. For a full appreciation of Maxwell's subtle references, one needs to control his various universes of discourse; one cannot expect him to explain it all, as Ludwig Boltzmann observed in a comment on Maxwell's gas theory: "There is not time to say why this or that substitution was made; he who cannot sense this should lay the book aside, for Maxwell is no writer of programme music obliged to set the explanation over the score."[17]

By this point in the discussion, many of my colleagues may have come to the conclusion that my approach to reading texts is wrong for the history of science, violating the norms and canons of our discipline in a variety of ways. First, translating nineteenth-century equations into a modern mathematical dialect or idiolect, in order to fill in intermediate steps and render allusions into an apprehensible form, goes against some of the most basic commitments of the modern history of science discipline. To the extent that we believe, with Benjamin Whorf, that language constrains cognition; to the extent that we follow Thomas Kuhn in judging that successive paradigms are incommensurable; and to the extent that we worry, with Herbert Butterfield, about taking a Whig approach to history – we will be disturbed about such translation into modern terms.[18]

Now I sincerely wish that I could muster the insight, flexibility of mind, technical ability, and sheer power of intellect that would be required to think myself into a completely nineteenth-century frame of mind, and fill in the equations and catch the allusions completely within that framework. Unfortunately, I am not always up to that task; nor have I found anyone else who is. The task of understanding Maxwell is a difficult one, and we need all the intellectual tools we can muster, including the tool of translation, in order to make headway.[19] Intellectual tools, like other tools, have dangers associated with them. Using a hammer carries the danger of a smashed finger; using a power saw carries the danger of a severed limb. Nevertheless, we use these tools, in order to get the job done, and we do the best we can to guard against the risks. So also, I would argue, with translation into modern terms in the history of science: we must use the tool, but use it with caution. Well used, the tool of translation can be enlisted against Whig history: We inevitably carry with us, as wanted or unwanted baggage, what we do know of modern science, and this does distort our view of the past. It will not help to try to somehow suppress our knowledge of modern science, because it will then continue to lurk below the surface, influencing our view of the past at a subliminal level. Better to confront the prejudices inherent in our knowledge of modern science openly, through an exercise of explicit comparison and contrast between present and past ideas. And what better tool for such an exercise of comparison and contrast than translation, which involves an explicit and detailed confrontation between the two languages.[20]

To the objector who says that, given incommensurability, translation simply cannot be done, I would reply, let the experiment be tried. To the objector who counters that the very attempt to translate will contaminate our minds, so that we will never again be good historians, I would reply that individuals who do not have the agility and flexibility of mind to deal with a variety of languages and worldviews, and who therefore feel that they may somehow become trapped forever in a linguistic or paradigmatic limbo, if they should so much as attempt to translate from one framework to another, should perhaps not be doing history of science. In any case, the contamination argument – that one must not think certain thoughts because one will be forever after contaminated by that experience – has been used by the enemies of intellectual freedom through the ages. We cannot, must not, buy into that game.[21]

The maxims of Whorf, Kuhn, and Butterfield, then, hinder the proper reading of Maxwell tests, by denying us the intellectual tools that we need in order to unpack the texts and perceive their meaning. There are yet other commitments of the modern history of science discipline that hinder the appropriate reading of Maxwell texts. We have come to view our discipline, more and more, as a subdiscipline of history; we have accordingly taken the membership of the history discipline to be our audience. But this is not an audience that is interested in the details of Maxwell's equations, and this brings obvious problems. There are other audiences for Maxwell scholarship, who are interested in the mathematical and physical details, including scientists, engineers, science teachers, philosophers of science, and various others. This is not a "narrow" audience; there are in fact many more of these people than there are historians. If the history of science discipline is construed as one that reaches out to a variety of audiences and has connections with a variety of other disciplines, then there is incentive for reading mathematical physics texts properly and reporting on this; if the discipline is construed restrictively, as a subdiscipline of history, then the incentives are diminished. If, however, the hard work of reading the mathematical physics texts is not done, then a part of the proper foundation for solid work in the history of physics will be missing. In particular, if the Maxwell texts are not read properly, then we are in danger of talking nonsense about the broader issues in Maxwell studies.[22]

Another element in the recent practice of history and philosophy of science that has a tendency to hinder the proper reading of Maxwell texts derives from a certain debunking trend in the treatment of science and scientists: "During the past twenty-five years, the so-called 'historical school,' or 'post-positivist school,'" "has been attempting to redraw the image of science, replacing the positivist or logical empiricist image," of "an objective, distinct, value-free, and cumulative science," utilizing "*the* scientific method," with a picture of the functioning of science that "den[ies] the central theses of the positivist image of science," finding science less rational, less "objective, [methodologically] distinct, value-free, and cumulative."[23] This trend, coupled with a historical tradition stemming from the criticism of Maxwell's work by his Continental opponents,[24] and reinforced by some persistent difficulties in the interpretation of Maxwell's various electromagnetic formalisms, has resulted in a rhetoric in which we are told that the symbols Maxwell used were "ambiguous," that he had "double vision," that he wrote "confused equation[s]," and that he "paid frustratingly little attention to [the algebraic] signs" in the equations; these equations "cannot ... be correct."[25] His account of electric charge has been characterized as "highly ambiguous and sometimes confused," and yet again as "both vague and confusing."[26] Now the existence of errors and confusions in parts of Maxwell's various articulations of electromagnetic theory cannot be denied; proceeding from these instances, however, to such global judgments as that Maxwell in general "paid . . . little attention to signs," so that attention to these kinds of details of his work is unwarranted, has the effect of hindering the attempt to read Maxwell carefully and understand him.[27]

Indeed, I suspect that there is not a little bit of Whig history at work here: one of Maxwell's sins is that his various formulations of electromagnetic theory differ in important respects – including the irritating signs in the equations – from modern electromagnetic theory. The proper response to this is not to lose patience with Maxwell and throw up one's hands. Instead, I think we owe it both to ourselves and to Maxwell to be restrained in our negative judgments concerning his thought processes and workmanship, so as not to hinder communication unnecessarily. Consider, for example, a student coming to us with a certain vision of a physics problem or a historical situation; even if one were to find the student's perspective a bit strange, one would hesitate to tell the student that the problem was that he or she had "paid [too] little attention," and that his or her ideas were therefore "ambiguous" and "confused." Instead, one would try to understand how the student was thinking about the problem, and what internal coherence the student's position might have. I think we should be careful to extend the same courtesy to Maxwell. This is not "hagiography," this is not to say that Maxwell could do no wrong; rather, it is to say that, whoever is one's partner in the communication process, too facile a judgment that he or she is not paying proper attention and is therefore confused, will be a hindrance rather than a help to communication. In particular, I think it is premature to give up on trying to make sense of Maxwell.

The hesitancy to use translation as a tool of analysis; the reluctance to bother an audience of general historians with scientific detail; and the quickness to dismiss the ratiocinations of historical scientific figures as confused – all of these converge in encouraging a certain superficiality in approaching the history of mathematical physics. A final impetus in the direction of superficiality comes from the Baconian strain in the history discipline. One favorite kind of historical success story is that of the scholar who, "by luck or by pluck," has discovered some very old documents – with much dust on them – which, without much cerebration by the historian, lead to great new historical insights.[28] It is a very empiricist success story, and, to the extent that the extreme empiricism which it celebrates has caught the professional imagination of historians of science, it converges with the other tendencies discussed

TABLE I

Parts of "Physical Lines"	Part I	Part II	Part III	Part IV
1. "Molecular vortices" in subtitle?	Yes	Yes	Yes	Yes
2. Symbols α, β, γ used?	Yes	Yes	Yes	Yes
3. α, β, γ identified as vortex rotation?	Yes	Yes	No	Yes

to hinder proper reading of texts in the history of mathematical physics, which necessarily involves a great deal of reading between the lines – leading to a texture of historical discourse in which even the appearance of Baconian empiricism cannot be maintained. Thus, in order to make better progress in Maxwell studies – especially as concerns Maxwell's electromagnetic theory – the Baconian and other forces that lead to superficiality must be overcome, and we must give more full and careful attention to the texts themselves, using all of the intellectuals tools at our disposal, fully aware of the dangers involved in using those tools, but also aware of the help they can render us.

And suppose that one does try to read the texts in the proper manner – rendering the mathematics appropriately and being alert to the allusions – what does one find? My own labors in this direction have centered on a particular text – Maxwell's paper entitled "On Physical Lines of Force," of 1861–62, in which two of his signal innovations in electromagnetic theory – the displacement current and the electromagnetic theory of light – made their first appearances. The paper was published in four parts, over a period of twelve months, and one of the questions that has exercised Maxwell scholars is the question of whether this paper presents a unified and coherent theory, or whether it instead presents a series of mutually inconsistent models.[29] The question of continuity between the successive parts of the paper is illustrated in Table I.

Each of the four parts of the paper had a subtitle of the form "The Theory of Molecular Vortices Applied to . . . " as indicated in row 1 of the table; as indicated in row 2, the symbols α, β, γ were used in each of the four parts of the paper to denote some vector; and, as indicated in row 3, the vector denoted by α, β, γ was explicitly identified as the angular velocity of the vortex rotations in Parts I, II, and IV, but not Part III. The question is, what shall we say about the meaning of α, β, γ in Part III, in which Maxwell does not explicitly identify that vector? One response among students of Maxwell's electromagnetic theory has been to refrain from assuming that the vector α, β, γ represents rotations in Part III; this in turn inevitably leads to the conclusion that Part III is at best weakly connected with the rest of the paper, and is at worst inconsistent with the rest of the paper.[30] Ludwig Boltzmann in the 1890s, and I in my work, have assumed that, in the absence of any indications on Maxwell's part to the contrary, he most probably intended the vector α, β, γ to have the same meaning in Part III as in Parts I, II, and IV.[31]

Even an arch-positivist such as Ernst Mach would agree with this kind of interpolation between the explicit statements in Parts II and IV. Mach discussed the question of a projectile passing behind a post: the complete positivist would have to say that we can make no statement about what the projectile was doing while it was out of sight, behind the post; the individual interested in really doing science, however, would not be able to write off the question in this manner, and, in the absence of contrary evidence, would have to be willing to interpolate between the visible trajectories before and after the projectile went behind the post.[32] If we really want to do history, I think we have to make similar kinds of interpolations, as in concluding that the vector α, β, γ represents vortex rotations in Part III of Maxwell's paper. (This interpolation is facilitated by the discussion above of the spherical shapes of the vortices, which provides a further continuity between Parts II and III of the paper.) This conclusion, in turn, necessitates some further reading between the lines, as appropriate in dealing with texts in mathematical physics: Maxwell was not explicit as to how the rotations represented by the vector α, β, γ were to be composed with certain other motions of the vortices that were discussed in Part III; once again, in Boltzmann's words, Maxwell did not "set the explanation over the score," and it was left to the reader to fill in the blanks.

I have attempted the interpretation of Maxwell's paper in the indicated manner in my own work and have arrived at an account of Maxwell's vortex model of the electromagnetic field. By interpolating the vortex rotations into Part III of Maxwell's paper – this undertaken in opposition to Baconian scruple; and by inferring the shapes of the vortices on the basis of a modern rendering of Maxwell's component equations – this undertaken in opposition to the maxims of Whorf, Kuhn, and Butterfield; I arrive, finally, at a vision of Maxwell's paper "On Physical Lines of Force" as a coherent whole – this in opposition to the debunking trend. (In the process, I engage in detailed analysis of Maxwell's mathematics and physics – this in opposition to the history-subdiscipline trend.) What results is an account of the emergence of the displacement current, and the electromagnetic theory of light, not from confusion, ambiguity, and incoherence, but rather from Maxwell's disciplined, rational, and coherent attempt to fashion a mechanical model of the electromagnetic field. It is my hope that my analysis of this text will be sustained by future historical scholarship and will eventually be seen as part of a cumulative effort in the Maxwell industry.[33] Thus, I do not think it is time to throw in the towel and give up on reading Maxwell texts; I think the battle for an understanding of Maxwell's work, based on careful and appropriate reading of the relevant texts, has just begun.[34]

Department of the History of Science
University of Wisconsin
U.S.A.

NOTES

[*] Department of the History of Science, Helen C. White Hall, University of Wisconsin, Madison, Wisconsin 53706. An early form of this paper was presented at the Annual Meeting of the History of Science Society, Seattle, October 1990.

[1] Paul Forman, talk presented at the History of Science Society meeting in Madison, Wis., October 1978, as reported in M. Norton Wise, "The Maxwell Literature and British Dynamical Theory," *Historical Studies in the Physical Sciences* 13 (1982): 175–201, on p.175. Cf., e.g., *ibid.*, pp. 175, 195–196, and Daniel M. Siegel, "The Origin of the Displacement Current," *Historical Studies in the Physical Sciences* 17 (1986): 99–146, esp. pp. 99–100, and *idem, Innovation in Maxwell's Electromagnetic Theory: Molecular Vortices, Displacement Current, and Light* (Cambridge: Cambridge University Press, 1991), pp. 1–2.

[2] See Siegel, *Innovation*.

[3] Emphasizing breadth and context, and explicitly criticizing excessive absorption in detail, are, e.g., Wise, "Maxwell Literature," esp. pp. 175, 195–196, and Ben Marsden, review of Siegel, *Innovation*, in *British Journal for the History of Science* 26 (1993): 116–117. Emphasizing detailed analysis are Siegel, "Displacement Current," esp. pp. 99–100, and *idem, Innovation*; Alan Chalmers, review of *Innovation*, in *Metascience* [New Series Issue Four 1993]: 17–27, esp. pp. 26–27, agrees.

[4] Lewis Campbell and William Garnett, *The Life of James Clerk Maxwell*, Robert H.

294 DANIEL M. SIEGEL

Kargon, ed. (1882; New York: Johnson Reprint, 1969), pp. 612–617, on p. 612.
[5] Edgar Allen Poe, "The Raven," in *An Anthology of Famous English and American Poetry*, William Rose Benét and Conrad Aiken, eds. (New York: Modern Library, 1944), pp. 556–559, on p. 556.
[6] On Maxwell's reading and poetry, see Campbell and Garnett, *Life, passim* and pp. 577–651; on the process of Maxwell's parodizing, see p. 279.
[7] The church in question is commonly referred to as "Great St. Mary's" – Frederick Brittain, *Illustrated Guide to Cambridge*, 15th ed. (Cambridge: W. Heffer & Sons, 1971), p. 46. Maxwell's "Deep St. Mary's . . ." thus preserves the trochaic meter of a common usage, while setting up the intended physical allusion through the substitution of "Deep" for "Great"; the juxtaposition of "deep . . . notes" with "gently rounded" then carries forward the allusion. In the 1990s, it might be the high-fidelity sound reproduction enthusiast who would catch the allusion with greatest facility.
[8] Another poem of Maxwell's – Campbell and Garnett, *Life*, following p. 70 – suggests, by its orthography, that it should be rendered out loud in archaic Scottish dialect.

The literary theory used here is eclectic and heuristic: concern with allusions is characteristic of traditional literary criticism; concern with the way in which the interpreter of the text must render or perform it, thereby adding something, is more characteristic of contemporary theory. There is, of course, some tension between the idea of catching an allusion, which may suggest an all-or-nothing process, and the notion of rendering a text, which suggests a spectrum of interpretation and implies a measure of destabilization. Keeping this in mind, I would, nevertheless, suggest that the two examples just considered – Maxwell's allusions to "The Raven" and to the diffraction of sound waves – are reasonably stable, as are the further examples to be considered in connection with Maxwell's scientific writings.
[9] Edson Ruther Peck, *Electricity and Magnetism* (New York: McGraw-Hill, 1953), pp. 214–217, 244, where Ampère's circuital law is applied to a long, thin, straight wire. Peck leaves it to "the reader [to] verify the agreement of signs" in the equations; he calculates the magnetic field outside the wire, and leaves it to the reader to calculate the field inside the wire (problem 7.55). These are modest demands on the reader, appropriate to an intermediate-level textbook; in contradistinction to more advanced material, Peck has "ma[de] the demonstration of the theorems sufficiently full to be read easily, without imposing on the reader the necessity of filling in the essential steps of the reasoning" (p. vi). In an advanced-level textbook, such as Herbert Goldstein, *Classical Mechanics* (Reading, Mass.: Addison-Wesley, 1950), the reader is indeed called upon to fill in even "essential steps," as on p. 287, "solving . . . we obtain." Examples could be multiplied.
[10] Albert A. Michelson and Edward W. Morely, "On the Relative Motion of the Earth and the Luminiferous Ether," *American Journal of Science* 34 (1887): 333–341, as reprinted in *Relativity Theory: Its Origin and Impact on Modern Thought*, L. Pearce Williams, ed. (New York: John Wiley, 1968), pp. 24–34, on p. 28.
[11] Seminal for my understanding of the role of the reader in approaching an historical text in mathematical physics was a seminar at Yale University, under the direction of Asger Aaboe, and with the participation of Derek T. Whiteside, in which we read *Sir Isaac Newton's Mathematical Principles of Natural Philosophy and His System of the World*, trans. Andrew Motte, rev. and ed. Florian Cajori (Berkeley: University of California Press, 1946).

[12] Joella G. Yoder, *Unrolling Time: Christiaan Huygens and the Mathematization of Nature* (Cambridge: Cambridge University Press, 1988), pp. 77–82, on pp. 78, 81. Cf. Thomas S. Kuhn, *The Structure of Scientific Revolutions*, 2nd ed., enl. (Chicago: University of Chicago Press, 1970), p. 76, on the difficulty of "retooling"; a reader of mathematics will translate into his or her own idiolect in order to avoid the "retooling" necessary to fully assimilate an argument in another mathematical dialect.

[13] See, e.g. (and current), David P. Stern, "All I Really Need to Know . . . ," *Physics Today* **46** (1993): 63. Stern's compendium of good maxims includes the following: "[When reading in physics] write down in your own words [notation] key sections and calculations."

[14] See, e.g., Richard P. Feynman, Robert B. Leighton, and Matthew Sands, *The Feynman Lectures on Physics*, 3 vols. (Reading: Addison-Wesley, 1964), vol. I, pp. 9-1–9-4; F = ma used as an example in Stern, "Need to Know."

[15] James Clerk Maxwell, "On Physical Lines of Force [1861–62]," in *The Scientific Papers of James Clerk Maxwell*, 2 vols., W. D. Niven, ed. (1890; New York: Dover, 1965 [2 vols. in 1]), vol. I, pp. 451–513, on p. 469.

[16] Siegel, *Innovation*, pp. 69–70. Independent evidence based on manuscript materials supports this conclusion concerning the shapes of the vortices – *ibid.*, pp. 177–179. The textbook from which my favored notation for rigid body motion stems is Goldstein, *Classical Mechanics*, p. 144.

[17] *A Random Walk in Science: An Anthology*, E. Mendoza, ed., comp. R. L. Weber (London: Institute of Physics; New York: Crane, Russak, 1973), p. 43.

[18] Kuhn, *Scientific Revolutions*, esp. pp. 111–135, 144–159, 198–204; *Language, Thought, and Reality: Selected Writings of Benjamin Lee Whorf*, John B. Carroll, ed. (New York: Technology Press [MIT] and John Wiley, 1956), esp. pp. 212–219. Herbert Butterfield, *The Whig Interpretation of History* (1931; New York: Scribner's, 1951).

[19] As concerns the particular case discussed above, discerning the equations for rigid-body rotation per se in component form is something that I and my readers might have been able to handle, with reference to an appropriate older textbook, such as Horace Lamb, *Higher Mechanics* (Cambridge: Cambridge University Press, 1920), pp. 20–24, 75, where the notation bears some similarities to Maxwell's. (Cf., from Maxwell's time in its early editions, Edward John Routh, *The Elementary Part of a Treatise on the Dynamics of a System of Rigid Bodies* . . . , 4th ed. [London: Macmillan, 1882], pp. 202–206.) Given the added complexity introduced by the proportionality constant involved in α, β, γ, and the necessity for transforming this, I do not think I could have caught the allusion – and I do not think my readers would find it prima facie convincing – without translation into vector form. Negotiating Maxwell's various component forms of his electromagnetic equations without reduction to a common mnemonic notation is altogether hopeless, and almost all recent work on Maxwell's electromagnetic theory, whether by individuals who are primarily historians, philosophers, or physicists, makes use of translations into vector form.

[20] For variant forms of this argument, see Siegel, *Innovation*, pp. 3–4, and "Displacement Current," pp. 101–103. Cf. a recent discussion of the incommensurability issue – F. Rochberg, "Introduction [to 'The Cultures of Ancient Science: Some Historical Reflections']," *Isis* **83** (1992): 547–553, on p. 549 – which worries that, if the "incommensurability thesis" and related doctrines are strictly applied, "the gulf between us

and the subject of our inquiry becomes unbridgeable, and the chances of reaching any understanding become nil."

For the other side, in a classic exposition of the dangers attendant upon the translation of historical mathematical arguments into modern notation, see Sabetai Unguru, "On the Need to Rewrite the History of Greek Mathematics," *Archive for History of Exact Sciences* 15 (1975–76): 67–114; more recently, see Marsden, review of Siegel, *Innovation*. Contra, see Siegel, *Innovation*, p. 186, n8.

[21] An anonymous referee, addressing the metaphor of the power saw, suggested that one might not want to use power tools on a piece of antique furniture – in order to avoid possibly damaging the object grievously and irretrievably – and that one would correspondingly not use modern notation on Maxwell. The difference is, if one assumes that competent historians will have the flexibility of mind not to be irretrievably impaired or contaminated by entertaining certain thoughts, then the experiment of translation can be tried in all cases, without fear of doing permanent damage in cases where the exercise fails.

[22] Marsden, review of Siegel, *Innovation*, e.g., seems to regard the most appropriate target audience for Maxwell studies as consisting of "historians wishing to comprehend Maxwell within his own environment," who have not been exposed to modern electromagnetic theory. I am not sure that many in this target audience would be interested in the mathematical and physical details of Maxwell's work; Marsden himself, in the review, chooses not to engage with these kinds of details – with what he appears to regard as overly "painstaking numerical, symbolic, and graphological analyses."

[23] Rachel Laudan, Larry Laudan, and Arthur Donavan, "Testing Theories of Scientific Change," in *Scrutinizing Science: Empirical Studies of Scientific Change, idem*, eds. (1988; Baltimore: Johns Hopkins University Press, 1992), pp. 3–5.

[24] Pierre Duhem, *Les Théories Electriques de J. C. Maxwell: Etude Historique et Critique* (Paris: A. Hermann, 1902), see also *idem, The Aim and Structure of Physical Theory*, trans. Philip P. Wiener (1954; New York: Atheneum, 1962), esp. pp. 55–104, and, e.g., Alfred O'Rahilly, *Electromagnetic Theory: A Critical Examination of Fundamentals* (1938, as *Electromagnetics*; New York: Dover, 1965).

[25] Joan Bromberg, "Maxwell's Displacement Current and His Theory of Light," *Archive for History of Exact Sciences* 4 (1967): 218–234, on pp. 222, 227, 233. Wise, "Maxwell Literature," p. 196. John Hendry, *James Clerk Maxwell and the Theory of the Electromagnetic Field* (Bristol: Adam Hilger, 1986), p. 181.

[26] Mary Hesse, "Logic of Discovery in Maxwell's Electromagnetic Theory," in *Foundations of Scientific Method: The Nineteenth Century*, R. N. Giere and R. S. Westfall, eds. (Bloomington: Indiana University Press, 1973), pp. 86–114, on p. 102. A. F. Chalmers, "Maxwell's Methodology and His Application of It to Electromagnetism," *Studies in History and Philosophy of Science* 4 (1973): 107–164, on p. 142.

[27] Wise, "Maxwell Literature," esp. pp. 175, 194–195.

[28] Peter Smith, "Cliometrics" (Paper delivered to the Mathematics Department, University of Wisconsin-Madison, ca. 1973).

[29] Siegel, *Innovation*, esp. pp. 1–4. Maxwell, "Physical Lines."

[30] Bromberg, "Displacement Current," p. 224, e.g., finds profound discontinuity in the transition from Part II to Part III, with no rotations in Part III. M. Norton Wise has evidently found the argument for rotations in Part III unconvincing, as in his commentary on this paper when it was presented at the Annual Meeting of the History of Science

Society, Seattle, October 1990.

[31] *Ueber Physikalische Kraftlinien*, Ludwig Boltzmann, ed. and trans., Ostwald's Klassiker der exacten Wissenschaften, No. 102 (Leipzig: Wilhelm Engelmann, 1898), esp. pp. 126–127; Siegel, *Innovation*, esp. pp. 79–80.

[32] Ernst Mach, *The Science of Mechanics: A Critical and Historical Account of Its Development*, trans. Thomas J. McCormack, 3rd rev. and enl. ed. (Chicago: Open Court, 1907), pp. 490–492.

[33] Reviews of *Innovation* to date express various opinions, but are generally encouraging. Early reviews – Peter Harman, in *Nature* 356 (1992): 753–754, and Bruce Hunt, in *Physics Today* 45 (1992): 68 – have found the interpretation of "Physical Lines" as a coherent whole basically convincing, while not necessarily agreeing on all details. Two reviewers have paired *Innovation* with other recent contributions to Maxwell scholarship in optimistic assessments of the enterprise: Crosbie Smith – in *Times Higher Education Supplement* (Nov. 27, 1992): 23 – pairs *Innovation* with *The Scientific Letters and Papers of James Clerk Maxwell*, Peter Harman, ed., vol. I (Cambridge: Cambridge University Press, 1990), in judging them "valuable contributions to the growing corpus of Maxwellian scholarship"; David Cahan – in *American Historical Review* 98 (1993): 861–863 – pairs *Innovation* with Bruce J. Hunt, *The Maxwellians* (Ithaca: Cornell University Press, 1991), judging them "fine studies," which "together . . . show the complexity of topics, contexts, and analyses needed to understand the development of Maxwellian theory and electrotechnology." Marsden, in *British Journal for the History of Science* – in a less positive review, which invokes the current disciplinary *shibboleths* – objects to the emphasis of text over context, finds the translation of Maxwell's mathematics into modern notation improper, and is impatient with the amount of mathematical and physical detail. Others, e.g. Robert March – in *Foundations of Physics* 23 (1993): 1157–1159 – and Jed Z. Buchwald – in *Isis* 84 (1993): 395–396 – find the focus and treatment quite appropriate.

[34] Cf. Harman, review of Siegel, *Innovation*: "Siegel's thoroughly convincing study stands, I believe, at the beginning of a truly historical analysis of the development of Maxwell's field theory."

STEPHEN G. BRUSH

PREDICTION AND THEORY EVALUATION IN PHYSICS AND ASTRONOMY*

This is a progress report on a long-term project to study the dynamics of theory-change in science. I use the word dynamics, just as in mechanics, to denote the analysis of *causes* of change, as distinct from the *description* of change ("kinematics"). I want to find out *why* scientists accept or reject a theory.

The process of theory-change in science may be separated into, first, the origin and development of a theory, and, second, the decision of the scientific community to accept or reject that theory. In the past, most research in the history of science has addressed only the first part of the process; one commonly finds only brief and mostly undocumented assertions about the response of the community. Experiments that are now considered to support the theory are described, but not the evidence that scientists accepted the theory *because* of those experiments. Detailed studies of the "reception" of theories have been limited to a few major examples such as heliocentric astronomy, Darwinian evolution, and Einstein's theory of relativity. The collection of papers edited by Arthur Donovan, Larry Laudan and Rachel Laudan, *Scrutinizing Science*, is the first systematic large-scale attempt to investigate this kind of question.[1]

The role of empirical tests in the evaluation of theories is of special importance because philosophers and other writers on science usually claim that such tests have or should have high priority. In particular, Karl Popper and others have claimed that *prediction* of the results of experiments is an essential function of scientific theories, and that theories that refuse to take the risk of falsification by making testable predictions are only pseudoscientific. Moreover, Popper argues that confirmation of a prediction made *before* the empirical fact was known, especially if that fact is contrary to what might have been expected on other accepted or plausible theories, is stronger evidence for the theory than the *explanation* of a previously-known fact. Other philosophers have disputed this claim, primarily on logical grounds (how can the evidential value of a fact depend on when it was known?).[2]

299

A.J. Kox and D.M. Siegel (eds.), No Truth Except in the Details, 299–318.
© 1995 *Kluwer Academic Publishers.*

Against this view, it has been argued by Kuhn and others that there is no such thing as a "crucial experiment" that unambiguously forces scientists to abandon a theory or to choose one theory over another, especially if the theories are so different that they belong to different paradigms.[3]

BIG BANG VS. STEADY STATE COSMOLOGY

The Kuhnian claim is too extreme, as can be seen from a well-known case that I have recently examined: the switch from the Steady State to the Big Bang cosmology.[4] These theories may not belong to incommensurable paradigms, yet they are radically different in their basic assumptions about the world. The Big Bang theory, developed by Georges Lemaître and George Gamow, assumes that the universe began at a certain time with the sudden creation of a quantity of mass-energy that has been strictly conserved ever since. The Steady State theory, proposed by Fred Hoyle, Hermann Bondi, and Thomas Gold, assumes that the universe has always existed, and that matter is continuously created all the time. The Big Bang theory led to the prediction by Gamow's colleagues, Ralph Alpher and Robert Herman, that space is now filled with black-body radiation at a temperature a few degrees above absolute zero; the Steady State theory made no such prediction.[5]

In the 1950s and early 1960s, while the Steady State theory probably never commanded the support of a majority of astronomers, it was strongly advocated by a substantial minority, especially in England. Its advocates, especially Bondi and Gold, argued that it satisfied Karl Popper's criterion for a good scientific theory because it was vulnerable to empirical tests: if you look out into space and see that distant galaxies are different from nearby ones, "then the steady-state theory is stone dead."[6] The Big Bang theory was somewhat more popular but lacked strong empirical support.

Within a few years after the discovery of the cosmic microwave radiation by Arno Penzias and Robert Wilson in 1965,[7] the community had largely abandoned the Steady State theory in favor of the Big Bang (or some modification of it); this empirical observation is generally cited as the only or at least the major reason for the change in theories. (Other evidence, such as radio source counts, was widely regarded as unfavorable to the Steady State, but, with one or two exceptions, did not persuade its advocates to abandon it.) Six of the Steady State advocates explicitly (though sometimes reluctantly) declared their support for the Big Bang; 12 others simply stopped publishing on cosmology. For many years Hoyle was the only persistent defender of the Steady State cosmology; in 1990 he managed to rally some of the old team to try once more to revive it, but by then it was far too late to reverse the trend.

In this case we also have statistical evidence for the theory-change, in the form of opinion surveys (predominantly of American astronomers) conducted before and after the crucial experiment, respectively in 1959 and 1980:[8]

| | 1959 | | 1980 | |
%	favorable	unfavorable	favorable	unfavorable
Big Bang	33	36	69	7
Steady State	24	55	2	91

The conversion of astronomers from Steady State to Big Bang cosmology seems to be an example of the decisive role of a single crucial experiment, and thus, though probably quite untypical, serves as a useful benchmark for other cases. It shows that Popper's falsificationist methodology *may* describe how science works – especially, when some of the scientists involved have gone on record as practitioners of that methodology! (Hoyle, on the other hand, felt no need to follow Popper's rules and was quite willing to change his theory to avoid empirical refutation.)

The fact that scientists change theories because of empirical evidence does not, of course, exclude the possibility that they choose theories for philosophical, religious, or social reasons. A comprehensive treatment of the history of cosmology would have to take account of any extant evidence of the influence of such factors.

ORIGIN OF THE MOON[9]

Four years after the discovery of the cosmic microwave radiation predicted by the Big Bang cosmology, the American space program conducted a test of another prediction. Before the Apollo series of lunar landings beginning in 1969, planetary scientists had discussed three hypotheses about the origin of the Moon (selenogony): fission (the Moon was ejected from the Earth); capture (the Moon was formed elsewhere in the Solar System, then captured into orbit around the Earth); and co-accretion (the Moon developed from a ring of material left over from the Earth's own formation). Samples returned from the Apollo missions were expected to provide a definitive test of these three hypotheses.

Harold Urey, a well-known nuclear chemist who became a leader in planetary science during the 1950s and 1960s, advocated the capture hypothesis. He argued that if the capture hypothesis were correct, studying the Moon would yield valuable information about the early history of the solar system – not only about the region where it had been formed, but perhaps about the condition of the Earth at the time of capture. But it would hardly be worth the expense to retrieve samples from the Moon if the Moon were just a piece torn out of the Earth. He predicted from his theory that the Moon would be found to have water on or near its surface, and that it had been cold ever since its capture.

The co-accretion hypothesis implied that the Moon had been formed from the same kind of material as the Earth; the fact that it now ha significantly lower density had to be explained by an additional hypothesis. The fission theory, as formulated by John A. O'Keefe, implied that the heat involved in the ejection of the protolunar material would have vaporized it, leading to the loss of the more volatile components before recondensation. In particular, the abundances of "siderophile" elements such as gold, platinum, and nickel should be lower in the Moon than in the Earth's crust.

Urey's predictions were refuted by the analysis of rocks returned from the Apollo 1 mission in July 1969, and by results of later missions: there was no indication of surface water at the Mare Tranquillitatis site, and there was evidence for an earlier high-temperature stage. On the other hand there was strong depletion of nickel in some of the samples. The outcome of this episode was that Urey decided to abandon his capture hypothesis; he collaborated with O'Keefe in developing a more elaborate fission model that would specifically account for the depletion of nickel.

Other planetary scientists did not draw quite the same conclusion. In fact, for several years the consensus was that all three hypotheses had been refuted. This impasse made it possible for a new selenogonical hypothesis, of a kind that had previously been considered too unlikely for serious consideration, to come into prominence. According to this hypothesis, the collision of a Mars-sized planet with the early Earth had vaporized much of Earth's crust, and the Moon had eventually condensed from the vapor formed out of the Earth and the other planet. This "giant impact" hypothesis was proposed by W. K. Hartmann and D. R. Davis in 1974–75, and by A. G. W. Cameron and W. R. Ward in 1976; it became the favored hypothesis of the planetary science community in the mid-1980s.

While the behavior of Urey and O'Keefe seems to fit the prediction-testing model, it is not so clear that other scientists evaluated these theories on the basis of their predictions from before the first lunar landing. (An exception is A. E. Ringwood, who made detailed estimates of the moon's thermal history from a fission hypothesis in 1966.) The reasons for rejecting the three pre-Apollo hypotheses were based partly on the chemical composition of lunar samples, and partly on theoretical arguments involving angular momentum. None of these hypotheses could plausibly explain how the Earth–Moon system came to have its present angular momentum. The capture hypothesis was shown to be dynamically impossible unless the Moon had been formed at about the same heliocentric distance as the Earth, and even then it would be rather unlikely. Selenogonists broke the impasse by, in effect, working backwards from the present state to a previous state that was chosen to have the desired angular momentum, even though it would seem to have very low *a priori* probability. Thus a theoretical criterion played the dominant role in selecting the new hypothesis.

The implausibility of a collision with precisely chosen impact parameters was somewhat ameliorated by observations and theories pertaining to the formation of other planets in the inner solar system during this period. Photographs of Mercury's surface taken by the Mariner 10 spacecraft in 1974 showed that the planet, like the Moon, was heavily cratered, suggesting that it had suffered a heavy bombardment by medium-sized objects after its formation. George Wetherill's calculations on the accumulation of the terrestrial planets from small solid particles ("planetesimals"), based on a theory developed by V. S. Safronov, offered an explanation for the cratering of Mercury that was also relevant to the Earth–Moon system. Wetherill found that a substantial fraction of the total mass in each region of the nebula would reside in bodies only one order of magnitude smaller than the dominant planetary "embryo" at a fairly late stage of the accumulation process. It is, therefore, an essential feature of this process that a terrestrial planet will probably be hit by an object as large as Mars during the final stage of its growth.

Testing of selenogonical predictions played a significant but limited role in the story from the mid-1960s to the mid-1980s. It remains to be seen whether the giant impact model will survive the testing of its own predictions.

GRAVITATIONAL LIGHT BENDING

Although the cosmic background radiation had been quantitatively predicted by Alpher and Herman before its discovery by Penzias and Wilson, it is not clear from the history of that case whether the *novelty* of the prediction was really a factor in persuading scientists to accept the Big Bang theory.[10]

We have to distinguish between the way scientists evaluate theories in their own technical literature and the way they promote them in popular writings. As Geoffrey Cantor has pointed out, successful novel predictions have an important rhetorical value in compelling assent.[11] But this may be true primarily in presentations to those who are not considered capable of verifying the logical connection between hypothesis and prediction. For those who are, temporal novelty may be of less significance.

There is some confusion in the use of the term "prediction" by scientists: often it is used to refer to the deduction of a previously-known fact, while at other times the dictionary meaning seems to be intended. The ambiguity itself suggests that novelty is not considered significant. I propose to use the word "forecast" when it is necessary to specify the prediction of a previously-unknown fact.

Since Popper himself reported that he was led to his falsifiability criterion partly because of the spectacular confirmation of Einstein's forecast of the gravitational bending of light by the English eclipse expedition of 1919 – providing a stark contrast with the excessive flexibility (and hence untestability) of Alfred Adler's psychological theories with which Popper was working at the same time – this seemed to me to be an especially appropriate case to examine.[12]

In fact, light-bending has turned out to be a rather good example of the non-importance of novelty in the discourse of scientists. One can judge the weight attributed to the forecast of light-bending by comparison with two other tests that were discussed at the same time: the advance of the perihelion of Mercury and the gravitational redshift of spectral lines. The former was a well-known discrepancy that theorists had failed to explain successfully despite several decades of work; Einstein managed to calculate the observed effect within the observational error, without introducing any arbitrary parameters. The latter was, like light-bending, a forecast from general relativity theory, but its observational confirmation was still in doubt in the 1920s and remained so for several decades.

So one can inquire whether light-bending provided better evidence for Einstein's theory than Mercury's perihelion because it was a forecast. My research suggests that it did not, and raises the possibility that the opposite may even be true in some cases. A novel fact is not decisive evidence for the hypothesis that led to its prediction unless competing theories fail to explain that fact, and it takes some time to determine whether this is the case. But if rival theories have already had a chance to explain a fact and have not successfully done so (as in the case of Mercury's perihelion), a new theory that plausibly explains it will immediately have an advantage.

Advocates of rival theories did try to explain light-bending, arguing that if they could do so, their theories should be preferred to that of Einstein, despite the latter's undoubted success in forecasting the effect. It was only after they failed that light-bending was accepted as a confirmation of general relativity. Even then, it was still not considered as strong a confirmation as Mercury's perihelion advance, which can be measured more accurately and depends on a "deeper" part of the theory.

ALFVÉN PLASMA PROGRAMME

In the "methodology of scientific research programmes," proposed by the philosopher Imre Lakatos, one does not simply accept or reject a theory on the basis of an empirical test. Instead, one judges the track record of a series of theories ("research programme") based on the same fundamental assumption ("hard core"). The falsification of a prediction does not refute the programme but simply calls for the introduction of a new auxiliary hypothesis, generating another theory to be tested. The programme is "progressive" or "degenerating" depending on whether the new theories tend to be confirmed or refuted by further tests. It is rational (according to Lakatos) for scientists to pursue a programme as long as it is at least as progressive as its competitors, and to abandon it only when a more progressive programme is available.

The choice between research programmes, according to the Lakatos methodology, is still ultimately based on the success or failure of predictions, which therefore play an even more crucial role than in Popper's methodology (which actually gives considerable importance to factors such as theoretical coherence).[13]

A test of the Lakatos methodology is furnished by the research programme of Hannes Alfvén, who published a number of forecasts about the behavior of plasmas and electric and magnetic fields in the magnetosphere and interplanetary space:[14]

1. Magnetohydrodynamic waves, predicted in 1942; discovered in laboratory experiments in 1949–59, and in the ionosphere, generated by nuclear explosions, in 1958; now universally known as "Alfvén waves";

2. Electric currents flowing along magnetic field lines, suggested by Kristian Birkeland in 1908 and developed by Alfvén as part of a general theory of magnetic storms and aurorae in 1939; "field-aligned currents" were discovered in 1966–67 and are generally known as "Birkeland currents";

3. Critical ionization velocity for the interaction of a neutral gas with a plasma, predicted by Alfvén in 1954 and discovered by several laboratory experiments in 1961–70;

4. Rings of the planet Uranus, predicted from Alfvén's theory in 1972 by Bibhas De (but refused publication); discovered in 1977;

5. Electrostatic double layers, predicted to occur in the Earth's magnetosphere by Alfvén in 1958 on the basis of earlier laboratory experiments; evidence allegedly supporting the hypothesis was reported in 1970 and thereafter, from rocket experiments, but this interpretation is still controversial;

6. Magnetic braking of a rotating plasma, proposed by Alfvén in 1942 as an explanation for the fact that the Sun rotates more slowly than expected on the nebular hypothesis; assumes that magnetic field lines from a strongly-magnetized early Sun are trapped in the surrounding ionized gas; accepted by several cosmogonists in the 1960s, but now out of favor[15];

7. Partial corotation, a modified form of magnetic braking in which the field lines are not completely "frozen" in the plasma. Alfvén proposed in 1967 that condensation of grains from a partially corotating plasma around a planet like Saturn will be affected by small bodies so as to leave a gap in the rings at 2/3 the planetocentric

distance of the body. He used this hypothesis to predict two new satellites of Saturn in 1968, but neither exists at the predicted distances, according to observations from the Voyager missions. (In 1986, another scientist argued that the prediction itself was based on an erroneous theoretical argument.)

By Lakatosian criteria, Alfvén's programme clearly seems to be progressive, even though a few of its predictions have been falsified. By the same criteria, rival programmes are less progressive; some made no forecasts at all. But I could find no evidence that any other scientists adopted Alfvén's programme on this basis – or even that anyone rejected it because of its falsified predictions. Probably personal factors (Alfvén's style of interaction with other scientists) influenced the response of the community. Moreover, it appears that most plasma scientists disregarded Alfvén's later predictions, despite the confirmation of his earlier ones, on the grounds that they were derived from hypotheses deemed wrong on theoretical grounds. This attitude – which was apparently not changed by the award to Alfvén of the 1970 Novel Prize in Physics – is shown explicitly in the referees' reports on De's paper predicting the rings of Uranus, when he submitted it again *after* the discovery, only to have it rejected once more.

The case of Alfvén's programme shows that if one's basic assumptions and method are considered unacceptable by other scientists, no amount of empirical confirmation will force them to accept it. I say this not as a criticism of the scientific community, but simply as a fact about science which many philosophers of science ignore.

SUBATOMIC PARTICLES

A dominant feature of 20th-century physics has been the discovery of new particles, some of which were predicted in advance. Does such a successful prediction count as strong evidence in favor of the theory that led to the prediction? Does such a theory gain an advantage over other theories that do not make successful predictions, or that simply take the existence of the particle as given?

I have looked at three cases: P. A. M. Dirac's prediction of the positron from his relativistic quantum theory of the electron (supplemented by the "hole" hypothesis); Hideki Yukawa's prediction of the meson from his theory of forces between nucleons; and Murray Gell-Mann's prediction of the Ω^- particle from his SU(3) symmetry-group model. These were selected because they have been widely celebrated as examples of successful prediction, and because they occurred long enough ago that we can draw on correspondence, recollections of the participants, and retrospective evaluations.

THE POSITRON

Dirac attempted to construct a wave equation that would satisfy the requirements of special relativity theory; he found that his equation led to states of negative energy. He assumed that these were normally nearly all filled, and that a vacancy ("hole") in the space-filling sea of negative-energy particles would behave like a particle of positive energy but opposite electric charge. In 1931, he predicted that encounters of gamma rays could provide enough energy to raise an electron from a negative-energy state to a positive-energy state, thus creating both an ordinary negatively-charged electron and a (previously-unobserved) positively-charged electron ("anti-electron," later called "positron"). An obvious generalization of Dirac's proposal was that protons, neutrons, and other particles have corresponding "antiparticles," and that the universe as a whole may be electrically neutral, although the rarity of antiparticles in our part of the universe remains to be explained.

In 1932 Carl D. Anderson discovered, in cosmic rays, particles similar to electrons but with positive charge. Anderson later stated that he was not aware of Dirac's prediction when he made the discovery. P. M. S. Blackett and others subsequently showed that Anderson's particle is the one predicted by Dirac. (According to Dirac's later recollections, Blackett missed making the discovery himself because he was too cautious in interpreting his observations as indicating a positron, even though he was aware of Dirac's prediction.)

There is considerable evidence that physicists (e.g. Blackett, Walther Bothe, Louis de Broglie, Edward Condon, Enrico Fermi, Jacob Frenkel, George Gamow, Robert Oppenheimer, Rudolf Peierls, and Erwin Schrödinger) regarded the discovery of the positron as strong evidence in favor of Dirac's theory. He was awarded the 1933 Nobel Prize in Physics, with a presentation speech that twice mentioned this "brilliant confirmation" of his theory. Wolfgang Pauli urged (successfully) that Dirac be invited to speak at the 1933 Solvay Congress because the discovery of the positron had revived his hole theory and made it a focus of research. By the third year after its discovery, most papers on the positron associated it with Dirac; even earlier – by the second year – most papers discussing Dirac's equation associated it with the positron.

But relatively few (27) of the 239 papers on the Dirac equation and/or the positron published in the first three years after the discovery explicitly stated that it confirmed or supported Dirac's theory. None stated that the theory should receive any more credit because it had predicted the positron before its discovery.

Indeed, in spite of their fascination with the positron and their recognition that it had been predicted by Dirac, leading theorists – Pauli, Heisenberg, Oppenheimer and others – considered Dirac's theory unacceptable: the hole hypothesis necessitated a many-particle theory, implied an infinite energy density everywhere in space, and introduced an asymmetry between positive and negative charges that seemed to be refuted by the discovery itself. From the viewpoint of the quantum field theorists of the next generation, the natural way to resolve these difficulties was "second quantization," converting the Dirac wave function into an operator and taking the existence of antiparticles as a fundamental postulate. Dirac's hole theory was still praised as an important step toward the modern theory, but nevertheless regarded as a hypothesis that ultimately had to be discarded.

In the new quantum electrodynamics developed in the 1940s by Sinitiro Tomonaga, Julian Schwinger and Richard Feynman, one still had to subtract infinite quantities in order to get a finite result to compare with empirical data, but now there was a systematic, logically consistent way to do this subtraction ("renormalization"). The new theory did not *predict* the positron as Dirac's theory did; it simply assumed its existence, or at best deduced it from general principles, such as relativistic invariance, without providing anything like Dirac's mechanism for the creation and annihilation of antiparticles. The only answer to the question "why do antiparticles exist?" was offered by Feynman's proposal to regard them as particles moving backwards in time – a proposal not taken seriously by other physicists.

In this case, the ability to predict a new particle was not the most important criterion in choosing a theory; it could not override the strong theoretical objections to Dirac's hypothesis. Yet the successful prediction did force theorists to seek, and eventually to find, satisfactory ways to overcome those objections. The improved theory was not completely developed and accepted by the community until new empirical data (the "Lamb shift" in the hydrogen spectrum) showed that Dirac's theory was not quantitatively accurate. Physicists reluctantly gave up a theory that had made a sensationally successful prediction of a completely new phenomenon (antiparticles) in favor of a more consistent (though still imperfect) theory that could provide a more accurate, non-novel prediction of the details of the behavior of electrons, positions, and radiations.

MESONS

Yukawa proposed in 1935 that the force between protons and neutrons was transmitted by the exchange of a particle. According to his theory, the force depends exponentially on the distance r, falling off as $e^{-r/R}$, where R is the "range" of the force. Taking R to be approximately the size of a nucleus, he found that the mass of the particle should be about 200 times the mass of an electron. Yukawa's proposal, together with earlier explanations of electromagnetic forces by exchange of photons, implies that "force" is not a fundamental entity in nature but can be reduced to the exchange of particles.

Two years later, a new particle was discovered in cosmic rays by Seth Neddermeyer and Carl Anderson, and was subsequently found to have a mass on the order of two hundred electron masses. Oppenheimer and Robert Serber, and independently E. C. G. Stueckelberg, suggested that this particle (subsequently called the "μ meson") was the particle predicted by Yukawa. Without fully committing themselves to this position, a number of physicists in Europe and America started to develop meson theories of nuclear forces. Heisenberg, whose earlier papers on exchange forces had inspired the development of Yukawa's theory, became a strong advocate of that theory.

In addition to predicting the existence of a new particle, Yukawa also suggested that this particle would spontaneously decay into an electron and an antineutrino; this was confirmed by several observers in 1938. Yukawa's theory was also used to explain (semi-quantitatively) the anomalous magnetic moments of the proton and neutron, and the electric quadrupole moment of the deuteron. His exponential formula for the nucleon force law was found to be useful in fitting scattering data, quite apart from its possible connection with the meson.

As in the case of Dirac's positron, there were relatively few explicit statements, in physics articles published after the discovery of the μ

meson, that this discovery confirmed Yukawa's theory of nuclear forces. I found only 9 such statements in 255 papers published during the three years after the Neddermeyer–Anderson report. None claims that Yukawa's theory should get more credit because the prediction came before the discovery; only one physicist (Heitler), in a review article, noted this fact. But by the third year after the discovery, most papers on nuclear forces associated them with mesons, even if they did not explicitly credit Yukawa's theory; somewhat less than half of the papers on mesons mentioned Yukawa.

By 1940, many theorists had accepted the premise that nuclear forces could be explained by the exchange of particles, but had serious doubts about whether the μ meson was suited for that role. In particular, its lifetime was much too long for a particle that was supposed to interact strongly with nucleons. On the other hand, the meson theory of nuclear forces was encountering mathematical difficulties even more severe than those of quantum electrodynamics, so it was difficult to know just what predictions the theory really made.

After the discovery of the π meson in 1947, many physicists concluded that this particle was actually the one predicted by Yukawa. Yukawa was quickly awarded the 1949 Nobel Prize in Physics, "for his prediction of the existence of mesons on the basis of theoretical work on nuclear forces"; the π meson had about the right mass and lifetime, and its discovery was therefore considered "a brilliant vindication of Yukawa's fundamental ideas," despite the fact that "it has not yet been possible to give a theory for the nuclear forces, which yields results that are in good quantitative agreement with experiment."[16]

Strenuous efforts by theoretical physicists to develop a satisfactory meson field theory of nuclear forces, comparable to quantum electrodynamics, did not succeed. Yukawa's theory is still considered valid as a semi-empirical approximation for describing nuclear forces. Since π mesons themselves, as well as nucleons, are now believed to consist of quarks, the forces acting between nucleons may in principle be reducible to the exchange of "gluons" between quarks. On the other hand, physicists have to some extent abandoned the notion that particles are the fundamental building blocks of the world, and often speak instead of the four fundamental interactions (strong, weak, electromagnetic, and gravitational).

THE "LAST" PARTICLE

One of the most sensational events in elementary-particle physics in the early 1960s was the prediction and subsequent discovery of the Ω^- hyperon. The prediction, based on the "eight-fold way" version of the symmetry group known as SU(3), was announced by Murray Gell-Mann at a conference in July 1962. He proposed that nine known hyperons should be classified together in a "decimet," or a 'decuplet," whose 10th member had not yet been discovered. This "last" particle, which he called Ω^- (presumably because Ω, omega, is the last letter in the Greek alphabet), would have, in addition to negative electric charge, an isotopic spin value of 0 and strangeness of -3; its spin would be 3/2 and its parity positive. Its mass, by extrapolation from the masses of the other 9 particles, would be 1685 MeV. It would be produced by a beam of high-energy K^- particles hitting protons; it would be metastable, decaying through weak interactions by three different routes.

The discovery of the Ω^- particle was announced by a group at Brookhaven National Laboratory in February 1964. Its mass was 1686 ± 12 MeV, it decayed by one of the three routes suggested by Gell-Mann, and its other properties were precisely as predicted. This event is often credited with bringing about a major change in the thinking of theoretical physicists, by persuading them to use the abstract mathematical methods of group theory in a creative way to uncover the secrets of nature, rather than merely as a way of systematizing facts already discovered by other methods.

The immediate reaction of physicists was that the discovery provided strong evidence for a connection between the symmetry group SU(3) and the nature of elementary particles. But most of the published statements do not clearly indicate whether the evidence would have been any weaker if the particle had been discovered before the prediction was made. In a few cases, where Ω^- simply appears in a list of particles and is not distinguished from those that were previously known, one can infer that its novelty has no special value.

Statements about the value of *novel* prediction in science are more often found in textbooks and popular science magazines than in the technical literature. For example, the leaders of the discovery team wrote in *Scientific American*:

It is one thing . . . to devise a scheme describing a set of known facts and quite another to create a generalization that will bring to light new phenomena previously undreamed of. The test of any grandscale theory is its ability to predict what was previously unpredictable and to lead to new knowledge [as Gell-Mann's prediction did].

But Yuval Ne'eman, who independently proposed the SU(3) theory and has written extensively about the history of particle physics, disagrees:[18]

... the importance attached to a successful prediction is associated with human psychology rather than with the scientific methodology. It would not have detracted at all from the effectiveness of the eightfold way if the Ω^- had been discovered *before* the theory was proposed. But human nature stands in great awe when a prophecy comes true, and regards the realization of a theoretical prediction as an irrefutable proof of the validity of the theory.

There is no doubt that the prediction stimulated the search for the particle, which might not have been discovered until much later if Gell-Mann had not suggested how to produce and detect it. Thus, the correct forecast contributed to the progress of science. But we still need to ask: what theory benefitted from the confirmation of the prediction? Does the success of a hypothesis derived from group theory mean that a Platonic–Pythagorean approach to physics has been vindicated? Is nature simply a concrete realization of a system of mathematical forms? Does this portend a defeat for the atomistic worldview, i.e., for the doctrine that physicists should seek the smallest fundamental particles of matter, and that any evidence of pattern or structure in those particles is to be explained by looking for still-smaller constituents?

One argument in favor of that interpretation is that the Eight-Fold Way grew out of – but ultimately displaced – another theory that did attempt to explain particles in terms of smaller constituents: this was the "Sakata model," developed by Yukawa's collaborator Shoichi Sakata and his colleagues in Japan. Sakata proposed that all hadrons are combinations of six known particles: the neutron, the proton, the lambda particle, and their corresponding antiparticles. Sakata's physics was frankly based on Marxist–Leninist philosophy, in particular the doctrine that there exists in nature an infinite number of strata. Sakata himself regarded the Eight-Fold Way and the emphasis on group theory as incompatible with his own approach.

But the triumph of Gell-Mann's Eight-Fold way did not lead physicists to abandon the atomistic approach. On the contrary, Gell-Mann himself soon afterwards proposed that elementary particles are composed of a still smaller particle: the quark. The SU(3) hypothesis, despite its confirmation by the discovery of the Ω^- particle, was soon superseded by other theories that successfully predicted more new particles. The history of this episode has become somewhat blurred in the minds of physicists; one occasionally reads that the quark model was confirmed by the prediction of the Ω^-.[19]

CONCLUSIONS

In all the cases studied here, the discovery of the predicted particle or phenomenon had a major impact on theoretical and experimental research. With the exception of the Alfvén case, there was a significant increase in publications on the theory that led to the prediction, whether that theory had previously been well known (Big Bang cosmology, Dirac's relativistic wave equation for the electron) or almost completely unknown (Yukawa's meson theory of nuclear forces). Thus, our results mostly support the claims of Karl Popper and others that empirical confirmation of a prediction provides "corroboration" of the hypothesis that yielded the prediction – provided one does not confuse corroboration with "verification."[20] As I interpret Popper's use of the term, corroboration does not increase the probability that a hypothesis is true; it merely makes it more reasonable to pursue that hypothesis than one that has not been corroborated. In this minimalist sense, all the theories (again with the exception of Alfvén's) were certainly corroborated by the discovery of the particles or phenomena they predicted.

Despite his careful restriction of the meaning of "corroboration," Popper (like other philosophers) still argues that successful predictions are essential for the progress of science, and specifically mentions Dirac's prediction of antiparticles and Yukawa's meson theory as examples.[21] but the cases studied here provide little or no evidence for the claim of Popper and others that novelty increases the importance of a prediction. Because of the ambiguous use of the term "prediction" by scientists, it is impossible to determine how much weight they intended to give to novelty in saying, for example, that Dirac's electron theory was confirmed by his successful prediction of the positron, or that the Big Bang cosmology was confirmed by the discovery of the cosmic microwave background radiation, unless the novelty was explicitly mentioned.

In the case of gravitational light-bending, it was possible to show that novelty was clearly *not* a factor. The non-novel phenomenon of Mercury's perihelion motion and the novel phenomenon of gravitational light bending were generally enumerated, without distinction, as predictions of Einstein's general theory of relativity. Unfortunately, there is no such "control" easily available in other cases. Although Dirac's theory had also made non-novel predictions about the properties of electrons, the most spectacular being the existence of spin, those had already been absorbed into the mainstream of theoretical understanding; they were not usually enumerated along with the positron as "predictions of Dirac's theory." As for Yukawa's meson theory, we do not even know how physicists would have assessed it before they became aware of the discovery of the muon, since they had not heard of the theory; on the other hand we do know that – as of 1940 – whatever credit it had gained by this novel prediction was largely cancelled by its quantitative difficulties in accounting for known properties of nuclear forces, as well as by the failure of its predictions about the lifetime of the muon. In the case of Ω^- as for the positron and the muon, physicists may make statements to a non-physicists or lay audience about the importance of novel prediction, which do not reflect their actual practice in evaluating theories.

It is also clear that the confirmation of a prediction, whether novel or not, is only one factor governing the response to a theory. The case of Alfvén's predictions, and the case of the positron, show that theoretical objections to a hypothesis can prevent its full acceptance despite the strongest empirical support. Conversely, the refutation of a prediction (such as Urey's predictions about the Moon from his capture theory) may lead an individual scientist to abandon a theory, but in general the community bases its rejection of a theory on more than a single falsification.

Institute for Physical Science and Technology
University of Maryland
U.S.A.

NOTES

* This paper is based on research supported by the National Endowment for the Humanities and the National Science Foundation. Author's address: Department of History and Institute for Physical Science & Technology, University of Maryland, College Park, MD 20742–2431, U.S.A.
[1] Donovan, Laudan and Laudan, eds., *Scrutinizing Science: Empirical Studies of Scientific Change* (Dordrecht: Kluwer, 1988). Classic studies in the "reception" genre are:

Dorothy Stimson, *The Gradual Acceptance of the Copernican Theory of the Universe* (New York: Baker and Taylor, 1917); David L. Hull, *Darwin and His Critics: The Reception of Darwin's Theory of Evolution by the Scientific Community* (Cambridge, MA: Harvard University Press, 1973); Stanley Goldberg, *Understanding Relativity: Origin and Impact of a Scientific Revolution* (Boston: Birkhäuser, 1984); Thomas F. Glick, ed., *The Comparative Reception of Relativity* (Boston: Reidel, 1987).

[2] Karl R. Popper, *The Logic of Scientific Discovery* (London: Hutchinson, 1959; New York: Science Editions, 1961), pp. 33, 272; *Conjectures and Refutations* (New York: Basic Books, 1962), pp. 117, 339–340; (1972), pp. 349, 352–354. Popper does not claim that a theory is *confirmed* by successful prediction; it still remains hypothetical or conjectural, but it has been "corroborated" and deserves preference over competing theories.

The importance of prediction has been widely discussed by philosophers of science going back to Comte, Whewell, and J. S. Mill in the 19th century and including J. M. Keynes, N. Campbell, H. Margenau, C. Hempel, P. W. Bridgman, R. B. Braithwaite, I. Lakatos, M. Gardner, R. Giere, N. Rescher, S. Toulmin, F. Suppe, R. Rosenkrantz, P. Achinstein, P. Maher, and P. Lipton. With a few exceptions they treat the problem from a normative standpoint and give little attention to historical evidence about the actual behavior of scientists.

[3] T. S. Kuhn, *The Structure of Scientific Revolutions*, 2nd ed. (Chicago: University of Chicago Press, 1970), p. 146 and elsewhere; *The Essential Tension* (Chicago: University of Chicago Press, 1977), p. 327–328. More generally, many philosophers of science now agree that the logical-positivist ideal of a neutral observation language to describe empirical facts is unattainable, and that experiment is not independent of theory.

[4] S. G. Brush, "Prediction and Theory Evaluation: Cosmic Microwaves and the Revival of the Big Bang" *Perspective on Science* 1 (1993): 565–602; a revised and much shorter version has been published as "How Cosmology Became a Science," *Scientific American* 267, no. 2 (August 1992): 64–72.

[5] Did the Steady State theory explicitly predict *no* background radiation? I have not found any such prediction published before 1965, but in 1981 Hoyle recalled a 1956 meeting with Gamow in which Hoyle argued for zero degrees and Gamow for tens of degrees, as the temperature of the radiation. F. Hoyle, "The Big Bang in Astronomy," *New Scientist* 92 (1981): 521–527. After 1965 the Steady State advocates admitted that the background radiation was *prima facie* evidence that the universe was different in the past and thus refuted their theory unless they could explain it by some new hypothesis, such as thermalization of starlight by interstellar grains.

[6] H. Bondi, "The Steady-State Theory of the Universe," in *Rival Theories of Cosmology*, Bondi *et al.*, eds. (London: Oxford University Press, 1960), pp. 12–21, on p. 19.

[7] Newspaper reports on the COBE results in Spring 1992 showing small inhomogeneities in the microwave background stated that Penzias and Wilson made their discovery in 1964. This mistake apparently stems from overlooking the fact that until Robert Dicke and P. J. E. Peebles pointed out the theoretical significance of the observation, Penzias and Wilson had not made a "discovery" but only noticed some unexplained noise in their apparatus. As Kuhn and others have stressed, you cannot really discover something unless you understand (within some theoretical framework) what you have discovered.

[8] "Discuss origin of Universe: Astronomers polled in a Science Service Grand Jury

disagree on theories explaining the origin of the universe, as well as on the observations needed to answer the problem," *Science News Letter* **76**, no. 2 (July 11, 1959): 22. The article reports responses of 33 astronomers (26 from the U.S.) who responded to the survey out of 61 who were asked. They were asked whether or not the universe started with a big bang, and whether they agreed with the theory that matter is being continuously created.

For the 1980 survey see C. M. Copp, "Relativistic Cosmology I: Paradigm Commitment and Rationality," *Astronomy Quarterly* **4** (1982): 103–116. Three hundred and eight American astronomers responded out of 650 asked. The survey asked respondents to assign a probability of 0.0, 0.5, 0.1, 10^{-3} or $\leq 10^{-6}$ to the statements: "The 2.7° microwave background radiation is a relic of the fireball phase of a Friedmann universe" and "The 2.7° microwave background will eventually be accounted for in a steady state cosmology" among several others. I have counted the 0.9 response as "favorable" and the 0.1 and smaller probability responses as "unfavorable."

[9] S. G. Brush, "Nickel for Your Thoughts: Urey and the Origin of the Moon," *Science* **217** (1982): 891–898; "A History of Modern Selenogony: Theoretical Origins of the Moon, from Capture to Crash 1955–1984," *Space Science Reviews* **47** (1988): 211–273; "Theories of the Origin of the Solar System 1956–985," *Reviews of Modern Physics* **62** (1990), 43–112.

[10] The suggestion that space has a temperature of about 3 °K had been made as early as 1926 by Arthur Eddington, and spectroscopic evidence for such a temperature was pointed out about 15 years later by Andrew McKellar. Whether these estimates were known to Alpher and Herman before they made their prediction is important to philosophers of science who discuss "novel predictions," although it may have little relevance to the history of the case.

[11] G. N. Cantor, "The Rhetoric of Experiment," in *The Uses of Experiment*, D. Gooding and T. Pinch, eds. (New York: Cambridge University Press, 1989), pp. 159–180.

[12] Brush, "Prediction and Theory Evaluation: The Case of Light Bending," *Science* **246** (1989): 1124–1129.

[13] See, e.g., K. Popper, *Logic of Scientific Discovery*, pp. 32–33. I. Lakatos, *Philosophical Papers*, vol. I, *The Methodology of Scientific Research Programmes* (New York: Cambridge University Press, 1978).

[14] Brush, "Prediction and Theory Evaluation: Alfvén on Space Plasma Phenomena," *Eos: Transactions of the American Geophysical Union* **71** (1990): 19–33.

[15] Brush, "Theories."

[16] I. Waller, "Physics 1949," in *Nobel Lectures – Physics, 1942–1962* (Amsterdam: Elsevier, 1964), pp. 125–127.

[17] W. B. Fowler and N. P. Samios, "The Omega-Minus Experiment," *Scientific American* **211**, no. 4 (October 1964): 36–45, on 39. See similar statements by D. B. Lichtenberg, "Spectroscopy of the strongly interacting particles," in *Lectures on Particles and Field Theory*, S. Deser and K. W. Ford, eds. (Englewood Cliffs, NJ: Prentice-Hall, 1965), vol. 2; pp. 77–143, on p. 135; J. L. Emmerson, *Symmetry Principles in Particle Physics* (Oxford, 1972), p. 129.

[18] Y. Ne'eman and Y. Kirsh, *The Particle Hunters* (New York: Cambridge University Press, 1986), p. 202.

[19] For further details on the three predictions of subatomic particles, see S. G. Brush, "Prediction and Theory Evaluation: Subatomic Particles," *Rivista di Storia della Scien-

za, Serie II, **1**, n. 2 (Dicembre 1993): 47–152.
[20] Popper, *Logic*, p. 243 S. G. Brush, "Dynamics of Theory Change: The Role of Predictions," *PSA 1994* **2** (in press).
[21] Popper, *Conjectures*, pp. 220, 243.

ABRAHAM PAIS

THE POWER OF THE WORD

I. THAT LITTLE CONJECTURAL SCIENCE

I spent the years 1946–1963 at the Institute for Advanced Study in Princeton, first as a temporary member, then with a five-year longer term appointment which in 1950 was converted into a professorship. I have many fine memories of that period. Among the best was my close contact with Faculty members in the School of Historical Studies.

One of these was the distinguished Welsh historian Sir Ernest Llewellyn Woodward (1890–1971), who among other things had written the official history of Britain's foreign policy in the Second World War. One favorite topic in our conversations was the writing of history. I remember in particular two stories he told me.

One was a favorite quotation of his from the French historian of religion Ernest Renan (1829–1892), who had written somewhere that he considered history to be *cette petite science conjecturale*, that little conjectural science. The other was a story Woodward liked to tell to illustrate this point:

It is known, he said, that it rained heavily on the morning of the battle of Waterloo, but not when the rain stopped. It is quite important, he continued, to know when the rain ended. The reason is that Napoleon did not have cannon brought up to a certain area on his flank. The historically interesting question is: Did he not do that because he had failed to notice a weak point in his defence, or could he not do that because heavy rains had made the ground too soggy for such movement? Historians have tried in vain, Woodward told me, to find out when the rains ended, by pouring over military and meteorological records, letters, diaries found on soldiers who lost their lives on the battle field – all to no avail. He considered this a prime example of how the historical relevance of an event often becomes manifest only after it has occurred.

The Waterloo story leads to questions at various levels. The first of these is objective. There are two possibilities, (a) the ground was too soggy, (b) it was not. One may imagine that the historian next pursues

319

A.J. Kox and D.M. Siegel (eds.), No Truth Except in the Details, 319–332.
© 1995 *Kluwer Academic Publishers.*

the issue further à la Rashomon, making either objective assumption (a) or else (b). One may further well imagine that our historian will not yet be content – as indeed he should not be – and will next raise a further question, now at the subjective level: suppose (b) is the case, then *why* did Napoleon not move cannon? Or, put in modern terms borrowed from the Watergate hearings: How much did he know and when did he know it?

The answer to this new question is clearly elusive and deliciously conjectural. The question itself is distilled from an historical record, and I do not think it at all bold to call it part of history. Its answer lies, it seems to me, beyond history, however. Somewhere between the question and the answers lies what I like to call[1] "the edge of history," a term more easily definable than history itself. There is of course no objection to our historian indulging in a bit of extrahistorical speculation regarding the answers – as long as he marks clearly where he, ex post facto witness, speaks for himself.

We have now reached the level of inquiry, familiar to all who have tried their hand at writing history of one kind or another, at which one historian's work distinguishes itself from another, from superb to mediocre to poor: A historical account rests inevitably on a subjective selection of factors and facts, it is not, cannot be, an objective play-by-play reconstruction of a part of the past.

Anyone delving into the past will of course have to submit to a disciplined study of all relevant sources, after first having learned how and where to scout these out. These activities form an important part of the historian's labors. She or he must at the same time be prepared for the possibility if not probability that these will not reveal all he may like to know – as in the case of the rains at Waterloo. What he acquires in this way is history's skeleton, to be fleshed out to history's body by an element of choice.

Let us recall what the great historian and teacher Theodor Mommsen (1817–1903) had to say on this subject: "One can say with more justification of the historian than of the mathematician or the philologist that he is not trained but born, not educated but self-educated . . . The stroke which forges a thousand links, the insight into the uniqueness of men and peoples, evinces such high genius as defies all teaching and learning." To those who rely exclusively on learning, on critical analysis of sources, Mommsen had this to say: "Through sham accomplishments you will only increase that excessively long line of men who thought they could learn history as a craft and discovered later, to their horror, that it is an art."[2] Who has not read books on history that were ever so well documented, yet ever so boring?

Elsewhere[3] I have expressed my own views of the writing of history of science:

"The recitation of facts and dates and a handful of formulae will lend an objective touch to the account to be presented. History is highly subjective, however, since it is created after the fact, and after the date, by the inevitable process of the selection of events deemed relevant by one observer or another. Thus there are as many (overlapping) histories as there are historians. I keep before me two admonitions; one by Thomas Babington Macaulay (1800–1859): "He who is deficient in the art of selection may, by showing nothing but the truth, produce all the effects of the grossest falsehoods. It perpetually happens that one writer tells less truth than another, merely because he tells more truths;"[4] and one by Thomas Carlyle (1795–1881): "He who reads the inscrutable Book of Nature as if it were a Merchant's Ledger, is justly suspected of never having seen that Book, but only some school Synopsis thereof; from which, if taken for the real Book, more error than insight is to be derived."[5]

All that has been said thus far applies to any of the multifarious kinds of history, of man and his deeds, of his thoughts and mentalities, of his ideas and ideologies, of his politics and economics, of old style history and new style history such as analyses of the lives of common men, or of feminism. I have never been more than a dabbler in all these domains, however much they do interest me. During most of my life history has in fact been among my favorite topics for leisure reading, not just because of interest in the past but, perhaps even more so, because of the grand literary style of the great historians. Take for example the opening lines of *The Decline and Fall* by Edward Gibbon (1737–1794):

"In the second century of the Christian era, the Empire of Rome comprehended the fairest part of the earth, and the most civilized portion of mankind. The frontiers of that extensive monarchy were guarded by ancient renown and disciplined valour. The gentle but powerful influence of laws and manners had gradually cemented the union of the provinces. Their peaceful inhabitants enjoyed and abused the advantages of wealth and luxury."

To me there resides a majestic cadence in such lines. I cannot judge whether this author has all his facts right. Nor do I care. I read Gibbon, now a page here, now one there, less to inform myself about certain past events than to reinforce my awe for the power of the word. My taste, indeed my need, for reading history is best explained in the eloquent words of George Steiner, who I consider to be the greatest living literary critic: "The writing of history achieves classic status less by its documentary exactitude or its sobriety of judgment than its literary power. Thucydides, Tacitus, Gibbon, Michelet are the masters of history because they are very great writers, because they have made the language live and made remembrance eloquent."[6] It stimulates me to be in the presence of "an Artist of History [who] may be distinguished from the Artisan of History; for here, as in all other provinces, there are Artists and Artisans; men who labour mechanically in a department, without an eye for the Whole, not feeling that there is a Whole; and men who inform and ennoble the humblest department with an Idea of the Whole."[7]

The power of the written word is composed of a blend of precision and ambiguity. Take for example the opening words of the Bible: "In the beginning God created . . . ." Not at such or such a time so many years ago, no, in the beginning, less precise, ever so more beguiling. Or take the opening sentence of Isak Dinesen's *Out of Africa*: "I had a farm in Africa, at the foot of the Ngong Hills." A phrase like that draws one into the story.

Newspapermen, scribes of almost instant history, are particularly aware of the power of the opening line – "the lead" – as I had occasion to note during my own one and only brief stint as reporter. It happened in the summer of 1950. I was visiting my friend Steve White who at that time was the main European correspondent for the since defunct *New York Herald Tribune*. He was stationed in Paris, living with his wife in a nearby cottage on the estate of the late politician Léon Blum. One day, while I was there, Steve received a phone call from his main office ordering him to proceed to Brussels in order to cover the serious riots in Belgium centering on Léopold III.[8] Steve asked if I wanted to come along. Very much so, I replied, provided that I shall not be in your way. That will be no problem, he said. I shall get you reporters' credentials. You can in fact be of help in interviewing since you understand Flemish.

Off we went. It was all new and interesting to me: interviews at prime minister Devieussart's office – who would start proceedings with: "Messieurs, vos porte-plumes [gentlemen, your fountain pens]"; rowdy street scenes with crowds chanting: "Léopold au poteau [L. to the gallows]"; I interviewing the Home Secretary, who was Flemish.

The many reporters present were a lively group, but every day around three or four in the afternoon they all would become very quiet. When I asked Steve why that was so he told me that everybody was thinking about his lead, the crucial opening sentence that had to draw the reader into the article. By observing Steve I also learned something valuable about the distinction between gathering objective information and writing an article. Whenever there was an interview with a group of reporters, Steve would walk up and down and listen, but would not take notes, unlike most others. When I asked him about that he said that he could get all that was said from the wire services, and that meanwhile he was more interested in reflecting on the general flavor of events – like the better historian, who gathers his objective information from existing documents, and only then sits down to give his own subjective account. Information is necessary, but not sufficient for insight . . . .

As already said (and as the preceding may well confirm) I am a dilettante in matters of general history. There is one area, however, in which I consider myself rather more qualified, and in which I have been active: the history of science. Even in that limited area I do not consider myself a historian, however, but rather a physicist with genuine interest in historical development, as an amateur in the original sense of that word.

I now turn to the more specific problems history of science presents.

II. ON THE ROLE OF HISTORY IN SCIENCE

Santayana's often quoted dictum: Those who do not remember the past are condemned to relive it, does not apply to scientists insofar as their professional activities are concerned. Science is an a-historic enterprise. There is a saying among experimental physicists: Yesterday's sensation is today's calibration and tomorrow's background. Put in more everyday language: A new, exciting discovery made yesterday becomes a means of gauging still unexplored territory today, and by tomorrow fades into a secondary effect that only obscures further novelty. Or, put still differently: What yesterday was a wonderful rarity causing ripples of excitement becomes common today and more of a nuisance tomorrow. The time scale is of course meant jocularly, but the evolution implied is factual. The pain, struggles and bitter disappointments that so often accompanied the earlier search for the new are quickly forgotten. Only the final outcome remains.

Two examples may serve to illuminate how in science the past recedes rapidly.

Consider first the modern insight about atomic structure: All matter is built up out of some one hundred basic species, the chemical elements. Each element is built up out of smallest units, atoms. The mass of each atom is almost entirely localized in a central body, the atomic nucleus. An atom consists of a nucleus surrounded by small and lightweight particles, electrons. Each element is characterized by its own distinctive number of electrons. Consider next the modern insight about nuclear structure: All atomic nuclei are built up out of two basic constituents, the proton, the nucleus of hydrogen, the lightest element, and the neutron.

The best way to start out teaching a course on atomic structure is, it seems to me, to begin right away with the modern picture of an atom as a nucleus plus electrons. Likewise I believe it is most profitable to begin a nuclear physics course with the proton + neutron picture of nuclei. Proceeding this way one obviously skirts numerous historical developments in science. To mention but some: What happened, and what survived, of the early speculations on atomic structure of matter, beginning with the days, centuries BC, of men like Democritus or Epicurus? How did the reduction of matter in its uncountably many forms to the more basic chemical elements come about? How did one discover the nucleus and the electron and how did one find out that all atoms are composed of these? How did one come to understand the proton-neutron structure of nuclei?

My preceding suggestions on how to start teaching atomic or nuclear physics are not meant to imply that I think these questions are dull or irrelevant. On the contrary, I find them fascinating. I therefore do not propose that the teacher shall withhold this historical background, but rather that someone who comes fresh to the subject can only appreciate that fascination if he or she has first acquired some familiarity with the modern state of affairs. Is it not much more stirring to be told that Democritus and Epicurus believed that atoms exist, but in an infinite variety of sizes and shapes, and that any one variety would forever be incapable of transforming itself into any other, if the listener already knows that atoms come in any of a hundred chemically immutable varieties? Is it not much more interesting to learn that until late in the nineteenth century some of the greatest scientists thought that an atom cannot be broken up into smaller parts (atom means uncuttable!) if he knows first that electrons can be ejected from an atom and that even the nucleus can be broken apart?

Thus it seems to be that one should *begin* a course in the way indicated, then *end* it by going back to the past. Tell the students that, now that they know enough about atoms or nuclei or whatever else the subject may be, the time has come to explain by what road our forefathers arrived at what was the starting point of the course. I am deeply convinced that initial familiarity with current thinking is indispensable for a deepened appreciation of how tortuous that road actually was, that what now seems simple to a beginning student was only grasped by the best scientists of earlier days after great exertion and often after false starts. As James Clerk Maxwell said more than a hundred years ago: "The history of science is not restricted to successful investigations. It has to tell of unsuccessful inquiries, and to explain why some of the ablest men have failed to find the key of knowledge, and how the reputation of others has only given a further footing to the errors in which they fell."[9]

It has further to be made clear that progress is not a series of jumps from peaks to ever higher peaks, but that rather those peaks are separated by valleys of confusion. These confusions do not by any means reflect on poor quality of earlier thinking, but rather are an unavoidable concomitant of incomplete experimental data combined with an urge to penetrate ever further into new territory. Such insights will instill in the good student a sense of pride in what has been achieved, a sense of respect for what it takes to become a good scientist, and a sense of wonder at the way science evolves, not by straight progression, but rather by detours which appear inevitably to be followed by a return to the narrow path of true wisdom. All these insights will prepare the student for the temporary disillusions he will have to face if he is willing and able to persist in his own future research.

Having spent the concluding stage of a course on these historical aspects, I like to conclude by quoting a line from Marianne Moore's poem "The Steeple-Jack": "It is a privilege to see so much confusion." I may also tell in the final hour what Niels Bohr once said to me, after we had spent a fruitless evening trying to understand something: "Tomorrow will be a wonderful day because tonight I do not understand anything."

So, Dr. Santayana, not to remember the past does not condemn the scientist to anything – although to remember it will add to the enjoyment of his professional activities. My opinion in this matter is not universally accepted, however. Thus in the preface to a valuable collection of essays, the first of several arguments for incorporating historical developments in the teaching of science is stated to be: "Historical perspective can be valuable both in seeking new knowledge and in applying existing knowledge."[10] It is a point of view with a venerable history. Thus in the 1830s, William Whewell (1794–1866), professor of mineralogy and of moral theology and casuistical divinity, sometimes referred to as the first modern historian of science, wrote: "The examination of the steps by which our ancestors acquired our intellectual estate . . . may teach us how to improve and increase our store . . . and afford us some indication of the most promising mode of directing our future efforts to add to its extent and completeness."[11]

Having spent half a century at the frontiers of physics I have never found the slightest evidence for such uses of historical perspective, however, either in my own work, or in that of my colleagues, whether persons of distinction or those less endowed. To repeat, I find historical insights in the evolution of science enriching but, please, do not let us pretend that they can serve to help solve current scientific problems. "History repeats itself" is a statement with wide ranging validity, but simply does not apply to the evolution of science.

III. ON THE ROLE OF SCIENCE IN HISTORY[12]

The history of science is very different from the science of history. We are not studying or attempting to study the working of those blind forces which, we are told, are operating on crowds of obscure people, shaking principalities and powers, and compelling reasonable men to bring events to pass in an order laid down by philosophers (J. C. Maxwell).[13]

In the preceding I have tried to outline why and how, in my view, scientists should be informed about some of the early background of their field. I should like next to reflect on a different though related topic: the history of science as an autonomous area of study and research; history of science, in other words, as a branch of history.[14]

I approach that subject with awareness of my limitations. To put history of science in historical rather than in scientific context it would greatly help to have a much firmer grasp of general historical issues than I possess. Of whole historical subareas I know only their name, say numismatics or sphragistics. I do not know much of the history of science beyond that of physics, and that subject only well – beyond the level of general education – insofar as the late nineteenth and the twentieth centuries are concerned. I have taught physics for many years but have never given a course in the history of physics.

While broad knowledge of history would be a good thing for all of us, its importance for history of science should not be exaggerated as long as one restricts oneself to "pure" science (a useful though arrogant term), the history of the development of theoretical concepts and of experiments, those in support of theories as well as those that lead the way into as yet unexplored territory. Maxwell must have had that restriction in mind when he took distance from the "blind forces" which affect the destinies of "obscure people," meaning presumably society at large. Only with that limitation understood would I agree with him.

There is much more to the history of science than the contemplation of pure science, however; more than ever before does science play a role in problems of war and peace and of the environment, topics vital to all of us, obscure or not, about which valuable historical analyses have been written. Also sociological issues, such as the growth and spreading of scientific institutions and of the scientific literature as well as the funding of science, have for good reasons been found worthy of historical discussions. Nor should one forget that there have been times when quite different blind forces were deemed to affect science, to wit when history, including that of science, was seen as the record of the revelation of God's wisdom and power. Newton considered the Bible to be a literal historical source. Less frequently one finds the view that science is the way to God also in modern times.[15]

Closest to pure science is technology, a term under which I lump applications of science to such varied areas as, for example, electronics, computers, transportation, medicine, agriculture. Devoted pure scientists – those who cannot imagine what it would be like to live without science – tend to pursue their objectives without having technological applications in the forefront of their thoughts. Nuclear fission was not discovered in the course of a quest for new weapons. At the same time, scientists are aware that they cannot forge ahead without availability of new experimental data, which may crucially depend on progress in technology. Thus the histories of pure science and of technology are inextricably linked. For example, familiarity with the workings of the steam engine led to the theory of thermodynamics; the late nineteenth-century discoveries of X-rays and of electrons are closely linked to advances in high vacuum and high voltage techniques. These examples bring home that the distinction between pure science and technology is rather arbitrary, and that the study of history of science demands at least some acquaintance with the history of technology.

Who, then, should write the history of science? To answer this question it is essential to distinguish first between the internal history of science – how science *per se* developed in the course of time – and its external history, how the relations between science and society at large evolved. It will be evident that in regard to the second category history of science can greatly benefit from the wisdom of non-science historians. One of the latter, a distinguished American historian, has put the case eloquently:

"Since the historical student is not planning to become a specialist in any scientific field, his needs call for a different kind of instruction ... Here lies the peculiar function of the historian: not so much to write the internal history of science as to trace the external connection of history and society. This relationship works both ways. Sometimes society motivates scientific discovery. Sometimes a new advance of science motivates social change ... In the end, we may hope, the natural and social sciences, each contributing to the wisdom of the other, will stand united in forwarding the common welfare of this nation and the world."[16]

To my mind there is no doubt that scientists, on the other hand, have the best opportunities for contributing to the internal history – as long as they have an adequate sense of history.[17] Scientists have in particular the best chances, if not obligations, the more recent the focus lies. Indeed, the closer we come to modern times, the more experimental techniques as well as theoretical concepts have created an ever increasing distance from everyday experience of phenomena in the world around us. The term "two cultures," which has gained currency in recent decades, was coined to denote the separation between the culture of science and the culture of the humanistic tradition. Even that notion of a duality no longer does justice to the current predicament. There are few if any scientists today who will claim mastery of all disciplines of physics, chemistry, astronomy, biology. Even worse, there are few if any physicists who will be familiar with all subdisciplines of physics, from particles to plasmas, from chaos to cosmology, in order to write with authority about the historical evolution of all these topics. Specialization, as evil as it is inevitable, has made broad overview increasingly difficult. We no longer live in bicultural but rather in multicultural times.

This raises a further question: Which is the audience the historian of science shall address? "Shall the works in history of science be addressed to other historians of science? To historians, and to philosophers who are historically minded? Or to scientists? This question of audience is basic to our researches, the level and tone of our writings, the technical demands we make upon our students . . . "[18]

The ability to reach across the board, from the specialized scientist to the intelligent interested layman, is given to only very few historians of science. The difficulties of the subject and, even more so I think, the varied quality of the writing, have caused the literature to divide itself into subcategories depending on the chosen audience.

First there are those historians of science who in essence only reach others in that same field. Their main mode of publication is in books and history of science journals that are hardly ever read by any but specialists. I doubt whether there is more than a handful among the many thousands of physicists who have ever laid eyes on journals like *Historical Studies in the Physical and Biological Science*, or *Isis*.[19] Neither had I, until in more recent years I developed interest and activity in this field. Now I scan those journals for items bearing on my own writing. What has perhaps struck me most in the course of such perusals is that their writing style so often would not appeal to physicists' way of thinking and expression. It appears to me that the time to which one distinguished historian of science referred in 1963 has come upon us: "Not far off is the time when historians of science will be so numerous that they may produce scholarly works which need satisfy only the members of their own profession, the only requirement being that of high standards."[20] Like it or not, history of science has become a discipline to a considerable extent closed upon itself.

Which brings me to the second type of audience, the physicists themselves. Their ignorance about the professional literature should not be construed as lack of interest in the history of their field. Quite the contrary is true. Particularly in recent times I have noted, especially in the United States but also in Europe, what one might well call a hunger for hearing about physics' more recent past, the last hundred years, say.

Last, but certainly not least, there is the audience made up of the general public, which in recent times has shown a markedly increased interest in science, from the historical as well as from the current events point of view, especially so in the post World War II period. No doubt the arrival on the scene of atomic weapons, space explorations, and the uses and misuses of atomic energy for peaceful purposes have contributed to the arousal of curiosity in wide circles.

Really active interest in historical issues by the physics community postdates World War II. The first historical articles in *Physics Today*, the trade journal (founded in 1948 and widely distributed by the American Institute of Physics), appeared in 1952.[21] These were reminiscences by distinguished physicists. The first article written by a new generation of historians of science did not appear in that journal until 1966.[22] It was written by my good friend Martin Klein, one of the small group of men trained as physicists who had switched to history of science, to the benefit of the community. This essay is respectfully dedicated to him on the occasion of his 70th birthday. (Welcome to the club Martin!)

Meanwhile, in 1961 the Sources of History of Quantum Physics project had begun its activities with financial support of the U.S. National Science Foundation. It has gathered a most valuable collection of documents, including transcripts of interviews with people who had led the way during the early days of the quantum theory.[23] This project was followed by a similar one in nuclear physics; more have developed since. In 1965 the American Institute of Physics established a permanent Center for the History of Physics. In 1969 *Historical Studies in the Physical Science* was founded. (It has since included also biological studies.) Also from that decade dates the founding of new university departments in the history of science.

Finally, in 1980 the American Physical Society established a Division of History of Physics. Among its responsibilities is the organization of historical sessions at the principal Physical Society meetings. Just as is the case for other Divisions, such sessions are open to all Society members; they are well attended. The physicists were relatively late in making this move. The history of chemistry section of the American Chemical Society dates from 1921, the American Association for the history of medicine from 1924. The technologists were even later than the physicists, however. The Society for the History of Technology (SHOT) was founded in 1958, as was their separate journal, *Technology and Culture*.[24]

This synopsis of dates and events is meant to illustrate how historical activities have recently evolved within the bosom of the physics community. These developments have of course proceeded hand in hand with the work of the longer established and more broad-based group of what one might call the professional historians of science. It is a pleasure to note that distinctions between those two groups are beginning to blur.

IV. ENVOI

This is an old Chassidic tale.[25]

When the Baal Shem had a difficult task before him, he would go to a certain place in the woods, light a fire and meditate in prayer – and what he had set out to perform was done.

When a generation later the "Maggid," Rabbi Baer of Meseritz, was faced with the same task he would go to the same place in the woods and say: We can no longer light the fire, but we can still speak the prayers – and what he wanted done became reality.

Again a generation later Rabbi Moshe Leib of Sassov had to perform this task. And he too went into the woods and said: We can no longer light a fire, nor do we know the secret meditations belonging to the prayer, but we do know the place in the woods to which it all belongs – and that must be sufficient; and sufficient it was.

But when another generation had passed and Rabbi Israel of Rishin was called upon to perform the task, he sat down on his golden chair in his castle and said: We cannot light the fire, we cannot speak the prayers, we do not know the place, but we can tell the story of how it was done. And, the story-teller adds, the story which he told had the same effect as the actions of the other three.

The Rockefeller University
New York, New York
U.S.A.

NOTES AND REFERENCES

[1] See A. Pais, *'Subtle Is the Lord . . .': The Science and the Life of Albert Einstein* (Oxford: Oxford University Press, 1982), ch. 8, esp. p. 164, where I have discussed other such questions, answers and issues.

[2] Th. Mommsen, *Reden und Aufsätze* (Berlin: Weidmann, 1905), p. 10. Translated in *The Varieties of History*, F. Stern, ed. (New York: Vintage Books, 1973), p. 192.

[3] A. Pais, *Inward Bound* (Oxford: Oxford University Press, 1985), p. 3.

[4] T. Macauly, "History." Reprinted in *Essays* (New York: Sheldon, 1860), vol. 1, p. 387,

and in Stern, *Varieties*, p. 71.

[5] T. Carlyle, "On History." Reprinted in *A Carlyle Reader*, G. B. Tennyson, ed. (New York: Random House, 1969) and in Stern, *Varieties*, p. 90.

[6] G. Steiner, *The New Yorker*, 11 March 1991.

[7] Carlyle, "On History," p. 90.

[8] King of Belgium since 1934. The root of the unrest was his having stayed in Belgium after May 1940, when that country capitulated to the Germans, rather than joining the Belgian government in exile in London. On August 11, 1950, Léopold renounced his sovereignty.

[9] W. D. Niven, ed., *The Scientific Papers of James Clerk Maxwell* (Cambridge: Cambridge University Press, 1890), p. 241. Reprinted: New York: Dover, 1965.

[10] S. G. Brush and A. L. King, eds., *History in the Teaching of Physics* (Hanover, NH: University Presses of New England, 1972), p. vii.

[11] W. Whewell, *History of the Inductive Sciences* (London: Parker & Son, 1834), vol. 1, p. 42. Reprinted: London: Cass, 1967. See also H. Kragh, *Introduction to the Historiography of Science* (Cambridge: Cambridge University Press, 1987), ch. 1, for still earlier expressions of similar views by men of distinction.

[12] See also J. D. Bernal, *Science in History* (London: Watts & Co., 1954), for more on this subject.

[13] *The Scientific Papers of James Clerk Maxwell*, p. 251.

[14] See also Pais, *Subtle Is the Lord*, ch. 8, and Pais, *Inward Bound*, ch. 7.

[15] Cf. S. L. Jaki, *The Road of Science and the Ways to God* (Edinburgh: Scottish Academic Press, 1978).

[16] A. M. Schlesinger, *Isis* **36** (1946): 162.

[17] This is not a generally shared view. For example, it has been written: "All history of science – internal or external, technical and popular – is social history. The scientists study things; the historians study the scientists" (A. Hunter Dupree, *American Historical Review* **71** (1966): 863). I find this opinion short-sighted, to put it politely.

[18] I. B. Cohen in *Scientific Change*, A. C. Crombie, ed. (London: Heinemann, 1963), p. 773.

[19] More information on this class of journals is found in A. G. Brush, *American Journal of Physics* **55** (1987): 683.

[20] See note 18.

[21] See K. T. Compton, *Physics Today* (February 1952): 4, and E. U. Codon, *ibid.* 6.

[22] M. J. Klein, "Thermodynamics and Quanta in Planck's Work," *Physics Today* (November 1966): 23–32.

[23] T. S. Kuhn, J. L. Heilbron, P. Forman, and L. Allen, *Sources for History of Quantum Physics* (Philadelphia: American Philosophical Society), 1967).

[24] J. M. Staudenmaier, *American Historical Review* **95** (1990): 715.

[25] See Gershom Scholem, *Jewish Mysticism*, 9th ed. (New York: Schocken, 1977), p. 349.

ROGER H. STUEWER

THE SEVENTH SOLVAY CONFERENCE: NUCLEAR PHYSICS AT THE CROSSROADS

INTRODUCTION

The seventh Solvay Conference, held at the Free University of Brussels from October 22–29, 1933,[1] occupies a place in the history of nuclear physics similar to the one that the first Solvay Conference, held twenty-two years earlier in the Hotel Métropole in Brussels, occupies in the history of quantum physics. Both were the first to be devoted to their respective areas of physics; both were held soon after fundamental discoveries had opened up and transformed their fields; and both served to consolidate knowledge that had been gained and to expose problems that awaited solutions. Martin J. Klein has discussed the first Solvay Conference in his writings.[2] In this paper, I shall examine the seventh Solvay Conference, using it as a vantage point from which to view the social and political currents and the theoretical and experimental developments that were buffeting and transforming nuclear physics in the fall of 1933.

ORGANIZATION AND PARTICIPANTS

The Statutes of the International Solvay Institute of Physics vested the responsibility for organizing each Solvay Conference in its Scientific Committee.[3] As adopted in 1930 after earlier modifications, the Statutes required the Scientific Committee to be composed of eight ordinary members to whom could be added one extraordinary member with the same rights as the ordinary members. The Scientific Committee elected Paul Langevin, professor of physics in the Collège de France, to succeed H.A. Lorentz as its president following Lorentz's death on February 4, 1928. Langevin had been an invited member of every Solvay Conference beginning with the first in 1911, and had been a member of the Scientific

333

A.J. Kox and D.M. Siegel (eds.), No Truth Except in the Details, 333–362.
© 1995 Kluwer Academic Publishers.

Committee since the fourth in 1924.[4] In addition to Langevin, the current Scientific Committee consisted of Niels Bohr (Copenhagen), Blas Cabrera (Madrid), Peter Debye (Leipzig), Théophile de Donder (Brussels), Albert Einstein (Berlin), Charles-Eugène Guye (Geneva), Abram F. Ioffe (Leningrad), and Owen W. Richardson (London). Langevin convened the Scientific Committee in Brussels in April 1932 to plan the program for the seventh Solvay Conference to be held one and one-half years later.[5]

At that time the most striking new development in physics was James Chadwick's discovery of the neutron, first reported in *Nature* on February 27, 1932.[6] Chadwick's discovery promised to open up entirely new avenues of research in experimental and theoretical nuclear physics, and the Scientific Committee did not doubt that this was the area in physics in which the "most important problems" lay.[7] Still, Langevin recalled, the Scientific Committee hesitated to choose nuclear physics as the subject for the seventh Solvay Conference, because only six months earlier, from October 11–18, 1931, an international conference on nuclear physics had been organized by Enrico Fermi and held in Rome, and a two-year interval between major conferences on the same subject seemed to be too short a time.[8] Nevertheless, because of the great interest in the subject, the Scientific Committee decided in the end to go ahead and choose nuclear physics as the subject for the seventh Solvay Conference.

The Scientific Committee sensed correctly that nuclear physics was in a period of rapid transformation. Chadwick's discovery of the neutron had been preceded by Harold C. Urey's discovery of deuterium in December 1931,[9] and it was soon followed by Carl D. Anderson's discovery of the positron in August 1932.[10] Moreover, in the same month of February 1932 that Chadwick reported his discovery, Ernest O. Lawrence in Berkeley and John D. Cockcroft and Ernest T.S. Walton in Cambridge reported the inventions of their new particle accelerators.[11] These discoveries and inventions in that *annus mirabilis* of nuclear physics stimulated still further developments, so that more time than usual was required to finalize the program and participants of the seventh Solvay Conference. Ultimately, the Scientific Committee decided to invite as many people as permitted under the Statutes,[12] and some did not receive their invitations until just a few months before the conference opened.[13] The end result, however, Langevin was pleased to note, was that the participants differed from those attending earlier Solvay

Conferences in two respects. First, "as we have expressly sought," they were divided equally between experimentalists and theorists "to confront very intimately the efforts of the one with the other." Second, a large number of young people had been invited to present lectures. "A young physics," said Langevin, "requires young physicists." "Nothing justifies better our hope in international collaboration," he declared, than "the appearance in all countries of these young people in whom we place our hope."[14]

The Scientific Committee had indeed arranged a "truly international meeting." Except for the conspicuous absence of anyone from the Institut für Radiumforschung in Vienna, whose scientific reputation had suffered in recent years,[15] all of the majors centers of nuclear research were represented.[16] From the Cavendish Laboratory in Cambridge came its director, Ernest Rutherford, and joining him were Chadwick, Cockcroft, Walton, and Charles D. Ellis. Paul A.M. Dirac attended from St. John's College, Cambridge. Patrick M.S. Blackett, who had just left the Cavendish, came from Birkbeck College, London,[17] joining O.W. Richardson from King's College, London. Nevill F. Mott and Rudolf Peierls too had just left Cambridge and came from the University of Bristol and University of Manchester, respectively.[18]

From the Institut du Radium in Paris came its director, Marie Curie, and her daughter Irène Joliot-Curie, son-in-law Frédéric Joliot, and protégé Salomon Rosenblum. Joining Langevin from the Collège de France was Edmond Bauer. Maurice de Broglie, member of the Institut de France, his younger brother Louis de Broglie and Francis Perrin, respectively professor and maître de conférences in the Faculté des Sciences, rounded out the Parisian participants.

Attending from Berlin was Lise Meitner from the Kaiser-Wilhelm Institut für Chemie and Erwin Schrödinger from the University of Berlin. Werner Heisenberg joined Peter Debye from the University of Leipzig, and Walther Bothe, who had left Giessen the preceding year, attended from the University of Heidelberg.[19]

From Russia, joining Ioffe from Leningrad was George Gamow. Other countries were represented by only a single participant. Niels Bohr attended from Copenhagen, Hendrik A. Kramers from Utrecht, Enrico Fermi from Rome, Wolfgang Pauli from Zurich (Charles-Eugène Guye was ill and could not attend from Geneva[20]), Blas Cabrera came from Madrid, and Ernest O. Lawrence, the only American who was invited, came from Berkeley, California. Finally, there was strong representation from Belgium. Jules Emile Verschaffelt, Secretary of the Conference, came from the University of Gent, and Léon Rosenfeld came from the University of Liège. From the University of Brussels, joining de Donder, were Emile Henriot, Auguste Piccard, Ernest Stahel, Max Cosyns, and Jacques Errera.[21] Edouard Herzen attended from the École des Hautes Études in Brussels as the representative of the Solvay family. In all, there were forty-one participants between the ages of twenty-six and sixty-five from eleven countries. Fully one-third of the participants, however, did not contribute to the published proceedings (see Table I).

Deeply missed by all was Paul Ehrenfest, who took his own life in Amsterdam on September 25, 1933,[22] just one month before the conference opened. With a heavy heart Langevin recalled Ehrenfest's participation in the third and fifth Solvay Conferences of 1921 and 1927 where he was "so to speak the soul of these meetings." At the former, he gave an exposition of Bohr's correspondence principle, filling in for Bohr who was absent owing to illness. At the latter, where Heisenberg's uncertainty principle was much discussed, he contributed his characteristic clarity of thought.[23] But now death had "destroyed the great spirit and heart of Ehrenfest," and Langevin took it to be his "pious duty" to evoke the memory of Ehrenfest and "to relate how much he will be missed during the course of this meeting."[24]

INTELLECTUAL MIGRATION

Langevin's words perhaps touched no one more than Ehrenfest's oldest friend at the conference, Abram Ioffe. Ioffe had received his Ph.D. degree *summa cum laude* under Wilhelm Conrad Röntgen in Munich in 1905, and two years later, when Ehrenfest arrived in St. Petersburg, Ioffe established a close friendship with him and became a regular participant in his stimulating physics discussion club.[25] Ehrenfest's career in Russia languished, however, while Ioffe's flourished. Both were of the same age and both were Jews, but Ehrenfest was a foreigner as well and in 1912 he left for Leiden. The following year Ioffe received a professorial appointment in St. Petersburg's Polytechnical Institute, and after the October 1917 revolution he became one of the leaders in restructuring physical research in that city (Petrograd after 1914, Leningrad after 1924). In 1921, as part of a broad institutional reorganization, he became director of the Physico-Technical Institute. Ioffe's institute became the

TABLE I

Seventh Solvay Conference participants (age in parenthesis).

Cambridge
  *E. Rutherford (62)
  **J. Chadwick (41)
  *C.D. Ellis (38)
  **J.D. Cockcroft (36)
  **P.A.M. Dirac (31)
  *E.T.S. Walton (30)

London
  O.W. Richardson (54)
  *P.M.S. Blackett (35)

Bristol
  *N.F. Mott (28)

Manchester
  *R. Peierls (26)

Paris
  *M. Curie (65)
  *P. Langevin (61)
  *M. de Broglie (58)
  E. Bauer (53)
  L. de Broglie (41)
  *S. Rosenblum (37)
  **I. Joliot-Curie (36)
  **F. Joliot (33)
  *F. Perrin (32)

Berlin
  *L. Meitner (54)
  E. Schrödinger (46)

Leipzig
  *P. Debye (49)
  **W. Heisenberg (31)

Heidelberg
  *W. Bothe (42)

Leningrad
  A.F. Ioffe (53)
  **G. Gamow (29)

Copenhagen
  *N. Bohr (48)

Utrecht
  H.A. Kramers (38)

Rome
  *E. Fermi (32)

Zurich
  *W. Pauli (33)

Madrid
  B. Cabrera (55)

Gent
  J.E. Verschaffelt (63)

Liège
  L. Rosenfeld (29)

Brussels
  *T. de Donder (61)
  E. Herzen (ca. 56)
  A. Piccard (49)
  E. Henriot (48)
  *E. Stahel (37)
  J. Errera (37)
  M. Cosyns (ca. 30)

Berkeley
  *E.O. Lawrence (32)

*Participated in discussion
**Presented paper

source of generation after generation of accomplished physicists, making Ioffe the "founder and organizer of modern physics in the Soviet Union."[26]

One whose career took root in Leningrad was George Gamow, whose life repeatedly intersected with Ioffe's until the two would part forever at the seventh Solvay Conference. Gamow arrived in Leningrad (then Petrograd) from Odessa in 1922.[27] Six years later, in June 1928, frustrated with his thesis work, he left to spend the summer in Max Born's institute in Göttingen, where immediately after his arrival he conceived his new quantum-mechanical theory of alpha decay.[28] Emboldened with his success, on July 21, 1928, he wrote to Niels Bohr, enclosing a letter of reference from Ioffe, proposing to visit Copenhagen before returning to Leningrad in the fall.[29] When Gamow arrived in Bohr's institute, however, Bohr was so impressed with Gamow's work and personality that he arranged fellowship support for him to spend the entire academic year 1928–29 in Copenhagen. Subsequently, Gamow spent the academic year 1929–30 at the Cavendish Laboratory in Cambridge on a Rockefeller Fellowship, and the academic year 1930–31 again Copenhagen. During the intervening summers he returned home to Russia, as he did again in the summer of 1931 after attending a meeting in May on nuclear physics in Zurich. That fall he was scheduled to present a paper at the conference in Rome organized by Fermi, but this time he was denied permission to leave Russia, and Max Delbrück had to read his paper in Rome for him.[30] Forced to remain in Leningrad, Gamow taught and worked on his own, married, and made several attempts to escape with his wife. Their plans finally came to fruition when Gamow received an invitation to attend the seventh Solvay Conference.

Gamow later recalled (in his inimitable English) the circumstances surrounding his and his wife's trip to Brussels as "something like a dubble-miracle."[31] Knowing that an invitation sent directly to Gamow to attend the Solvay Conference would be insufficient, as in the past, for him to obtain an exit permit, Bohr persuaded Langevin, who was well known for his Communist sympathies and was chairman of the Franco-Russian Scientific Cooperation Committee,[32] to write to the Soviet government requesting that Gamow be officially designated as a Soviet delegate to the Solvay Conference. Whether Ioffe, as a member of the Scientific Committee, played any role in Langevin's request was never clear to Gamow. In his opinion, Ioffe's role could have been positive, neutral, or even negative, as Gamow felt that Ioffe never really liked him very much.[33]

In any case, Langevin's request succeeded. After much uncertainty and "psychologikal warfare,"[34] Gamow, and quite mysteriously his wife as well, were permitted to leave Russia together. Later, Bohr became quite upset when Gamow told him that they did not intend to return to Russia, since on Bohr's initiative Langevin had given his personal guarantee that Gamow would return. Only after Gamow explained his situation to Marie Curie, and after she then intervened with Langevin, obtaining his acquiescence to Gamow's decision, did both Bohr and Gamow feel ethically comfortable with it.[35] After the Solvay Conference, Gamow spent successive two-month periods in Paris, Cambridge, and Copenhagen before leaving for the United States in the early summer of 1934 to participate in the University of Michigan summer school in Ann Arbor. While there he received an offer of a professorship at George Washington University in Washington, D.C., beginning that fall.[36]

Gamow was motivated to leave Russia because on his return in 1931 he saw that the position of scientists had deteriorated greatly through increased political interference: "proletarian science" was now supposed to combat "erring capitalistic science."[37] Gamow thus decided to leave his homeland to seek greater political freedom in Europe and the United States. In complete contrast to this voluntary decision of Gamow's was the forced expulsion of scholars from Germany as a result of the brutal racial policies of Adolf Hitler. The political upheaval in Germany was on the mind of everyone at the Solvay Conference, and some of the participants had been directly affected by it.

Event followed event in Germany with breathtaking rapidity in 1933: Hitler became Chancellor on January 30; the Reichstag building in Berlin was torched on February 27; the Enabling Act, empowering the Nazi regime to govern without a constitution for four years, was passed by the new Reichstag on March 24; the Nazi Law for the Restoration of the Career Civil Service went into effect on April 7; and the infamous book burning in a square opposite the University of Berlin (and in many other university cities as well) took place on the evening of May 10 – a scene, said one observer, "which had not been witnessed in the Western world since the late Middle Ages."[38] By April 15 a correspondent for the *New York Evening Post* could sum up:

An indeterminate number of Jews have been killed. Hundreds of Jews have been beaten or tortured. Thousands of Jews have fled. Thousands of Jews have been, or will be, deprived of their livelihood. All of Germany's 600,000 Jews are in terror.[39]

The Nazi Civil Service Law of April 7 had an immediate and devastating effect on Jewish teachers and scholars, precipitating an unprecedented intellectual migration.[40] Hitler, unlike Stalin, found no compelling reasons to confine scientists and other scholars, but moved to ostracize them.[41] On May 19, *The Manchester Guardian* published a long list of nearly two hundred scholars who were dismissed from over thirty institutions of higher learning throughout Germany between April 14 and May 4,[42] and that number would climb significantly in succeeding months.

To assist the exiled scholars, refugee organizations were established rapidly in England and other European countries and the United States. In England, Sir William Beveridge, Director of the London School of Economics and Political Science, became the prime mover in establishing the Academic Assistance Council (AAC), and in early May of 1933 he persuaded Lord Rutherford to serve as its president.[43] The formation of the AAC was announced in newspapers throughout Britain on May 24; it held its first meeting on June 1;[44] and *Nature* publicized its work on June 3 and in subsequent issues.[45] In Denmark, Niels Bohr's Institute of Theoretical Physics in Copenhagen became a way-station or haven for many refugees from Nazi Germany.[46] In the United States, Oswald Velben, head of the School of Mathematical Sciences of the Institute for Advanced Study in Princeton, played a leading role in the establishment of the Emergency Committee for Aid to Displaced German (later Foreign) Scholars,[47] whose guiding light became Stephen Duggan, Director of the Institute of International Education in New York.[48] The work of the Emergency Committee began in early June 1933,[49] and its formation was announced in *Science* on July 21.[50] These rescue efforts were all the more remarkable because of the severe economic depression at the time. Edward R. Murrow, second-in-command of the Emergency Committee, noted in a memorandum that by October 1933 – just at the time of the Solvay Conference – more than 2000 out of a total of 27,000 teachers had been dropped from the faculties of some 240 colleges and universities in the United States.[51]

The most prominent physicist to be caught in the maelstrom in Germany was Albert Einstein, who had been Ehrenfest's closest friend.[52] At the Solvay Conference Langevin only noted that Einstein was not in attendance because he had left Europe to fulfill a call to the United States.[53] Everyone present, however, knew that Langevin's bland remark masked Einstein's true fate. For on March 28, 1933, returning from a trip to the United States, Einstein disembarked with his wife Elsa from the *Belgenland* at Antwerp, Belgium, wrote a letter of resignation to the Prussian Academy of Sciences, and then was driven to Brussels where he surrendered his German citizenship at the German embassy.[54] Moving to Le Coq-sur-Mer, a small resort near Ostend, he also severed his ties with the Bavarian Academy of Sciences on April 21.[55] During the summer he traveled twice to Britain to lecture and to meet with dignitaries. In early September he left Belgium for good and resided close to London as a guest of a British Member of Parliament.[56]

On October 3, 1933, Einstein greatly advanced the cause of the Academic Assistance Council when it joined with several other refugee organizations to sponsor a meeting in the Royal Albert Hall in London. Lord Rutherford was in the chair, and along with Einstein on the platform were Sir Austen Chamberlain, Sir William Beveridge, and Sir James Jeans.[57] Einstein was the featured speaker of the evening, and he spoke to a packed audience of over 10,000 people, delivering his first public address in the English language. Announcing himself as "a man, a good European and a Jew," he praised the refugee agencies for their work, and he spoke vigorously in defense of "intellectual and individual freedom," without which "there would be no Shakespeare, Goethe, Newton, Faraday, Pasteur or Lister."[58] This was the last address Einstein delivered in Europe. On October 7 he boarded the *Westernland* at Southampton for New York, arriving on October 17 with his wife Elsa, his secretary Helen Dukas, and his collaborator Walther Meyer to take up an appointment in the Institute for Advanced Study in Princeton.[59] A few days later, his chairman in London, Ernest Rutherford, left Cambridge for Brussels to attend the seventh Solvay Conference.

Rutherford thus was personally familiar with the painful circumstances surrounding Einstein's absence in Brussels. But he needed only to look around himself to see that Einstein's case was not unique. Every German present was fully aware of the devastation that had been wrought in Germany. Werner Heisenberg, for example, had written to Bohr from Leipzig on June 30, 1933,[60] reporting that his Solvay lecture was nearly finished, and then telling Bohr that he had often spoken or written to Max Planck and Max von Laue, trying with them to retain James Franck and Max Born in Germany, quite likely unsuccessfully, leaving "the future completely uncertain." Heisenberg also mentioned Felix Bloch. Bloch's case, in fact, illustrates well the loss of gifted young physicists from Germany and shows how his fate, even though he himself had not been invited to the Solvay Conference, had become intertwined with the lives of a number of those who actually were in attendance.

After switching from engineering to physics as a student in the Eidgenössische Technische Hochschule (ETH) in Zurich, Bloch took Peter Debye's elementary physics course at the ETH and was inspired by it.[61] Debye, in turn, was impressed with Bloch's abilities, and when Debye decided to move to Leipzig in 1927 he suggested to Bloch that he should also transfer there to study further under Heisenberg. At Leipzig Bloch met Rudolf Peierls and became Heisenberg's first Ph.D. student, completing his degree in 1928. He then became Wolfgang Pauli's assistant in Zurich (1928–29), received a Lorentz Fund Fellowship for further study in Utrecht with H.A. Kramers and in Haarlem with A.D. Fokker (1929–30), returned to Leipzig as Heisenberg's assistant (1930–31), spent six months with Bohr in Copenhagen (1931–32), and completed his *Habilitationsschrift* under Heisenberg in Leipzig in the spring of 1932. As others had,[62] Bloch too witnessed the Nazi storm troopers in the streets of Leipzig and experienced classroom disruptions, but, being a Swiss citizen, he was unaffected by the Nazi Civil Service Law of April 7, 1933. As a Jew and human being, however, he found that law intolerable, so he simply quit his position in Leipzig as *Privatdozent* and went home to Zurich, refusing to return to Leipzig despite Heisenberg's urging him to do so – a stance that Bloch considered to be thoughtless and naive. While in Zurich, Bloch received an invitation to lecture for two or three weeks at the Institut Henri Poincaré in Paris, where he lived in the home of Paul Langevin. Then, in August or September of 1933, while again visiting Bohr's institute in Copenhagen, he received an offer of a position from Stanford University, which he soon accepted. By the time of the Solvay Conference, therefore, Bloch had been associated with no less than seven participants, Debye, Heisenberg, Peierls, Pauli, Kramers, Bohr, and Langevin, and it would have been natural for them, even though Bloch was not present, to have Bloch's odyssey on their minds.

Debye was destined to remain in Germany as long as possible, and Heisenberg would never leave, nor would another Solvay participant, Walther Bothe, from the University of Heidelberg. Lise Meitner, too, protected by her Austrian citizenship, and hoping and believing that the excesses of the Nazi regime would pass, would remain in her position at the Kaiser-Wilhelm Institut für Chemie in Berlin-Dahlem until she was placed in immediate danger of incarceration four and one-half years later when Hitler extended his domain to Austria.

Not so Erwin Schrödinger, who since 1927 had been at the University of Berlin as Max Planck's successor in the chair of theoretical physics. Although not Jewish, Schrödinger was repelled by the Nazi racial policies, and when the Oxford professor of physics Frederick A. Lindemann visited Schrödinger in Berlin in the middle of April 1933, Schrödinger voiced his willingness to leave and accept a position in Oxford.[63] Lindemann promised to approach the Imperial Chemical Industries for financial support, and by June Schrödinger and his wife Anny were quietly making arrangements to give up their house in Berlin and to send their furnishings to England. On July 21 Lindemann learned that Schrödinger would be elected to a senior fellowship in Magdalen College, Oxford.[64] By then the Schrödingers had left Berlin by car and were in Zurich visiting Wolfgang Pauli, after which they summered in the South Tirol. In September Lindemann again visited Schrödinger at Lake Garda, bringing details of Schrödinger's appointment in Oxford. Planck, learning of Schrödinger's decision not to return to Berlin, was shaken by it,[65] but Heisenberg was simply angry, writing to his mother on September 17 that Schrödinger had no reason to leave, "since he was neither Jewish nor otherwise endangered."[66] In fact, the Nazis had classified Schrödinger as "politically unreliable,"[67] a ground for dismissal under the Nazi Civil Service Law. On October 3, Lindemann, writing from Oxford, informed Schrödinger that he had that day been elected as a Fellow of Magdalen College.[68] Two weeks later Schrödinger attended a conference in Paris, and then went on to Brussels to attend the seventh Solvay Conference. On October 24, two days after the conference opened, the Berlin *Deutsche Zeitung* carried an article regretting the loss of Schrödinger to German science.[69]

Another physicist permanently lost to German science was Rudolf Peierls, at age 26 the youngest participant to be invited to the Solvay Conference. After studying at the Universities of Berlin (1925–26) and Munich (1926–28), Peierls transferred to the University of Leipzig where like Bloch he studied under Heisenberg.[70] He took his Ph.D. degree at Leipzig in July 1929, with a thesis on a problem suggested to him by Wolfgang Pauli while he was visiting Zurich that spring, when

Heisenberg was away on a trip to the United States. Peierls then worked for three years as Pauli's assistant at the ETH in Zurich (1929–32), during which time he also enlarged his circle of friends and associates through visits to Holland, Denmark, and the Soviet Union (where he met his future wife Genia in Odessa in the summer of 1930).[71] He completed his *Habilitationsschrift* in Zurich and then, on Pauli's recommendation, he received a Rockefeller Fellowship which he decided to split between Rome (fall 1932–spring 1933) and Cambridge (spring–fall 1933). While in Rome working with Fermi, he received an offer of an appointment in Hamburg to begin at Easter 1933, which, even though it meant giving up the second half of his fellowship, he decided to accept.[72] When the time came to take up the appointment, however, the Nazis were in power, and Peierls declined the Hamburg offer, going to Cambridge instead. By the time he left for Brussels in October to attend the Solvay Conference, he had received a two-year grant from a refugee organization to support him at the University of Manchester, where another refugee, Hans A. Bethe, whom Peierls had first met in Munich as a student, lived with him and his family for a year in a spare room in their house.[73]

The Nazi racial policies thus impinged directly or indirectly on the lives of all of the physicists from Germany who had been invited to attend the Solvay Conference – Einstein, Debye, Heisenberg, Bothe, Meitner, Schrödinger, and Peierls. But their impact was also felt by many of the other participants, such as Bohr and Langevin, who sheltered refugees for longer or shorter periods of time. Frédéric Joliot, also, found fellowship support for the refugee Walter Elsasser and a place for him to work in the Institut Henri Poincaré in Paris, where he and yet another refugee from Berlin, K. Guggenheimer, independently carried out the earliest shell-model studies on nuclei in 1933–34.[74] The list of helpers and helped could easily be extended.[75] No one at the Solvay Conference, in fact, was left entirely untouched by the plight of the refugees, for by 1933 physics had become a truly international enterprise, with numerous ties among physicists having been forged through traveling fellowships, visiting lectureships and professorships, participation in the Michigan summer schools, and attendance at professional meetings. These included a series of international conferences in the early 1930s that helped to build an increasingly large community of nuclear physicists.

The first international conference on nuclear physics was held from May 20–24, 1931, at the ETH in Zurich.[76] This meeting, a *Physika-*

lische Vortragswoche, was organized by Egon Bretscher and Eugene Guth. Gamow delivered the opening lecture, which was followed by lectures by Guth (Zurich and Vienna) and Theodor Sexl (Vienna), Walther Bothe (Giessen), P.M.S. Blackett (Cambridge), Hendrik B.G. Casimir (Leiden), Hermann J.J. Schüler (Potsdam), and Immanuel Estermann, Otto Robert Frisch, and Otto Stern (Hamburg). Also attending and participating in the discussions were Wolfgang Pauli from Zurich, Frédéric Joliot, Maurice de Broglie, and Louis Leprince-Ringuet from Paris, Hans Kopfermann from Berlin-Dahlem, and Derek A. Jackson from Oxford. Gamow, Bothe, Blackett, Pauli, Joliot, and de Broglie would also attend the seventh Solvay Conference.

That fall, from October 11–18, 1931, a much larger international conference on nuclear physics was organized by Enrico Fermi and held in Rome.[77] Guglielmo Marconi served as honorary president and Orso M. Corbino as effective president. Lectures were delivered by Nevill F. Mott (Cambridge), Samuel Goudsmit (Ann Arbor), Bruno Rossi (Florence), Walther Bothe (Giessen), Charles D. Ellis (Cambridge), Niels Bohr (Copenhagen), Léon Rosenfeld (Liège), Arnold Sommerfeld (Munich), Emil Rupp (Berlin), Ralph H. Fowler (Cambridge), and Guido Beck (Leipzig). Gamow's lecture was read by Max Delbrück (Zurich). Thirty-two other physicists also attended from England, Germany, France, Switzerland, Italy, and the United States. The participants who also would attend the seventh Solvay Conference were Fermi, Mott, Bothe, Ellis, Bohr, Rosenfeld, P.M.S. Blackett, Marie Curie, Peter Debye, Werner Heisenberg, Lise Meitner, Wolfgang Pauli, O.W. Richardson, and the absent George Gamow.

Two years later, from September 24–30, 1933, a third international meeting devoted to nuclear physics, the fifth All-Union Conference in Physics, was held in Ioffe's Physico-Technical Institute in Leningrad.[78] The hosts for this meeting were Vladimir A. Fock, Dmitri D. Ivanenko, Igor Y. Tamm, and Dmitry V. Skobelzyn.[79] Lectures were delivered by Frédéric Joliot (Paris), Francis Perrin (Paris), Louis H. Gray (Cambridge), P.A.M. Dirac (Cambridge), Franco Rasetti (Rome), and the Soviet physicists Ivanenko, Skobelzyn, S.E. Frisch, K.D. Simelidov, A.I. Leipunski, and George Gamow, although Gamow's was excluded from the published proceedings "for technical reasons."[80] Joliot, Perrin, Dirac, and Gamow also would attend the seventh Solvay Conference just one month later.

The seventh Solvay Conference thus became the fourth international conference devoted to nuclear physics in less than two and one-half years, with no less than seventeen of the Solvay participants having attended one or more of the earlier three. The personal and professional bonds that had been formed among them before coming to Brussels had created a sense of community in a time of crisis and diaspora.

NUCLEAR QUESTIONS

The discoveries of deuterium, the neutron, and the positron, and the inventions of the Cockcroft–Walton accelerator and cyclotron all found a place in the papers and discussions at the seventh Solvay Conference. Some questions pertaining to them had been settled, while others were still open and awaited solutions. I shall now discuss some of these questions to illustrate how the seventh Solvay Conference stood at the crossroads in nuclear physics in the fall of 1933.

Cockcroft opened the conference with a long paper describing in detail the proton accelerator that he and Walton, with Rutherford's strong support, had invented and constructed at the Cavendish Laboratory.[81] A crucial stimulus, too, was provided by Gamow, who visited Cambridge in early 1929, taking with him calculations indicating that protons of relatively low energy should be able to tunnel quantum mechanically through the potential barriers of light nuclei.[82] Cockcroft now reviewed Gamow's theory and its predictions for the disintegration of lithium, boron, and fluorine by protons of energy 100, 300, and 600 keV, which he compared with experiment, finding good agreement. At the end of his paper, Cockcroft also reported on experiments in which accelerated "deutons," the nuclei of Urey's heavy hydrogen, were used to disintegrate the nuclei of lithium, beryllium, carbon, and nitrogen.

Rutherford opened the discussion by calling attention to recent experiments that he and M.L.E. Oliphant, using their own low-energy 200-keV accelerator, had carried out at the Cavendish,[83] from which Rutherford concluded that when lithium is disintegrated by protons, one possible outcome is that a hitherto unknown isotope of helium of mass three is produced. Lawrence followed with a lengthy description of the new 27-inch cyclotron that he and his colleagues at Berkeley had recently put into operation,[84] accelerating deutons and finding controversial results – as others soon made clear. For the moment, however, Marie Curie made a brief remark pertaining to the oldest fundamental issue to surface at the Solvay Conference. She noted that:

It is interesting to remark that the reaction

$$_3Li^7 + {}_1H^1 = {}_2{}_2He^4$$

is, to my knowledge, the first nuclear reaction in which one can verify with precision and without any uncertainty the relation of Einstein between mass and energy, providing one uses for Li the atomic mass determined by Bainbridge.[85]

Curie's brief comment spoke volumes. Einstein's mass–energy relationship, $E = mc^2$, had remained inaccessible to a precise experimental test for almost three decades after he had derived it in 1905, because, as Einstein himself showed,[86] it required the determination of isotopic masses to an accuracy of 1 part in 10^5, whereas even F.W. Aston's most precise mass spectograph, which became operational in 1925, was capable of measuring isotopic masses to an accuracy of only 1 part in 10^4.[87] Nevertheless, Einstein's mass–energy relationship had been widely accepted as valid and used as a tool in analyzing nuclear reactions, which is precisely what Cockcroft and Walton did in early 1932: They calculated the loss of mass in the above reaction and compared it to the energy gained by the product α particles, finding reasonable agreement. Only one year later did Kenneth T. Bainbridge, working at the Bartol Research Foundation of the Franklin Institute in Philadelphia, point out that this reaction could be used as a *test* of Einstein's relationship.[88] Employing a more precise mass spectrograph that he himself had developed, Bainbridge determined the mass of lithium-7 to higher precision, used Aston's values for the masses of helium and hydrogen, inserted the kinetic energies of the incident proton and product α particles, and in June 1933 reported that, "Within the probable error of the measurements the equivalence of mass and energy is satisfied."[89] Four months later, Marie Curie emphasized the fundamental significance of Bainbridge's conclusion at the seventh Solvay Conference.

Cockcroft's paper was followed by ones by James Chadwick and by Frédéric Joliot and Irène Joliot-Curie.[90] Both intersected with certain remarks that Lawrence had made following Cockcroft's paper, pertaining to a second fundamental issue discussed at the Solvay Conference, namely, the question of the mass of the neutron.[91]

When Chadwick had submitted his full report on his discovery for publication in May 1932,[92] he had calculated the mass of the neutron from the reaction

$$_5B^{11} + _2He^4 \rightarrow _7N^{14} + _0n^1$$

by using Aston's values for the atomic masses of boron, helium, and nitrogen, and by measuring the kinetic energies of the incident α particle and product nitrogen nucleus and neutron. By simple arithmetic he found that the mass of the neutron was 1.0067 amu (atomic mass units), which, since the sum of the proton and electron masses was 1.0078 amu, he took to mean that the neutron consists of a proton-electron compound with a binding energy of 1 to 2 MeV (million electron volts).[93] As early as 1920 Rutherford had speculated on the existence of the neutron as a proton and electron in close combination,[94] and Chadwick now took his discovery to vindicate not only Rutherford's speculation but his specific model as well. In subsequent papers, Chadwick maintained his belief in the neutron as a complex particle, even though he realized that that picture entailed serious difficulties pertaining to its spin and statistics.[95]

Lawrence was the first to challenge Chadwick's value. His senior colleague in the chemistry department at Berkeley, G.N. Lewis, succeeded in producing a substantial quantity of deuterium in early 1933, and by June, Lawrence, Lewis, and M. Stanley Livingston were accelerating deutons of energy from 0.6 to 1.33 MeV onto targets of carbon, gold, platinum, lithium fluoride, and several other compounds.[96] In each case they found that protons of about 18-cm range in air were produced. The only interpretation that Lawrence could find for "this group of protons common to all targets" was that "the deuton itself is breaking up, presumably into a proton and a neutron."[97] On this assumption, Lawrence could calculate the mass of the neutron. Thus, he and his colleagues had found in every case that 1.2 MeV deutons produced 3.6 MeV protons (18 cm in range), indicating that they had gained 2.4 MeV in the break-up process. Therefore, the neutrons too, assuming they flew off in an opposite direction, had gained an equal amount of energy, for a total released energy of 4.8 MeV. The mass-energy balance of the break-up process would then be

$$m_d = m_p + m_n + 4.8 \text{ MeV},$$

from which Lawrence calculated "that the mass of the neutron is about 1.0006 [amu] rather than 1.0067 as estimated by Chadwick."[98] Lawrence did not revise his value during the following months even though researchers in the Cavendish Laboratory expressed strong skepticism about it.[99]

Irène Joliot-Curie and Frédéric Joliot, too, challenged Chadwick's value. In a note presented to the Académie des Sciences on July 17, 1933,[100] they suggested that the recent discovery of the positron had to be taken into account, and that instead of the reaction assumed by Chadwick in which the heavy isotope of boron was bombarded by α particles and a neutron was produced, the α particles were actually interacting with the light isotope of boron, and either a neutron and a positron, or a proton, appeared as products. The mass-energy balances of these two new reactions were

$$_5B^{10} + _2He^4 + T_\alpha = _6C^{13} + _0n^1 + _1e^+ + T_1$$

and

$$_5B^{10} + _2He^4 + T_\alpha = _6C^{13} + _1p^1 + T_2,$$

where T_α is the kinetic energy of the incident α particle and T_1 and T_2 are the total kinetic energies of the products in the first and second reactions, respectively. Subtracting the first equation from the second yielded

$$_0n^1 = (T_2 - T_1) + _1p^1 - _1e^+,$$

and by measuring the two kinetic energies and inserting the masses of the proton and positron, Joliot-Curie and Joliot calculated that the mass of the neutron was 1.011 amu, a substantially higher value than Chadwick's. Because the sum of the neutron and positron masses (1.011 + 0.0005 amu) would now exceed the proton mass (1.0073 amu), their value implied that the proton consisted of a neutron-positron compound with a rather high binding energy.

Chadwick, Lawrence, Joliot-Curie and Joliot thus arrived in Brussels prepared to argue for three very different values for the mass of the neutron. Lawrence got his word in first. He was more than ever convinced of the correctness of his low value of 1.0006 amu, because just before leaving Berkeley he learned about some experiments of H. Richard Crane, Charles C. Lauritsen, and A. Soltan at the California Institute of Technology in Pasadena, which he himself then repeated, and which he took to confirm his deuton break-up hypothesis.[101] He therefore confidently repeated his neutron-mass calculation in Brussels.[102] He also argued that the break-up occurs when the deuton strikes the potential barrier of the target nucleus. Heisenberg was not convinced. He remarked that the intensity of the effect then should depend on the atomic number of the target nucleus, contrary to experiment.[103] Bohr agreed that this discrepancy presented "grave difficulties" for Lawrence's interpretation.[104]

Lawrence's replies evidently were unpersuasive, since no one took his side in the debate. Chadwick, in particular, did not budge.[105] He reiterated his belief that the boron reaction he had assumed (employing the heavy isotope of boron) was correct, and he argued that only that reaction and the reaction

$$_3Li^7 + _2He^4 \rightarrow _5B^{10} + _0n^1$$

permitted the determination of the neutron mass, the former yielding 1.0066 ± 0.001 amu, the latter an upper limit of 1.0072 ± 0.0005 amu. Since both values were less than the sum of the proton and electron masses, Chadwick still felt that the neutron should be viewed as "the result of an intimate union of a proton with an electron,"[106] although he again recognized the troublesome spin and statistics difficulties implied in that model.

Joliot-Curie and Joliot again challenged Chadwick's value by again questioning his basic assumptions.[107] They pointed out that Chadwick's calculation assumed that no energy was emitted in the form of γ rays, and they then cited a number of reactions for which this was known to be untrue. They also displayed cloud-chamber photographs showing the creation of electron-positron pairs through the direct materialization of γ rays. Thus, new reactions were possible, in particular the ones they had assumed earlier involving the light isotope of boron. They therefore again focussed on these reactions and presented the same calculation as before for the neutron mass, arriving however at a slightly different value, 1.012 amu, because they used a slightly different figure for the kinetic energy of the products of one of the reactions. In any case, their value was, as before, much higher than Chadwick's, even "noticeably greater" than the mass of the proton.[108]

Both Lawrence and Chadwick once again defended their values, the upshot being that none of the protagonists persuaded any of the others to accept a different value for the mass of the neutron. The issue was a fundamental one, because its outcome would determine whether the neutron should be regarded as a proton-electron compound or as a new elementary particle. If Chadwick's or Lawrence's values were correct, the former should be the case; if Joliot-Curie and Joliot's value was correct, the latter should be true, and the proton should be regarded as a neutron-positron compound. Walther Bothe soon summarized the situation when he noted that the "important question" of whether the neutron or the proton was "the actual elementary particle . . . still cannot be answered with certainty."[109]

That question was not discussed in Dirac's paper on the theory of the positron,[110] nor was it of central concern to Gamow, whose subject was the origin of γ rays and nuclear energy levels.[111] Instead, it entered directly into Heisenberg's paper, entitled "General Theoretical Considerations on the Structure of the Nucleus."[112] Heisenberg's paper, as seen in its broadest perspective, represented a milestone in the history of the liquid-drop model of the nucleus.[113]

Gamow first proposed the liquid-drop model of the nucleus during a discussion on the structure of atomic nuclei at a meeting of the Royal Society in London on February 7, 1929, which he attended on Rutherford's invitation while he was visiting Cambridge.[114] Gamow envisioned the nucleus as a collection of α particles possessing kinetic and potential energy and having short-range attractive forces among them so that the nucleus could be treated like "a small drop of water in which the particles are held together by surface tension."[115] A year later, while in Cambridge on a Rockefeller Foundation fellowship, he developed this model quantitatively, incorporating a Coulomb repulsive term for the α particles at the surface of the nucleus.[116] He calculated the total energy E of the nucleus as a function of the total number N of its constituent α particles, in other words, as a function of the mass of the nucleus. He found that E was given by the sum of an attractive (negative) term varying as the cube root of N and a repulsive (positive) term varying as the 5/3 root of N, so that a plot of E versus N possessed a minimum – the most distinctive feature of the nuclear mass-defect curve.

Shortly after Chadwick's discovery of the neutron in early 1932, Heisenberg published his seminal theory of nuclear structure in which he introduced the concept of charge exchange as the origin of the nuclear force between protons and neutrons.[117] Still, he was uncertain at that time whether the neutron was a proton-electron compound or a new elementary particle. Only after Ettore Majorana in early 1933, while visiting Heisenberg's institute in Leipzig, proposed a new nuclear force in which both charge and spin are exchanged,[118] did Heisenberg begin to become more and more convinced that the neutron was indeed a new elementary particle. Moreover, Majorana's nuclear force bound two protons to two neutrons, while Heisenberg's bound only one proton to one neutron. In other words, Majorana's nuclear force saturated at the α particle while Heisenberg's saturated at the deuteron, and since the α particle was known to be a highly stable particle while the deuteron was not, that was a compelling reason to adopt Majorana's nuclear force over Heisenberg's. Majorana's also provided a direct connection to Gamow's liquid-drop model of the nucleus, since Gamow had assumed that α particles were the basic constituents of nuclei.

These were the fundamental physical ideas that served as the foundation for Heisenberg's Solvay paper. He succinctly noted that Majorana's theory could be "considered as corresponding to a form of Gamow's drop model made precise by the neutron hypothesis."[119] He assumed that nuclei were composed of n_1 neutrons and n_2 protons bound together in α-particle subunits by Majorana's exchange force, and in a long quantum-mechanical calculation he derived expressions for the total kinetic and potential energies, to which he added a term for the total Coulomb repulsive energy of the protons at the nuclear surface. In this way, he derived an expression for the total energy E of a nucleus as a function of the total number of neutrons n_1 and protons n_2 comprising it. Plotting E versus $(n_1 + n_2)$, he found, as Gamow had earlier, the distinctive minimum in the mass-defect curve, which he compared with Aston's mass-spectrographic data, finding qualitatively reasonable agreement.[120] Heisenberg discussed a number of other topics as well in his extensive paper, but his basic message was clear: Under the assumption that nuclei were composed only of protons and neutrons, he had shown that Majorana's exchange force provided a new and deeper theoretical foundation for Gamow's liquid-drop model of the nucleus.

In the discussion following Heisenberg's paper, Fermi pointed out that there was a great theoretical tension between conceiving the nucleus as an agglomeration of particles (liquid-drop model) and as a system of individual particles occupying various energy states (shell structure).[121] Bohr took even a more skeptical stance. He argued that Gamow's liquid-drop model was "very schematic," because even the heaviest nuclei contained only about fifty α particles, and even with the densest packing imaginable only about ten would reside in the interior of the nucleus, with the rest at its surface.[122] But most of the discussion was on another topic, β decay, because the first to speak was Wolfgang Pauli, and Pauli chose this opportunity to advance – for the first time – his neutrino hypothesis for publication,[123] thus making the Solvay Conference a milestone in the history of that hypothesis as well.

Experimentally, the history of β decay took a decisive turn in 1927 when Charles D. Ellis and William A. Wooster in the Cavendish Laboratory found that the average energy per disintegration of the β particles emitted from RaE ($_{83}Bi^{210}$) was close to the average energy of the continuous β-ray spectrum for that element and not at its upper limit, thereby proving that RaE β particles actually emerged from the nucleus with a continuous distribution of energies and not with some maximum energy that was then dissipated in passing through the electronic distribution.[124] Ellis and Wooster's result was confirmed by Lise Meitner and Wilhelm Orthmann in Berlin in 1930,[125] and at the end of that year, on December 4, 1930, Pauli wrote a letter addressed principally to Meitner and Hans Geiger at a conference in Tübingen in which he first proposed the "possibility that there could exist in the nuclei electrically neutral particles that I wish to call neutrons, which have spin 1/2 and obey the exclusion principle, and additionally differ from light quanta in that they do not travel with the velocity of light."[126] These "neutrons" would have a small mass (Pauli then thought it would be of the same order of magnitude as the mass of the electron) and would be emitted along with an electron in β decay, thereby preserving the conservation laws.

Pauli discussed his hypothesis again at a conference in Pasadena in June 1931, in Ann Arbor a few weeks later, and yet again in Rome that October at the conference organized by Fermi – all the while declining to publish it, although it did appear in print after the Rome conference because Samuel Goudsmit discussed it there.[127] Pauli himself, however, only offered his hypothesis for publication two years later at the Solvay Conference, by which time he had adopted Fermi's name, "neutrino," for his hypothetical new particle. He strongly opposed Bohr's alternative interpretation of β decay – which Bohr had repeatedly advocated since 1929 and which he again advanced at the Solvay Conference – that the conservation laws were being violated and, in general, should be regarded as only statistically valid in the nuclear realm.[128] Neither Bohr nor Pauli relinquished his position at the Solvay Conference.

EPILOGUE

Less than two weeks after the close of the Solvay Conference, on November 9, 1933, Heisenberg in Leipzig, Dirac in Cambridge, and Schrödinger in Oxford learned that they had been awarded the Nobel Prize for Physics, the 1932 award going to Heisenberg alone, the 1933 award being shared by Dirac and Schrödinger.[129] They joined five of the other Solvay participants (Marie Curie, Rutherford, Bohr, Richardson, Louis de Broglie) and the absent Einstein as Nobel Laureates.[130] That Schrödinger was now in England and Einstein in the United States symbolized the intellectual decapitation of Germany that had begun a few months prior to the Solvay Conference and would continue apace in the months and years to follow.[131]

In nuclear physics further progress occurred shortly after the Solvay Conference. Joliot-Curie and Joliot returned to Paris, where in January 1934 they discovered artificial radioactivity, [132] the last major discovery in the Institut du Radium to be witnessed by its founder, Marie Curie, who died six months later in July. In Rome, Enrico Fermi, also in January 1934, published his celebrated theory of β decay,[133] revealing the potency of Pauli's neutrino hypothesis and providing a firm theoretical foundation for excluding electrons from the nucleus. Then, during the following months, Fermi and his team in Rome followed up Joliot-Curie and Joliot's discovery of artificial radioactivity, discovering the efficacy of slow neutrons in producing nuclear reactions,[134] and opening up the study of neutron physics.

In Berkeley, Lawrence found to his chagrin in March 1934 that his deuton break-up hypothesis had to be discarded – he and his colleagues actually had been observing protons and neutrons from deuton–deuton reactions produced when deutons in their beam struck deuterium contaminants in their targets.[135] He therefore was forced to admit that his low value for the mass of the neutron was in error. That question, in general, was soon settled definitively when Chadwick in Cambridge joined forces with Maurice Goldhaber, yet another refugee and former student of Schrödinger's in Berlin, to carry out entirely new measurements of the neutron mass based on the photodisintegration of the deuton (or diplon, as it was called in Cambridge at the time).[136] They found, much to Chadwick's surprise, that the mass of the neutron was 1.0080 amu, much greater than Chadwick's earlier value of 1.0067 amu, and even greater than the mass of the hydrogen atom (1.0078 amu). That result left no doubt that the neutron was a new elementary particle, and it offered conclusive experimental support for excluding electrons from the neutron, and hence from the nucleus.

All of the above developments occurred by the time the fifth international conference on nuclear physics took place in London and Cambridge from October 1–6, 1934,[137] just one year after the Solvay Conference. In 1935 yet another development followed when C.F. von Weizsäcker, who was working on his *Habilitationsschrift* under Heisenberg in Leipzig, built upon his mentor's Solvay paper and proposed a semi-empirical nuclear-mass formula that again displayed the distinctive minimum found earlier by Gamow and Heisenberg.[138] Weizsäcker's formula, when somewhat refined the following year by Hans A. Bethe,[139] became a basic tool in the analysis of nuclear binding energies. It represented the culmination of the line of research opened up by Gamow when he applied his liquid-drop model to an understanding of the nuclear mass-defect curve.[140] In 1936 Bohr would initiate another line of development that would display the fruitfulness of the liquid-drop model in understanding nuclear reactions.[141]

The seventh Solvay Conference was the last one that was held before the outbreak of war in Europe. It stood at the crossroads both in nuclear physics and in human history. By the time the eighth Solvay Conference was held in 1948 the world had been fundamentally changed by the political and scientific forces so evident in October 1933.

Program in History of Science and Technology
School of Physics and Astronomy
University of Minnesota
U.S.A.

NOTES

[1] Institut International de Physique Solvay, *Structure et Propriétiés des Noyaux Atomiques. Rapports et Discussions du Septième Conseil de Physique tenu à Bruxelles du 22 au 29 Octobre 1933* (Paris: Gauthier-Villars, 1934); hereafter cited as Solvay *Rapports*.
[2] Martin J. Klein, "Einstein, Specific Heats, and the Early Quantum Theory," *Science* **148** (1965): 173–180; *Paul Ehrenfest*, vol. I, *The Making of a Theoretical Physicist* (Amsterdam: North-Holland, 1970), pp. 251–252, 257–258.
[3] Solvay *Rapports*, pp. 347–353, esp. pp. 348–350.
[4] Jagdish Mehra, *The Solvay Conferences on Physics: Aspects of the Development of Physics since 1911* (Dordrecht: Reidel, 1975), lists the invited participants and mem-

bers of the Scientific Committees for the various Solvay Conferences as given in their published proceedings.

[5] Langevin, Solvay *Rapports*, p. ix, stated that the Scientific Committee met eighteen months before the conference, i.e., in April 1932.

[6] James Chadwick, "Possible Existence of a Neutron," *Nature* **129** (1932): 312.

[7] Langevin, Solvay *Rapports*, p. ix.

[8] *Ibid.*

[9] Harold C. Urey, F.G. Brickwedde, and G.M. Murphy, "A Hydrogen Isotope of Mass 2," *Physical Review* **39** (1932): 164–165.

[10] Carl D. Anderson, "The Apparent Existence of Easily Deflectable Positives," *Science* **76** (1932): 238–239; "The Positive Electron," *Physical Review* **39** (1933): 491–494; latter reprinted in *Foundations of Nuclear Physics: Facsimiles of Thirteen Fundamental Studies as They Were Originally Reported in the Scientific Journals*, Robert T. Beyer, ed. (New York: Dover, 1949), pp. 1–4.

[11] E.O. Lawrence and M.S. Livingston, "The Production of High Speed Light Ions without the Use of High Voltages," *Physical Review* **40** (1932): 19–35; J.D. Cockcroft and E.T.S. Walton, "Experiments with High Velocity Positive Ions. I. Further Developments in the Method of Obtaining High Velocity Positive Ions," *Proceedings of the Royal Society* A **136** (1932): 619–630; both papers reprinted in *The Development of High-Energy Accelerators*, M. Stanley Livingston, ed. (New York: Dover, 1966), pp. 118–134 and 11–23.

[12] Langevin, Solvay *Rapports*, p. x.

[13] E.O. Lawrence, for example, did not receive his invitation until July 1933, and he did not accept it until September; see C. Lefebure to Lawrence, July 8, 1933, and Lawrence to Lefebure, September 22, 1933, Lawrence Correspondence, The Bancroft Library, University of California, Berkeley. Langevin apologized for the delay in his opening remarks; see Solvay *Rapports*, p. x.

[14] Langevin, Solvay *Rapports*, p. x.

[15] Roger H. Stuewer, "Artificial Disintegration and the Cambridge–Vienna Controversy," in *Observation, Experiment, and Hypothesis in Modern Physical Science*, Peter Achinstein and Owen Hannaway, eds. (Cambridge, Mass. and London: The MIT Press, 1985), pp. 239–307.

[16] The participants are listed on unpaginated pages at the end of the Solvay *Rapports*.

[17] Bernard Lovell, *P.M.S. Blackett: A Biographical Memoir* (London: The Royal Society, 1976), p. 22.

[18] Nevill Mott, *A Life in Science* (London and Philadelphia: Taylor & Francis, 1986), p. 46; Rudolf Peierls, *Bird of Passage: Recollections of a Physicist* (Princeton: Princeton University Press, 1985), pp. 99–100.

[19] See for example H. Maier-Leibnitz, "Walther Bothe (1891–1957)," *Physikalische Blätter* **47** (1991): 62–64.

[20] Langevin, Solvay *Rapports*, p. vii.

[21] Cosyns was identified only as coming from Brussels. Errera was not listed as an invited participant, but he appeared on the official photograph of the conference; see Mehra, *Solvay Conferences*, pp. 208–209.

[22] Hendrik B.G. Casimir, *Haphazard Reality: Half a Century of Science* (New York: Harper & Row, 1983), pp. 148–150.

[23] Langevin, Solvay *Rapports*, p. viii. Langevin remembered that one day after a break

in the discussions the participants returned to the meeting room to find a quotation from the Bible that Ehrenfest had written on the blackboard "recalling that at the Tower of Babel the people spoke in different languages and no longer understood each other" – a "picturesque and lively" characterization of their own confusion at the time.

24 Ibid., p. ix.

25 Klein, Ehrenfest, pp. 83–86; V.Ya. Frenkel, "Fiftieth Anniversary of the A.F. Ioffe Physicotechnical Institute, USSR Academy of Sciences (Leningrad)," Soviet Physics Uspekhi 11 (1969): 831–854, esp. 831–833; Paul R. Josephson, Physics and Politics in Revolutionary Russia (Berkeley: University of California Press, 1991), pp. 28–38; Horst Kant, Abram Fedorovič Ioffe: Vater der sowjetischen Physik (Leipzig: BSB B.G. Teubner Verlagsgesellschaft, 1989), pp. 17, 26–29.

26 Quoted in Arnold Kramish, Atomic Energy in the Soviet Union (Stanford: Stanford University Press, 1960), p. 15. For a full discussion of the establishment of Ioffe's institute, see Josephson, Physics and Politics, pp. 82–96.

27 George Gamow, My World Line: An Informal Autobiography (New York: Viking, 1970), pp. 28–54.

28 George Gamow, "Zur Quantentheorie des Atomkernes," Zeitschrift für Physik 51 (1928): 204–212; Roger H. Stuewer, "Gamow's Theory of Alpha-Decay," in The Kaleidoscope of Science, Edna Ullmann-Margalit, ed. (Dordrecht: Reidel, 1986), pp. 147–186.

29 Gamow to Bohr, July 21, 1928; Ioffe to Bohr, [1928], Bohr Scientific Correspondence, Archive for History of Quantum Physics.

30 George Gamow, "Quantum Theory of Nuclear Structure," in Reale Accademia d'Italia, Convegno di Fisica Nucleare Ottobre 1931-IX (Roma: Reale Accademia d'Italia, 1932), pp. 65–81; see also p. 5 for Delbrück's contribution.

31 George Gamow, "The facts concerning my getaway from Russia ...," (October 1950), Gamow papers, Library of Congress, p. 2.

32 Gamow, My World Line, p. 128. Pierre Biquard, Frédéric Joliot-Curie: The Man and His Theories (New York: Paul S. Eriksson, 1966), p. 126, implies that Langevin did not actually join the Communist Party until after 1946.

33 Gamow, "facts concerning my getaway," p. 3.

34 Ibid.

35 Ibid., p. 4; Gamow, My World Line, pp. 129–130. Where these conversations took place is not clear: The first source implies at the Solvay Conference, the second in Paris.

36 Gamow, My World Line, pp. 131–132.

37 Gamow, "facts concerning my getaway," p. 1.

38 William L. Shirer, The Rise and Fall of the Third Reich: A History of Nazi Germany (New York: Simon and Schuster, 1960), p. 241. For the other events see Shirer, ch. 7, pp. 188–230, and Edward Yarnall Hartshorne, Jr., The German Universities and National Socialism (London: Allen & Unwin, 1937), pp. 14–17.

39 Quoted in Norman Bentwich, The Refugees from Germany April 1933 to December 1935 (London: Allen & Unwin, 1936), pp. 28–29. The number of persons affected in Germany actually may have been as high as 875,000; see Herbert A. Strauss, "The Movement of People in a Time of Crisis," in The Muses Flee Hitler: Cultural Transfer and Adaptation 1930–1945, Jarrell C. Jackman and Carla M. Borden, eds. (Washington: Smithsonian Institution Press, 1983), p. 47.

40 For the refugee physicists, see Laura Fermi, Illustrious Immigrants: The Intellectual

Migration from Europe 1930–41 (Chicago: University of Chicago Press, 1968), pp. 174–214; Charles Weiner, "A New Site for the Seminar: The Refugees and American Physics in the Thirties," in *The Intellectual Migration: Europe and America, 1930–1960*, Donald Fleming and Bernard Bailyn, eds. (Cambridge, Mass.: Harvard University Press, 1969), pp. 109–234; Daniel J. Kevles, *The Physicists: The History of a Scientific Community in Modern America* (New York: Knopf, 1978), pp. 279–283; Gerald Holton, "The Migration of Physicists to the United States," in Jackman and Borden, *Muses*, pp. 169–188; Roger H. Stuewer, "Nuclear Physicists in a New World: The Émigrés of the 1930s in America," *Berichte zur Wissenschaftsgeschichte* **7** (1984): 23–40.

[41] Alan Beyerchen, "Anti-Intellectualism and the Cultural Decapitation of Germany under the Nazis," in Jackman and Borden, *Muses*, pp. 38–41.

[42] Reproduced in Weiner, "New Site," p. 234.

[43] Lord Beveridge, *Power and Influence* (London: Hodder and Stoughton, 1953), pp. 167–184, 236; *A Defence of Free Learning* (London: Oxford University Press, 1959), p. 2.

[44] Beveridge, *Defence*, pp. 3–5.

[45] "Academic Assistance Council," *Nature* **131** (1933): 793; "Nationalism and Academic Freedom," *ibid.* 853–855; "Science and Intellectual Liberty," *ibid.* **133** (1934): 701–702.

[46] Finn Aaserud, *Redirecting Science: Niels Bohr, Philanthropy, and the Rise of Nuclear Physics* (Cambridge: Cambridge University Press, 1990), pp. 105–164; Abraham Pais, *Niels Bohr's Times, in Physics, Philosophy, and Polity* (Oxford: Clarendon Press, 1991), pp. 381–386.

[47] Weiner, "New Site," p. 213.

[48] Stephen Duggan and Betty Drury, *The Rescue of Science and Learning: The Story of the Emergency Committee in Aid of Displaced Foreign Scholars* (New York: Macmillan, 1948).

[49] *Ibid.*, pp. 173–177.

[50] "The Emergency Committee in Aid of Displaced German Scholars," *Science* **78** (1933): 52–53.

[51] Nathan Reingold, "Refugee Mathematicians in the United States, 1933–1941: Reception and Reaction," in Jackman and Borden, *Muses*, p. 206.

[52] Klein, *Ehrenfest*, pp. 293–323.

[53] Langevin, Solvay *Rapports*, p. vii.

[54] Albert Einstein, *Ideas and Opinions* (New York: Bonanza, 1954), pp. 205–209; Otto Nathan and Heinz Norden, eds., *Einstein on Peace* (New York: Avenel, 1960), pp. 215–216; Ronald W. Clark, *Einstein: The Life and Times* (New York: World, 1971), pp. 463–464.

[55] Einstein, *Ideas and Opinions*, pp. 210–211; Nathan and Norden, *Einstein on Peace*, p. 216.

[56] Clark, *Einstein*, pp. 479–497.

[57] Nathan and Norden, *Einstein on Peace*, pp. 236–243; Clark, *Einstein*, pp. 502–504.

[58] Nathan and Norden, *Einstein on Peace*, pp. 237–238.

[59] Clark, *Einstein*, pp. 505–506.

[60] Heisenberg to Bohr, June 30, 1933, Bohr Scientific Correspondence, Archive for History of Quantum Physics (AHQP). In addition to the University of Minnesota, there currently are sixteen other Libraries of Deposit worldwide for the AHQP.

[61] This account is based on two interviews with Bloch, one with Thomas S. Kuhn, May 14, 1964, Archive for History of Quantum Physics, another with Charles Weiner, August 15, 1968, Center for History of Physics, American Institute of Physics, College Park, Maryland.

[62] For example, Eugene Feenberg; see his interview with Charles Weiner, April 13, 1973, Center for History of Physics, American Institute of Physics, College Park, Maryland, p. 23.

[63] Walter Moore, *Schrödinger: Life and Thought* (Cambridge: Cambridge University Press, 1989), p. 269.

[64] *Ibid.*, p. 271.

[65] David C. Cassidy, *Uncertainty: The Life and Science of Werner Heisenberg* (New York: Freeman, 1992), p. 310.

[66] Quoted in Cassidy, *Uncertainty*, p. 310.

[67] Moore, *Schrödinger*, p. 273.

[68] *Ibid.*, p. 276.

[69] *Ibid.*

[70] Peierls, *Bird of Passage*, pp. 16–45.

[71] *Ibid.*, pp. 46–81.

[72] *Ibid.*, p. 90.

[73] *Ibid.*, p. 99.

[74] Walter M. Elsasser, *Memoirs of a Physicist in the Atomic Age* (New York: Science History Publications and Bristol: Adam Hilger, 1978), pp. 161–164, 187–188.

[75] Numerous documents on the refugees are preserved in the archive of the Society for the Protection of Science and Learning, the successor organization of the Academic Assistance Council, in the Department of Western Manuscripts, Bodleian Library, University of Oxford.

[76] E. Bretscher and E. Guth, "Zusammenfassender Bericht über die Physikalische Vortragswoche der Eidg. Technischen Hochschule Zürich vom 20.–24. Mai 1931," *Physikalische Zeitschrift* **32** (1931): 649–674.

[77] Reale Accademia d'Italia, *Convegno di Fisica Nucleare.*

[78] M.P. Bronshtein, V.M. Dukelsky, D.D. Ivanenko, and Yu.B. Khariton, ed., *Atomic Nuclei: A Collection of Papers for the All-Union Nuclear Conference* (Leningrad and Moscow: State Technical-Theoretical Publishing House, 1934) [in Russian]. I am grateful to Morton Hamermesh for translating substantial portions of these proceedings for me. For a list of the All-Union Conferences in Physics, 1931–1939, see Josephson, *Physics and Politics*, p. 333.

[79] Helge Kragh, *Dirac: A Scientific Biography* (Cambridge: Cambridge University Press, 1990), p. 139.

[80] "Preface," in *Atomic Nuclei*, Bronshtein et al., eds. Gamow's paper probably was not included because he did not return to Leningrad after the Solvay Conference.

[81] J.D. Cockcroft, "La Désintégration des Éléments par des Protons accélérés," Solvay *Rapports*, pp. 1–56.

[82] Stuewer, "Gamow's Theory of Alpha-Decay," pp. 177–178.

[83] Solvay *Rapports*, pp. 57–61.

[84] *Ibid.*, pp. 61–70.

[85] *Ibid.*, p. 76.

[86] A. Einstein, "Über das Relativitätsprinzip und die aus demselben gezogenen Fol-

gerungen," *Jahrbuch für Radioaktivität und Elektronik* **4** (1907): 443. For a full discussion, see Daniel M. Siegel, "Classical-Electromagnetic and Relativistic Approaches to the Problem of Nonintegral Atomic Masses," *Historical Studies in the Physical Sciences* **9** (1978): 323–360.

[87] F.W. Astron, *Mass-Spectra and Isotopes* (London: Arnold, 1933), p. 73.

[88] Bainbridge was quoted to that effect in a *Science News* article, "Energy Turned into Mass for the First Time in History," *Science Supplement* **77** (April 7, 1933): 9.

[89] Kenneth T. Bainbridge, "The Equivalence of Mass and Energy," *Physical Review* **44** (1933): 123.

[90] J. Chadwick, "Diffusion anomale des Particules α. Transmutation des Éléments par des Particules α. Le Neutron," Solvay *Rapports*, pp. 81–112; F. and I. Joliot, "Rayonnement pénétrant des Atomes sous l'Action des Rayons α," *ibid.*, pp. 121–156.

[91] Roger H. Stuewer, "Mass-Energy and the Neutron in the Early Thirties," *Science in Context* **6** (1993): 195–238.

[92] James Chadwick, "The Existence of a Neutron," *Proceedings of the Royal Society* A **136** (1932): 692–708; reprinted in Beyer, *Foundations*, pp. 5–21.

[93] *Ibid.*, p. 702.

[94] E. Rutherford, "Nuclear Constitution of Atoms," *Proceedings of the Royal Society* A **97** (1920): 374–400; reprinted in *The Collected Papers of Lord Rutherford of Nelson O.M., F.R.S.*, James Chadwick, ed. (London: Allen and Unwin, 1965), vol. 3, pp. 14–38, esp. 34.

[95] See for example, James Chadwick, "The Neutron," *Proceedings of the Royal Society* A **142** (1933): 1–25.

[96] E.O. Lawrence, M.S. Livingston, and G.N. Lewis, "The Emission of Protons from Various Targets Bombarded by Deutons of High Speed," *Physical Review* **44** (1933): 56.

[97] *Ibid.*

[98] M.S. Livingston, Malcolm C. Henderson, and E.O. Lawrence, "Neutrons from Deutons and the Mass of the Neutron," *Physical Review* **44** (1933): 781–782; quote on p. 782.

[99] See for example, Rutherford to Lewis, July 27, 1933, Lewis Correspondence, The Bancroft Library, University of California, Berkeley.

[100] I. Curie and F. Joliot, "La complexité du proton et la masse du neutron," *Comptes rendus de l'Académie des Sciences* **197** (1933): 237–238; reprinted in *Oeuvres Scientifiques Complètes* (Paris: Presses Universitaires de France, 1961), pp. 417–418.

[101] H.R. Crane, C.C. Lauritsen, and A. Soltan, "Production of Neutrons by High Speed Deutons," *Physical Review* **44** (1933): 692–693; M. Stanley Livingston, Malcolm C. Henderson, and Ernest O. Lawrence, "Neutrons from Deutons and the Mass of the Neutron," *ibid.* 781–782; "Neutrons from Beryllium Bombarded by Deutons," *ibid.* 782.

[102] Lawrence, Solvay *Rapports*, pp. 67–69.

[103] Heisenberg, *ibid.*, p. 71.

[104] Bohr, *ibid.*, p. 72.

[105] Chadwick, *ibid.*, pp. 100–103.

[106] *Ibid.*, p. 102.

[107] F and I. Joliot, *ibid.*, pp. 121–156.

[108] *Ibid.*, p. 156.

[109] W. Bothe, "Das Neutron und das Positron," *Die Naturwissenschaften* **21** (1933):

825–831; quote on p. 830.

[110] P.A.M. Dirac, "Théorie du Positron," Solvay *Rapports*, pp. 203–212. Dirac's paper, however, was a significant contribution in its own right; see Kragh, *Dirac*, pp. 145–150.

[111] G. Gamow, "L'Origins des Rayons γ et les Niveaux d'Energie nucléaires," Solvay *Rapports*, pp. 231–260.

[112] W. Heisenberg, "Considérations théoriques générales sur la Structure du Noyau," *ibid.*, pp. 289–323.

[113] Roger H. Stuewer, "The Origin of the Liquid-Drop Model and the Interpretation of Nuclear Fission," *Perspectives on Science* 2 (1994): 76–129.

[114] G. Gamow, "Discussion on the Structure of Atomic Nuclei," *Proceedings of the Royal Society* A **123** (1929): 386–387.

[115] *Ibid.* p. 386.

[116] G. Gamow, "Mass Defect Curve and Nuclear Constitution," *ibid.* **126** (1930): 632–644.

[117] W. Heisenberg, "Über den Bau der Atomkerne. I," *Zeitschrift für Physik* **77** (1932): 1–11; "II," *ibid.* **78** (1932): 156–164; "III," *ibid.* **80** (1933): 587–596.

[118] Ettore Majorana, "Über die Kerntheorie," *ibid.* **82** (1933): 137–145.

[119] Heisenberg, Solvay *Rapports*, p. 316.

[120] *Ibid.*, p. 318.

[121] Fermi, Solvay *Rapports*, p. 334.

[122] *Ibid.*

[123] Pauli, *ibid.*, pp. 324–325. For discussions see Laurie M. Brown, "The Idea of the Neutrino," *Physics Today* **31** (September 1978): 23–28; Roger H. Stuewer, "The Nuclear Electron Hypothesis," in *Otto Hahn and the Rise of Nuclear Physics*, William R. Shea, ed. (Dordrecht: Reidel, 1983), pp. 19–67, esp. pp. 39–42; Abraham Pais, *Inward Bound: Of Matter and Forces in the Physical World* (Oxford: Clarendon Press, 1986), pp. 313–320.

[124] C.D. Ellis and W.D. Wooster, "The Average Energy of Disintegration of Radium E," *Proceedings of the Royal Society* A **117** (1927): 109–123. For discussions see Brown, "Idea," p. 24; Pais, *Inward Bound*, pp. 303–309. A detailed analysis has been given by Carsten Jensen, "A History of the Beta Spectrum and Its Interpretation, 1911–1934," Ph.D. dissertation (University of Copenhagen, 1990).

[125] L. Meitner and W. Orthmann, "Über eine absolute Bestimmung der Energie der primären β-Strahlen von Radium E," *Zeitschrift für Physik* **60** (1930): 143–155.

[126] Pauli to Meitner, Geiger, and others, December 4, 1930, in Wolfgang Pauli, *Wissenschaftlicher Briefwechsel mit Bohr, Einstein, Heisenberg u.a.*, Karl von Meyenn, ed. (Berlin: Springer, 1985), vol. 2, p. 39. For translations see Brown, "Idea," p. 27; Pais, *Inward Bound*, p. 315.

[127] S. Goudsmit, "Present Difficulties in the Theory of Hyperfine Structure," in Reale Accademia d'Italia, *Convegno di Fisica Nucleare*, p. 41.

[128] Pauli, Solvay *Rapports*, p. 324; Bohr, *ibid.*, pp. 327–328. For discussions see Joan Bromberg, "The Impact of the Neutron: Bohr and Heisenberg," *Historical Studies in the Physical Sciences* **3** (1971): 307–341, esp. 310–322; Roger H. Stuewer, "Niels Bohr and Nuclear Physics," in *Niels Bohr: A Centenary Volume*, A.P. French and P.J. Kennedy, eds. (Cambridge, Mass.: Harvard University Press, 1985), pp. 197–220, esp. pp. 199–201; Rudolf Peierls, "Introduction," *Niels Bohr Collected Works*, vol. 9, *Nuclear Physics (1929–1952)* (Amsterdam: North-Holland, 1986), pp. 3–89, esp. pp. 4–14.

[129] Cassidy, *Uncertainty*, p. 323; Kragh, *Dirac*, p. 115; Moore, *Schrödinger*, p. 280.

[130] Subsequently, twelve more of the Solvay participants, Chadwick, Curie-Joliot, Joliot, Debye, Fermi, Lawrence, Pauli, Blackett, Cockcroft, Walton, Bothe, and Mott would also receive the Nobel Prize in Physics or Chemistry. Only two earlier Solvay Conferences, the third (1921) and fifth (1927), had higher percentages of past and future Nobel Laureates among their participants.

[131] The felicitous term, "intellectual decapitation," is Beyerchen's; see Jackman and Borden, *Muses*, p. 41.

[132] Irène Curie and F. Joliot, "Un nouveau type de radioactivité," *Comptes rendus de l'Académie des Sciences* **198** (1934): 254–256; reprinted in *Oeuvres Scientifiques Complètes*, pp. 515–516, and Beyer, *Foundations*, pp. 39–41.

[133] E. Fermi, "Versuch einer Theorie der β-Strahlen. I," *Zeitschrift für Physik* **88** (1934): 161–177; reprinted in Enrico Fermi, *Collected Papers (Note e Memorie)*, vol. I, *Italy 1921–1938* (Chicago: University of Chicago Press, 1962), pp. 575–590, and Beyer, *Foundations*, pp. 45–61.

[134] E. Fermi, "Radioattività indotta da Bombardamento di Neutroni. – I," *Ricerca. Scientifica* **5** (1934): 283; reprinted in *Collected Papers*, vol. 1, pp. 645–646.

[135] G.N. Lewis, M.S. Livingston, M.C. Henderson, and E.O. Lawrence, "On the Hypothesis of the Instability of the Deuton," *Physical Review* **45** (1934): 497. For discussions see Stuewer, "Mass-Energy and the Neutron" and J.L. Heilbron and Robert W. Seidel, *Lawrence and His Laboratory: A History of the Lawrence Berkeley Laboratory*, vol. 1 (Berkeley: University of California Press, 1989), pp. 153–175.

[136] J. Chadwick and M. Goldhaber, "A 'Nuclear Photo-Effect': Disintegration of the Diplon by γ-Rays," *Nature* **134** (1934): 237–238; Maurice Goldhaber, "The Nuclear Photoelectric Effect and Remarks on Higher Multipole Transitions: A Personal History," in *Nuclear Physics in Retrospect: Proceedings of a Symposium on the 1930s*, Roger H. Stuewer, ed. (Minneapolis: University of Minnesota Press, 1979), pp. 83–106.

[137] The Physical Society, *International Conference on Physics London 1934: A Joint Conference Organized by the International Union of Pure and Applied Physics and The Physical Society. Papers & Discussions*, vol. 1 (Cambridge: Cambridge University Press, 1935).

[138] C.F. v. Weizsäcker, "Zur Theorie der Kernmassen," *Zeitschrift für Physik* **96** (1935): 431–458.

[139] Hans A. Bethe and Robert F. Bacher, "Nuclear Physics. A. Stationary States of Nuclei," *Reviews of Modern Physics* **8** (1936): 165–168.

[140] Stuewer, "Origin of the Liquid-Drop Model."

[141] Niels Bohr, "Neutron Capture and Nuclear Constitution," *Nature* **137** (1936): 344–348; reprinted in *Collected Works*, vol. 9, pp. 152–156.

LIST OF PUBLICATIONS OF MARTIN J. KLEIN

1. "Diffraction of Sound around a Circular Disk," *Journal of the Acoustical Society of America* **19** (1947): 132–142 (with H. Primakoff, J. B. Keller, and E. L. Carstensen).
2. "Theory of Critical Fluctuations," *Physical Review* **76** (1949): 1861–1868 (with L. Tisza).
3. "A Note on the Classical Spin Wave Theory of Heller and Kramers," *Physical Review* **80** (1950): 1111 (with Robert S. Smith).
4. "Thin Ferromagnetic Films," *Physical Review* **81** (1951): 378–380 (with Robert S. Smith).
5. "On Order Parameters," *American Journal of Physics* **19** (1951): 153–158.
6. "On a Degeneracy Theory of Kramers," *American Journal of Physics* **20** (1952): 65–71.
7. "Isolated and Adiabatic Susceptibilities," *Physical Review* **86** (1952): 807.
8. "The Ergodic Theorem in Quantum Statistical Mechanics," *Physical Review* **87** (1952): 111–115.
9. "Order, Organization, and Entropy," *British Journal for the Philosophy of Science* **4** (1953): 158–160.
10. "A Note on Wild's Solution of the Boltzmann Equation," *Proceedings of the Cambridge Philosophical Society* **50** (1954): 293–297.
11. "The Principle of Minimum Entropy Production," *Physical Review* **96** (1954): 250–255 (with Paul H. E. Meijer).
12. "Principle of Detailed Balance," *Physical Review* **97** (1955): 1446–1447.
13. "Overhauser Nuclear Polarization Effect and Minimum Entropy Production," *Physical Review* **98** (1955): 1736–1739.
14. "Entropy and the Ehrenfest Urn Model," *Physica* **22** (1956): 569–575.
15. "Generalization of the Ehrenfest Urn Model," *Physical Review* **103** (1956): 17–20.
16. "Negative Absolute Temperatures," *Physical Review* **104** (1956): 589.
17. "Thin Ferromagnetic Films II," *Physical Review* **109** (1958): 288–291 (with S. J. Glass).
18. "A Note on a Problem Concerning the Gibbs Paradox," *American Journal of Physics* **26** (1958): 80–81.
19. "A Note on the Domain of Validity of the Principle of Minimum Entropy Production," in *Proceedings of the International Symposium on Transport Processes in Statistical Mechanics, Brussels, 1956* (New York: Interscience Publishers, 1958), pp. 311–318.
20. "Ehrenfest's Contributions to the Development of Quantum Statistics, I, II," *Proceedings of the Dutch Academy of Sciences* **B62** (1959): 41–50, 51–62.

363

A.J. Kox and D.M. Siegel (eds.), *No Truth Except in the Details*, 363–367.
© 1995 *Kluwer Academic Publishers*.

21. "Grüneisen's Law and the Third Law of Thermodynamics," *Philosophical Magazine* **3** (1958): 538 (with S. J. Glass).
22. "Sublimation and the Third Law of Thermodynamics," *Physica* **25** (1959): 277–280 (with S. J. Glass).
23. "Remarks on the Gibbs Paradox," *Nederlands Tijdschrift voor Natuurkunde* **25** (1959): 73–76.
24. *Collected Scientific Papers of Paul Ehrenfest*, Martin J. Klein, ed. (Amsterdam: North-Holland Publishing Company, 1959).
25. "Thermal Expansion Coefficient of Solid He^3," *Physical Review Letters* **5** (1960): 363 (with R. D. Mountain).
26. "The Laws of Thermodynamics," in *Rendiconti della Scuola Internazionale di Fisica "Enrico Fermi," Corso X* (Bologna: Zanichelli, 1960), pp. 1–22.
27. "The Principle of Minimum Entropy Production," in *Rendiconti della Scuola Internazionale di Fisica "Enrico Fermi," Corso X* (Bologna: Zanichelli, 1960), pp. 198–204.
28. "Pressure Fluctuations," *Physica* **26** (1960): 1073–1079.
29. "Max Planck and the Beginnings of the Quantum Theory," *Archive for History of Exact Sciences* **1** (1962): 459–479.
30. "A Note on Negative Thermal Expansion Coefficients," *Journal of Physics and Chemistry of Solids* **23** (1962): 425–427 (with R. D. Mountain).
31. "The Onsager Conference," *Physics Today* **16**, no. 1 (1963): 45–60.
32. "Ehrenfest Comes to Leiden: Fifty Years After," *Delta* **5**, no. 4 (1962): 5–14. Dutch translation: *De Gids* **125** (1962): 394–405.
33. "The Origins of Ehrenfest's Adiabatic Principle," in *Proceedings of the Xth International Congress on History of Science* (Ithaca, 1962), pp. 801–804.
34. "Planck, Entropy, and Quanta, 1901–1906," *The Natural Philosopher* **1** (1963): 83–108.
35. "Einstein's First Paper on Quanta," *The Natural Philosopher* **2** (1963): 59–86.
36. "Einstein and the Wave-Particle Duality," *The Natural Philosopher* **3** (1964): 1–49.
37. "Adiabatic Processes and the Equation of State," *Physica* **30** (1964): 818–824 (with R. D. Mountain).
38. "Einstein and Some Civilized Discontents," *Physics Today* **18**, no. 1 (1965): 38–44.
39. "Lord Rutherford and the Origins of Nuclear Physics at Manchester," Essay Review, *Scientific American* **212**, no. 3 (1965): 129–134.
40. "Einstein, Specific Heats, and the Early Quantum Theory," *Science* **148** (1965): 173–180.
41. "Thermodynamics and Quanta in Planck's Work," *Physics Today* **19**, no. 11 (1966): 23–32.
42. "Thermodynamics in Einstein's Thought," *Science* **157** (1967): 509–516.
43. *Letters on Wave Mechanics by Einstein, Schrödinger, Planck, and Lorentz*, K. Przibram, ed., Martin J. Klein, transl. and introduction (New York: Philosophical Library, 1967).
44. "Gibbs on Clausius," *Historical Studies in the Physical Sciences* **1** (1969): 127–149.

45.   "Maxwell, His Demon, and the Second Law of Thermodynamics," *American Scientist* **58**, no. 1 (1970): 84–97.

46.   *Paul Ehrenfest.* Volume 1, *The Making of a Theoretical Physicist* (Amsterdam: North-Holland Publishing Company, 1970).

47.   "Boltzmann, Monocycles, and Mechanical Explanation," George Sarton Memorial Lecture 1969, in *Philosophical Foundations of Science*, Boston Studies in the Philosophy of Science, vol. 11, R. J. Seeger and R. S. Cohen, eds. (Dordrecht: Reidel, 1974), pp. 155–175.

48.   "Max Born on His Vocation," *Science* **169** (1970): 360–361. (Review of three books by Max Born.)

49.   "The First Phase of the Bohr–Einstein Dialogue," *Historical Studies in the Physical Sciences* **2** (1970): 1–39.

50.   "Mechanical Explanation at the End of the Nineteenth Century," *Centaurus* **17** (1972): 58–82.

51.   "The Use and Abuse of Historical Teaching in Physics," in *History in the Teaching of Physics*, S. G. Brush and A. L. King, eds. (Hanover NH: The University Press of New England, 1972), pp. 12–27.

52.   "Paul Ehrenfest," in *Dictionary of Scientific Biography*, C. C. Gillispie, ed. (New York: Charles Scribner's Sons, 1970–1980), vol. 4, pp. 292–294.

53.   "Albert Einstein" (in part), in *Dictionary of Scientific Biography*, C. C. Gillispie, ed. (New York: Charles Scribner's Sons, 1970–1980), vol. 4, pp. 312–319, 332–333.

54.   "Josiah Willard Gibbs," in *Dictionary of Scientific Biography*, C. C. Gillispie, ed. (New York: Charles Scribner's Sons, 1970–1980), vol. 5, pp. 386–393.

55.   "Biography of a Reluctant Subject," *Science* **174** (1971): 1315–1316. (Review of R. W. Clark, *Einstein*).

56.   "Aspects of a Revolution," *Science* **177** (1972): 418–419. (Review of A. Hermann, *The Genesis of Quantum Theory (1899–1913)*.)

57.   "The Development of Boltzmann's Statistical Ideas," in *The Boltzmann Equation: Theory and Applications*, E. G. D. Cohen and W. Thirring, eds. (Vienna: Springer-Verlag, 1973), pp. 53–106.

58.   "Einstein on Scientific Revolutions," in *Vistas in Astronomy*, vol. 17, A. Beer and K. A. Strand, eds. (Oxford: Pergamon Press, 1975), pp. 113–120.

59.   "The Maxwell–Boltzmann Relationship," in *Transport Phenomena – 1973*, J. Kestin, ed. (New York: American Institute of Physics, 1973), pp. 297–308.

60.   "The Historical Origins of the Van der Waals Equation," *Physica* **73** (1974): 28–47.

61.   "Carnot's Contribution to Thermodynamics," *Physics Today* **27**, no. 8 (1974): 23–28.

62.   "Closing the Carnot Cycle," in *Sadi Carnot et l'essor de la thermodynamique* (Paris: CNRS, 1976), pp. 213–219.

63.   "The Beginnings of the Quantum Theory," in *History of Twentieth Century Physics. Proceedings of the International School of Physics "Enrico Fermi," Course LVII*, C. Weiner, ed. (New York: Academic Press, 1977), pp. 1–39.

64.   "A Memoir in Physics," *Science* **186** (1974): 342–343. (Review of K. Mendelssohn, *The World of Walther Nernst*.)

65. "Einstein, Boltzmann's Principle, and the Mechanical World View," in *Proceedings of the XIVth International Congress of the History of Science* (Tokyo: Science Council of Japan, 1975), vol. 1, pp. 183–194.

66. "Some Unnoticed Publications by Einstein," *Isis* **68** (1977): 601–604 (with Allan Needell).

67. "The Early Papers of J. Willard Gibbs: A Transformation of Thermodynamics" (presented at the XVth International Congress of the History of Science, Edinburgh, 1977), in *Human Implications of Scientific Advance*, E. G. Forbes, ed. (Edinburgh: Edinburgh University Press, 1978).

68. "Physics in Nazi Germany," *Science* **199** (1978): 871. (Review of A. D. Beyerchen, *Scientists under Hitler*.)

69. "An Autobiography," *Science* **202** (1978): 622. (Review of W. M. Elsasser, *Memoirs of a Physicist in the Atomic Age*.)

70. "Einstein and the Development of Quantum Physics," in *Einstein: A Centenary Volume*, A. P. French, ed. (Cambridge, Mass.: Harvard University Press, 1979), pp. 133–151.

71. "Einstein and the Academic Establishment," in *Einstein: A Centenary Volume*, A. P. French, ed. (Cambridge, Mass.: Harvard University Press, 1979), pp. 209–213. (Reprint of no. 38).

72. "A Bit of Light," *The Sciences* **19**, no. 3 (1979): 10–13.

73. "No Firm Foundation: Einstein and the Early Quantum Theory," in *Some Strangeness in the Proportion: A Centennial Symposium to Celebrate the Achievements of Albert Einstein*, H. Woolf, ed. (Reading, Mass.: Addison-Wesley, 1980), pp. 161–185.

74. "Fluctuations and Statistical Physics in Einstein's Early Work," in *Albert Einstein: Historical and Cultural Perspectives. The Centennial Symposium in Jerusalem*, G. Holton and Y. Elkana, eds. (Princeton: Princeton University Press, 1982), pp. 39–58.

75. Review of T. S. Kuhn, *Black-Body Theory and the Quantum Discontinuity, 1894–1912, Isis* **70** (1979): 429–434.

76. "Not by Discoveries Alone: The Centennial of Paul Ehrenfest," *Physica* **106A** (1981): 3–14.

77. "Some Turns of Phrase in Einstein's Early Papers," in *Physics as Natural Philosophy: Essays in Honor of Laszlo Tisza on His Seventy-Fifth Birthday*, A. Shimony and H. Feshbach, eds. (Cambridge, Mass.: MIT Press, 1982), pp. 364–375.

78. "The Scientific Style of Josiah Willard Gibbs," in *Springs of Scientific Creativity*, R. Aris, H. T. Davis, and R. Stuewer, eds. (Minneapolis: University of Minnesota Press, 1983), pp. 142–162.

79. Review of A. Pais, *'Subtle Is the Lord . . .': The Science and the Life of Albert Einstein, Isis* **75** (1984): 377–379.

80. "Great Combinations Come Alive: Bohr, Ehrenfest and Einstein," in *The Lesson of the Quantum Theory. Niels Bohr Centenary Symposium October 3–7, 1985*, J. de Boer, E. Dal, and O. Ulfbeck, eds. (Amsterdam: North-Holland, 1986), pp. 325–342.

81. Review of J. L. Heilbron, *The Dilemmas of an Upright Man: Max Planck as Spokesman for German Science, Nature* **324** (1986): 190–191.

82. "Osservando L'America. La visita di P. Ehrenfest negli USA," in *La ristrut-turazione delle scienze tra le due guerre mondiali*, G. Battimelli, M. de Maria, and A. Rossi, eds. (Rome: La Goliardica Editrice Universitaria di Roma, 1986), vol. 2, pp. 9–24.

83. Introduction to Ernst Mach, *Principles of the Theory of Heat*, B. McGuinness, ed. (Dordrecht and Boston: Reidel, 1986), pp. ix–xx, 417–419.

84. "Some Historical Remarks on the Statistical Mechanics of Josiah Willard Gibbs," in *From Ancient Omens to Statistical Mechanics. Essays on the Exact Sciences Presented to Asger Aaboe*, J. Berggren and B. Goldstein, eds. (Copenhagen: University Library, 1987), pp. 281–289.

85. "Boldness and Caution: Aspects of Niels Bohr's Scientific Style," in *New Trends in the History of Science*, R. Visser, H. Bos, L. Palm, and H. Snelders, eds. (Amsterdam: Rodopi, 1989), pp. 97–114.

86. "Physics in the Making of Leiden: Paul Ehrenfest as Teacher," in *Physics in the Making. Essays on Developments in 20th Century Physics in Honour of H. B. G. Casimir on the Occasion of His 80th Birthday*, A. Sarlemijn and M. J. Sparnaay, eds. (Amsterdam: North-Holland, 1989), pp. 29–44.

87. "Duhem on Gibbs," in *Beyond History of Science. Essays in Honor of Robert E. Schofield*, E. Garber, ed. (Bethlehem: LeHigh University Press, 1990), pp. 52–66.

88. "The Physics of J. Willard Gibbs in His Time," in *Proceedings of the Gibbs Symposium, Yale University, May 15–17, 1989*, D. G. Caldi and G. D. Mostow, eds. (Providence: The American Mathematical Society, 1990), pp. 1–21.

89. *Physicists' Inaugural Lectures in History* (Amsterdam: Amsterdam University Press, 1993). (Lecture given on 3 May 1993 to inaugurate the Pieter Zeeman Visiting Professorship in the History of Modern Physics at the University of Amsterdam.)

90. *The Collected Papers of Albert Einstein. Volume 3, The Swiss Years: Writings, 1909–1911* (Princeton: Princeton University Press, 1993). Edited in collaboration with A. J. Kox, Jürgen Renn and Robert Schulmann.

91. *The Collected Papers of Albert Einstein. Volume 5, The Swiss Years: Correspondence, 1902–1914* (Princeton: Princeton University Press, 1993). Edited in collaboration with A. J. Kox and Robert Schulmann.

INDEX OF NAMES

Boston Studies in the Philosophy of Science

87. R.S. Cohen and T. Schnelle (eds.): *Cognition and Fact.* Materials on Ludwik Fleck. 1986   ISBN 90-277-1902-0
88. G. Freudenthal: *Atom and Individual in the Age of Newton.* On the Genesis of the Mechanistic World View. Translated from German. 1986
ISBN 90-277-1905-5
89. A. Donagan, A.N. Perovich Jr and M.V. Wedin (eds.): *Human Nature and Natural Knowledge.* Essays presented to Marjorie Grene on the Occasion of Her 75th Birthday. 1986   ISBN 90-277-1974-8
90. C. Mitcham and A. Hunning (eds.): *Philosophy and Technology II.* Information Technology and Computers in Theory and Practice. [*Also* Philosophy and Technology Series, Vol. 2] 1986   ISBN 90-277-1975-6
91. M. Grene and D. Nails (eds.): *Spinoza and the Sciences.* 1986
ISBN 90-277-1976-4
92. S.P. Turner: *The Search for a Methodology of Social Science.* Durkheim, Weber, and the 19th-Century Problem of Cause, Probability, and Action. 1986.
ISBN 90-277-2067-3
93. I.C. Jarvie: *Thinking about Society.* Theory and Practice. 1986
ISBN 90-277-2068-1
94. E. Ullmann-Margalit (ed.): *The Kaleidoscope of Science.* The Israel Colloquium: Studies in History, Philosophy, and Sociology of Science, Vol. 1. 1986
ISBN 90-277-2158-0; Pb 90-277-2159-9
95. E. Ullmann-Margalit (ed.): *The Prism of Science.* The Israel Colloquium: Studies in History, Philosophy, and Sociology of Science, Vol. 2. 1986
ISBN 90-277-2160-2; Pb 90-277-2161-0
96. G. Márkus: *Language and Production.* A Critique of the Paradigms. Translated from French. 1986   ISBN 90-277-2169-6
97. F. Amrine, F.J. Zucker and H. Wheeler (eds.): *Goethe and the Sciences: A Reappraisal.* 1987   ISBN 90-277-2265-X; Pb 90-277-2400-8
98. J.C. Pitt and M. Pera (eds.): *Rational Changes in Science.* Essays on Scientific Reasoning. Translated from Italian. 1987   ISBN 90-277-2417-2
99. O. Costa de Beauregard: *Time, the Physical Magnitude.* 1987
ISBN 90-277-2444-X
100. A. Shimony and D. Nails (eds.): *Naturalistic Epistemology.* A Symposium of Two Decades. 1987   ISBN 90-277-2337-0
101. N. Rotenstreich: *Time and Meaning in History.* 1987   ISBN 90-277-2467-9
102. D.B. Zilberman: *The Birth of Meaning in Hindu Thought.* Edited by R.S. Cohen. 1988   ISBN 90-277-2497-0
103. T.F. Glick (ed.): *The Comparative Reception of Relativity.* 1987
ISBN 90-277-2498-9
104. Z. Harris, M. Gottfried, T. Ryckman, P. Mattick Jr, A. Daladier, T.N. Harris and S. Harris: *The Form of Information in Science.* Analysis of an Immunology Sublanguage. With a Preface by Hilary Putnam. 1989   ISBN 90-277-2516-0

Boston Studies in the Philosophy of Science

105. F. Burwick (ed.): *Approaches to Organic Form*. Permutations in Science and Culture. 1987   ISBN 90-277-2541-1
106. M. Almási: *The Philosophy of Appearances*. Translated from Hungarian. 1989
                                                        ISBN 90-277-2150-5
107. S. Hook, W.L. O'Neill and R. O'Toole (eds.): *Philosophy, History and Social Action*. Essays in Honor of Lewis Feuer. With an Autobiographical Essay by L. Feuer. 1988   ISBN 90-277-2644-2
108. I. Hronszky, M. Fehér and B. Dajka: *Scientific Knowledge Socialized*. Selected Proceedings of the 5th Joint International Conference on the History and Philosophy of Science organized by the IUHPS (Veszprém, Hungary, 1984). 1988   ISBN 90-277-2284-6
109. P. Tillers and E.D. Green (eds.): *Probability and Inference in the Law of Evidence*. The Uses and Limits of Bayesianism. 1988   ISBN 90-277-2689-2
110. E. Ullmann-Margalit (ed.): *Science in Reflection*. The Israel Colloquium: Studies in History, Philosophy, and Sociology of Science, Vol. 3. 1988
                                    ISBN 90-277-2712-0; Pb 90-277-2713-9
111. K. Gavroglu, Y. Goudaroulis and P. Nicolacopoulos (eds.): *Imre Lakatos and Theories of Scientific Change*. 1989   ISBN 90-277-2766-X
112. B. Glassner and J.D. Moreno (eds.): *The Qualitative-Quantitative Distinction in the Social Sciences*. 1989   ISBN 90-277-2829-1
113. K. Arens: *Structures of Knowing*. Psychologies of the 19th Century. 1989
                                                        ISBN 0-7923-0009-2
114. A. Janik: *Style, Politics and the Future of Philosophy*. 1989
                                                        ISBN 0-7923-0056-4
115. F. Amrine (ed.): *Literature and Science as Modes of Expression*. With an Introduction by S. Weininger. 1989   ISBN 0-7923-0133-1
116. J.R. Brown and J. Mittelstrass (eds.): *An Intimate Relation*. Studies in the History and Philosophy of Science. Presented to Robert E. Butts on His 60th Birthday. 1989   ISBN 0-7923-0169-2
117. F. D'Agostino and I.C. Jarvie (eds.): *Freedom and Rationality*. Essays in Honor of John Watkins. 1989   ISBN 0-7923-0264-8
118. D. Zolo: *Reflexive Epistemology*. The Philosophical Legacy of Otto Neurath. 1989   ISBN 0-7923-0320-2
119. M. Kearn, B.S. Philips and R.S. Cohen (eds.): *Georg Simmel and Contemporary Sociology*. 1989   ISBN 0-7923-0407-1
120. T.H. Levere and W.R. Shea (eds.): *Nature, Experiment and the Science*. Essays on Galileo and the Nature of Science. In Honour of Stillman Drake. 1989
                                                        ISBN 0-7923-0420-9
121. P. Nicolacopoulos (ed.): *Greek Studies in the Philosophy and History of Science*. 1990   ISBN 0-7923-0717-8
122. R. Cooke and D. Costantini (eds.): *Statistics in Science*. The Foundations of Statistical Methods in Biology, Physics and Economics. 1990
                                                        ISBN 0-7923-0797-6

Boston Studies in the Philosophy of Science

123. P. Duhem: *The Origins of Statics*. Translated from French by G.F. Leneaux, V.N. Vagliente and G.H. Wagner. With an Introduction by S.L. Jaki. 1991
ISBN 0-7923-0898-0

124. H. Kamerlingh Onnes: *Through Measurement to Knowledge*. The Selected Papers, 1853-1926. Edited and with an Introduction by K. Gavroglu and Y. Goudaroulis. 1991
ISBN 0-7923-0825-5

125. M. Čapek: *The New Aspects of Time: Its Continuity and Novelties*. Selected Papers in the Philosophy of Science. 1991
ISBN 0-7923-0911-1

126. S. Unguru (ed.): *Physics, Cosmology and Astronomy, 1300-1700*. Tension and Accommodation. 1991
ISBN 0-7923-1022-5

127. Z. Bechler: *Newton's Physics on the Conceptual Structure of the Scientific Revolution*. 1991
ISBN 0-7923-1054-3

128. É. Meyerson: *Explanation in the Sciences*. Translated from French by M-A. Siple and D.A. Siple. 1991
ISBN 0-7923-1129-9

129. A.I. Tauber (ed.): *Organism and the Origins of Self*. 1991
ISBN 0-7923-1185-X

130. F.J. Varela and J-P. Dupuy (eds.): *Understanding Origins*. Contemporary Views on the Origin of Life, Mind and Society. 1992
ISBN 0-7923-1251-1

131. G.L. Pandit: *Methodological Variance*. Essays in Epistemological Ontology and the Methodology of Science. 1991
ISBN 0-7923-1263-5

132. G. Munévar (ed.): *Beyond Reason*. Essays on the Philosophy of Paul Feyerabend. 1991
ISBN 0-7923-1272-4

133. T.E. Uebel (ed.): *Rediscovering the Forgotten Vienna Circle*. Austrian Studies on Otto Neurath and the Vienna Circle. Partly translated from German. 1991
ISBN 0-7923-1276-7

134. W.R. Woodward and R.S. Cohen (eds.): *World Views and Scientific Discipline Formation*. Science Studies in the [former] German Democratic Republic. Partly translated from German by W.R. Woodward. 1991
ISBN 0-7923-1286-4

135. P. Zambelli: *The Speculum Astronomiae and Its Enigma*. Astrology, Theology and Science in Albertus Magnus and His Contemporaries. 1992
ISBN 0-7923-1380-1

136. P. Petitjean, C. Jami and A.M. Moulin (eds.): *Science and Empires*. Historical Studies about Scientific Development and European Expansion.
ISBN 0-7923-1518-9

137. W.A. Wallace: *Galileo's Logic of Discovery and Proof*. The Background, Content, and Use of His Appropriated Treatises on Aristotle's *Posterior Analytics*. 1992
ISBN 0-7923-1577-4

138. W.A. Wallace: *Galileo's Logical Treatises*. A Translation, with Notes and Commentary, of His Appropriated Latin Questions on Aristotle's *Posterior Analytics*. 1992
ISBN 0-7923-1578-2
Set (137 + 138) ISBN 0-7923-1579-0

Boston Studies in the Philosophy of Science

139. M.J. Nye, J.L. Richards and R.H. Stuewer (eds.): *The Invention of Physical Science*. Intersections of Mathematics, Theology and Natural Philosophy since the Seventeenth Century. Essays in Honor of Erwin N. Hiebert. 1992
ISBN 0-7923-1753-X

140. G. Corsi, M.L. dalla Chiara and G.C. Ghirardi (eds.): *Bridging the Gap: Philosophy, Mathematics and Physics*. Lectures on the Foundations of Science. 1992
ISBN 0-7923-1761-0

141. C.-H. Lin and D. Fu (eds.): *Philosophy and Conceptual History of Science in Taiwan*. 1992
ISBN 0-7923-1766-1

142. S. Sarkar (ed.): *The Founders of Evolutionary Genetics*. A Centenary Reappraisal. 1992
ISBN 0-7923-1777-7

143. J. Blackmore (ed.): *Ernst Mach – A Deeper Look*. Documents and New Perspectives. 1992
ISBN 0-7923-1853-6

144. P. Kroes and M. Bakker (eds.): *Technological Development and Science in the Industrial Age*. New Perspectives on the Science–Technology Relationship. 1992
ISBN 0-7923-1898-6

145. S. Amsterdamski: *Between History and Method*. Disputes about the Rationality of Science. 1992
ISBN 0-7923-1941-9

146. E. Ullmann-Margalit (ed.): *The Scientific Enterprise*. The Bar-Hillel Colloquium: Studies in History, Philosophy, and Sociology of Science, Volume 4. 1992
ISBN 0-7923-1992-3

147. L. Embree (ed.): *Metaarchaeology*. Reflections by Archaeologists and Philosophers. 1992
ISBN 0-7923-2023-9

148. S. French and H. Kamminga (eds.): *Correspondence, Invariance and Heuristics*. Essays in Honour of Heinz Post. 1993
ISBN 0-7923-2085-9

149. M. Bunzl: *The Context of Explanation*. 1993
ISBN 0-7923-2153-7

150. I.B. Cohen (ed.): *The Natural Sciences and the Social Sciences*. Some Critical and Historical Perspectives. 1994
ISBN 0-7923-2223-1

151. K. Gavroglu, Y. Christianidis and E. Nicolaidis (eds.): *Trends in the Historiography of Science*. 1994
ISBN 0-7923-2255-X

152. S. Poggi and M. Bossi (eds.): *Romanticism in Science*. Science in Europe, 1790–1840. 1994
ISBN 0-7923-2336-X

153. J. Faye and H.J. Folse (eds.): *Niels Bohr and Contemporary Philosophy*. 1994
ISBN 0-7923-2378-5

154. C.C. Gould and R.S. Cohen (eds.): *Artifacts, Representations, and Social Practice*. Essays for Marx W. Wartofsky. 1994
ISBN 0-7923-2481-1

155. R.E. Butts: *Historical Pragmatics*. Philosophical Essays. 1993
ISBN 0-7923-2498-6

156. R. Rashed: *The Development of Arabic Mathematics: Between Arithmetic and Algebra*. Translated from French by A.F.W. Armstrong. 1994
ISBN 0-7923-2565-6

Boston Studies in the Philosophy of Science

157. I. Szumilewicz-Lachman (ed.): *Zygmunt Zawirski: His Life and Work.* With Selected Writings on Time, Logic and the Methodology of Science. Translations by Feliks Lachman. Ed. by R.S. Cohen, with the assistance of B. Bergo. 1994   ISBN 0-7923-2566-4

158. S.N. Haq: *Names, Natures and Things.* The Alchemist Jābir ibn Ḥayyān and His *Kitāb al-Aḥjār* (Book of Stones). 1994   ISBN 0-7923-2587-7

159. P. Plaass: *Kant's Theory of Natural Science.* Translation, Analytic Introduction and Commentary by Alfred E. and Maria G. Miller. 1994
ISBN 0-7923-2750-0

160. J. Misiek (ed.): *The Problem of Rationality in Science and its Philosophy.* On Popper vs. Polanyi. The Polish Conferences 1988–89. 1995
ISBN 0-7923-2925-2

161. I.C. Jarvie and N. Laor (eds.): *Critical Rationalism, Metaphysics and Science.* Essays for Joseph Agassi, Volume I. 1995   ISBN 0-7923-2960-0

162. I.C. Jarvie and N. Laor (eds.): *Critical Rationalism, the Social Sciences and the Humanities.* Essays for Joseph Agassi, Volume II. 1995   ISBN 0-7923-2961-9
Set (161–162) ISBN 0-7923-2962-7

163. K. Gavroglu, J. Stachel and M.W. Wartofsky (eds.): *Physics, Philosophy, and the Scientific Community.* Essays in the Philosophy and History of the Natural Sciences and Mathematics. In Honor of Robert S. Cohen. 1995
ISBN 0-7923-2988-0

164. K. Gavroglu, J. Stachel and M.W. Wartofsky (eds.): *Science, Politics and Social Practice.* Essays on Marxism and Science, Philosophy of Culture and the Social Sciences. In Honor of Robert S. Cohen. 1995   ISBN 0-7923-2989-9

165. K. Gavroglu, J. Stachel and M.W. Wartofsky (eds.): *Science, Mind and Art.* Essays on Science and the Humanistic Understanding in Art, Epistemology, Religion and Ethics. Essays in Honor of Robert S. Cohen. 1995
ISBN 0-7923-2990-2
Set (163–165) ISBN 0-7923-2991-0

166. K.H. Wolff: *Transformation in the Writing.* A Case of Surrender-and-Catch. 1995   ISBN 0-7923-3178-8

167. A.J. Kox and D.M. Siegel (eds.): *No Truth Except in the Details.* Essays in Honor of Martin J. Klein. 1995   ISBN 0-7923-3195-8

168. J. Blackmore (ed.): *Ludwig Boltzmann* His Later Life and Philosophy, 1900–1906. 1995   ISBN 0-7923-3231-8

169. R.S. Cohen, R. Hilpinen and Qiu Renzong (eds.): *Realism and Anti-Realism in the Philosophy of Science.* Beijing International Conference, 1992. 1995 (forthcoming)   ISBN 0-7923-3233-4

170. I. Kuçuradi and R.S. Cohen (eds.): *The Concept of Knowledge.* The Ankara Seminar. 1995 (forthcoming)   ISBN 0-7923-3241-5

Boston Studies in the Philosophy of Science

Also of interest:
R.S. Cohen and M.W. Wartofsky (eds.): *A Portrait of Twenty-Five Years Boston Colloquia for the Philosophy of Science, 1960-1985.* 1985   ISBN Pb 90-277-1971-3

Previous volumes are still available.

KLUWER ACADEMIC PUBLISHERS – DORDRECHT / BOSTON / LONDON